City Development: Issues and Best Practices

Editors-in-Chief

Huhua Cao, Geography, Environment and Geomatics, University of Ottawa, Ottawa, ON, Canada

John Zacharias, College of Architecture and Landscape, Peking University, Beijing, China

Claude Ngomsi, United Nations Human Settlements Programme (UN-Habitat), Nairobi, Kenya

The current rate of urbanization is unprecedented and poses enormous challenges for governing bodies. New approaches to development and urban management are needed in the context of globalization and the need for local sustainability. While the developing world itself offers an abundance of lessons, case studies and best practices, these have rarely been positioned as cutting-edge contributions to reformed practices in city development. It is well recognized that the experience of the developed world is an incomplete guide to the new challenges posed by urbanization in the contemporary world.

The "City Development: Issues and Best Practices" book series includes academic research, comparative and applied research, and case studies at the scale of the neighborhood, city, region, nation and supranational levels. This series will offer an opportunity to present the latest academic research and best practices in urban development with the goal of promoting sustainable and inclusive development, learning from the diverse and complementary experiences of rapidly urbanizing areas of the world. Although this book series focuses primarily on the developing world, we intend to include the latest academic research and evolving best practices from developed countries.

The series is intended for geographers, planners, engineers, urban designers, architects, political scientists, sociologists, and economists, as well as policy makers and representatives from government, civil society, industry, etc. who are interested in the developing world. The series will also interest students seeking a foundation in the comparative analysis of key issues facing rapidly urbanizing areas of the world. It will include monographs, edited volumes and textbooks. Book proposals and final manuscripts will be peer-reviewed.

The areas to be covered in the series include, but are not limited to, the following:

- Urban Equity and Inclusivity
- Reforming Informal Settlements
- Climate Change and Adaptation of the Built Environment
- Health Crises and Urban Risk Management
- Privacy, Surveillance, Security and Collective Wellbeing
- Governance and Urban Resilience
- Participatory Policy and the Right to the City
- Indigenous and Ethnic Minority Urbanization
- Mobility and Urban Transformation
- Land Use and Urban Landscapes
- Urban Infrastructure and Transport Systems
- Urban Intelligence and Technologies
- Participatory Budgeting and Community Building
- Industrial Parks and Agro-Processing Zones
- Housing and Land Tenure Issues

Kh Md Nahiduzzaman
Editor

Making Sense of Planning and Development for the Post-Pandemic Cities

Springer

Editor
Kh Md Nahiduzzaman
Faculty of Applied Science
The University of British Columbia (UBC)
Okanagan
Kelowna, BC, Canada

ISSN 2731-7773 ISSN 2731-7781 (electronic)
City Development: Issues and Best Practices
ISBN 978-981-97-5480-9 ISBN 978-981-97-5481-6 (eBook)
https://doi.org/10.1007/978-981-97-5481-6

This Springer imprint is published by the registered company Springer Nature Singapore Pte Ltd.
The registered company address is: 152 Beach Road, #21-01/04 Gateway East, Singapore 189721, Singapore

Preface

The International Conference on Canadian, Chinese and African Sustainable Urbanization (ICCCASU) is pleased to present the second book of the series: "Making Sense of Planning and Development for the post-pandemic Cities", a cooperative venture with Springer Publishing. This book stands at the forefront of post-pandemic transformational planning, disseminating novel academic research and exemplary urban development practices to foster sustainable, inclusive, and resilient growth. By drawing insights from diverse, rapidly urbanizing regions, the series confronts the unprecedented challenges of urbanization, advocating for a re-evaluation of planning strategies in the globalization era while prioritizing local resilience and what does it mean for the local sustainability. Embracing a holistic perspective, it integrates lessons from both the developing and developed nations, showcasing research, analyses, and case studies across various scales. This inclusive approach invites scholars and practitioners from diverse disciplines to engage in collaborative dialogues aimed at enhancing territory-specific (urban, rural or regional) resilience.

The convergence of rapid urbanization and the deepening impact of the recent global pandemic has reinstated the need for a paradigm "shift" toward more resilient and inclusive planning approaches. As we navigated through the post-COVID era, transformational planning emerges as a pivotal domain to *re*-imagine urban and rural territorial landscapes and their interrelationships. Beyond the traditional scope of integration, this "transformative" wave of planning emphasizes the need to include health-centric infrastructure into mainstreaming urban planning, design, and governance domains. By fostering resilient, inclusive communities and prioritizing equitable access to healthcare and other essential services, transformational planning seeks to safeguard and strike a balance between sustainability of the built-environment and human well-being. This all-encompassing approach recognizes the intrinsic interplay between resilient planning, public health, and need for ecological preservation, advocating for innovations that harmonize the needs of people living under diverse social, economic, environmental, and cultural contexts.

Through collaborative efforts and forward-thinking initiatives, transformational planning endeavors to forge a future where cities thrive as livable, resilient hubs of prosperity, and sustainability.

Kelowna, BC, Canada

Professor Kh Md Nahiduzzaman

About ICCCASU

ICCCASU, established in 2015 as a collaborative venture between the United Nations Human Settlements Program (UN-Habitat) and the University of Ottawa, has since evolved into a consortium comprising numerous Canadian, Chinese, and African universities, NGOs, and partners worldwide. Serving as an international hub for urban research and dialogue, ICCCASU convenes diverse stakeholders—ranging from government representatives to civil society, industry leaders, and academics—to delve into the issues of urbanization and city development. It fosters a unique triangular dialogue among African nations, Canada, and China.

ICCCASU has organized five conferences and training programs thus far: ICCCASU I [2015] in Ottawa; ICCCASU II [2017] in Yaoundé, Cameroon; ICCCASU 3 [2019] in Chengdu, China; ICCCASU 4 [2021] in Montreal, Canada, and ICCCASU 5 [2023] in Nairobi, Kenya. Moreover, it has hosted various side events, including a roundtable at Habitat III in 2016 and workshops at the 10th World Urban Forum in Abu Dhabi in 2020. At the 11th World Urban Forum in Katowice, Poland, in 2022, ICCCASU organized a roundtable on city transformations in the context of the pandemic, underscoring its ongoing commitment to global urban dialogue and collaboration.

In addition to its renowned conferences, ICCCASU has expanded its activities to include training sessions, workshops, consultations, publications, research endeavors, and exemplary case studies. This multifaceted approach has solidified ICCCASU's status as a respected and influential international urban planning think tank, fostering extensive networks among researchers and practitioners worldwide. Recognizing cities as hubs of innovation, ICCCASU advocates for sustainable and inclusive urban planning practices that uplift marginalized communities, enhance educational opportunities, stimulate economic prosperity, and revolutionize land management practices, thus contributing significantly to global discussions on the sustainable urban development.

About ICOCASH

Contents

Contributors

Abue Jawfore Taufique Ahamed Ahade Department of Maritime Law and Policy, Bangabandhu Sheikh Mujibur Rahman Maritime University, Dhaka, Bangladesh

Md Moynul Ahsan Department of Real Estate Development and Management, Ankara University, Ankara, Turkey

Reazul Ahsan City and Metropolitan Planning Department, University of Utah, Incheon, South Korea

Omar Al-Mahadi Architecture and City Design Department, King Fahd University of Petroleum & Minerals, Dharan, Saudi Arabia

Muhammad Ahmad Al-Rashid Department of Urban and Regional Planning, Faculty of Built Environment, Universiti Malaya, Kuala Lumpur, Malaysia

Wail Ismail Bakhit Urban Planning and Policy Consultant, Khobar, Saudi Arabia

Johannes Bhanye African Climate and Development Initiative (ACDI), University of Cape Town, Cape Town, South Africa

Tiziana Campisi Department of Engineering and Architecture, University of Enna Kore, Cittadella Universitaria, Enna, Italy

Jayaprakash Chadchan School of Architecture, Christ University, Bangalore, India

Uchendu E. Chigbu Department of Land and Spatial Sciences, Namibia University of Science and Technology, Windhoek, Namibia

Soumaya Ezazaa Urban Design and Planner, Reichen et Robert & Associés International, Rabat, Morocco

Mareli Hugo Department of Urban and Regional Planning, University of the Free State, Bloemfontein, South Africa

Kazi Humayun Kabir Department of Urban Planning and Design, The University of Hong Kong, Hong Kong, China;
Development Studies Discipline, Khulna University, Khulna, Bangladesh

Md. Ayatullah Khan Development Studies Discipline, Khulna University, Khulna, Bangladesh;
Department of Geography, Hong Kong Baptist University, Hong Kong, China

Ewa Korcelli-Olejniczak Institute of Geography and Spatial Organization, Polish Academy of Sciences, Warsaw, Poland

E. M. Lentsoane University of Limpopo, Polokwane, South Africa

Cathrine Marenga Department of Land and Spatial Sciences, Namibia University of Science and Technology, Windhoek, Namibia

Abraham Matamanda Department of Urban and Regional Planning, University of the Free State, Bloemfontein, South Africa

Malcon L. Mazambani Department of Land and Spatial Sciences, Namibia University of Science and Technology, Windhoek, Namibia

Kgosi Mocwagae Department of Urban and Regional Planning, University of the Free State, Bloemfontein, South Africa

Seyed Navid Mashhadi Moghaddam University of Ottawa, Ottawa, ON, Canada

Muhammad Nadeem Graduate School of Urban Innovation, Yokohama National University, Yokohama, Japan

Kh Md Nahiduzzaman Faculty of Applied Science, The University of British Columbia (UBC) Okanagan, Kelowna, BC, Canada

George Onatu University of Johannesburg, Johannesburg, South Africa

Hassan Radoine School of Architecture, Planning and Design (SAP+D), Mohammed VI Polytechnic University (UM6P), Ben Guerir, Morocco

Sadia Binte Rahman Department of Law, Bangladesh University of Professionals, Dhaka, Bangladesh

Penehafo Ricardo Department of Land and Spatial Sciences, Namibia University of Science and Technology, Windhoek, Namibia

Bilal Saghir Department of City and Regional Planning, King Fahd University of Petroleum and Minerals, Dhahran, Saudi Arabia

Jiyon Shin World Bank, City and Metropolitan Planning, University of Utah, Incheon, South Korea

Sujit Kumar Sikder Leibniz Institute of Ecological Urban and Regional Development, Dresden, Germany

Talha Tasnim BRAC Urban Development Programme, BRAC Center, Dhaka, Bangladesh

Amena Jahan Urmy Department of Law, Bangladesh University of Professionals, Dhaka, Bangladesh

Arun Baby M. Wilson School of Architecture, Christ University, Bangalore, India

Introduction—Navigating Pandemics in Urban Spaces: Challenges, Strategies, and the Future of Urban Planning

Seyed Navid Mashhadi Moghaddam

Abstract The COVID-19 pandemic has exposed the vulnerabilities of urban environments, necessitating a reevaluation of urban planning and development to enhance city resilience, sustainability, and equity in the post-pandemic era. "Making Sense of Planning and Development for the Post-Pandemic Cities" provides a comprehensive analysis across four thematic areas: Urban Infrastructure & Resilience, Dynamics of Urban Transformation, Transformational Real Estate Market, and Urban Planning & Development. These themes collectively explore the impact of the pandemic on urban areas, with specific case studies from Poland and Bangladesh, the rapid urban transformations necessitated by global crises, the changing dynamics of real estate markets, and the critical role of urban planning in fostering economic growth and social development. The book highlights the importance of local governance, sustainable urban forms, and adaptive social practices in building resilient cities. It also discusses the challenges of managing crowd density in temporary events and emphasizes the need for clean drinking water and healthy indoor ventilation. Through varied geographical contexts and perspectives, the work aims to guide policymakers, urban planners, and academics towards creating more resilient, sustainable, and inclusive urban spaces in the aftermath of the pandemic.

Keywords Urban Planning · Pandemic · Post Pandemic Cities · Vulnerable cities · COVID-19 Urban Transformation

1 Introduction

The COVID-19 pandemic has highlighted the vulnerability of our cities and the pressing need to reassess urban planning and development. As the world emerges from the crisis, there is a growing consensus that we must reshape our urban environments to better withstand the challenges of the post-pandemic era. "Making Sense

S. N. M. Moghaddam (✉)
University of Ottawa, Ottawa, ON, Canada
e-mail: navid.mm@uottawa.ca; Smash091@uottawa.ca

K. M. Nahiduzzaman (ed.), *Making Sense of Planning and Development for the Post-Pandemic Cities*, City Development: Issues and Best Practices,
https://doi.org/10.1007/978-981-97-5481-6_1

of Planning and Development for the Post-Pandemic Cities" is a comprehensive exploration of the transformations and adaptations required to make our cities more resilient, sustainable, and equitable. The book is organized into three themes, each addressing a key aspect of post-pandemic urban life.

Theme 1: Urban Infrastructure and Resilience delves into the importance of urban size, functions, and resilience to external impacts. It examines the COVID-19 pandemic's effect on urban areas in Poland and Bangladesh, with a focus on how local governance, urban form, and social practices contribute to resilience.

Theme 2: Dynamics of Urban Transformation and Real Estate Market highlights the rapid changes in urban spaces brought about by the pandemic, and the need for cities to adapt and become more resilient to crises such as natural disasters, violent conflicts, and epidemics. The chapters in this section explore the challenges and complexities surrounding contemporary cities, urban sustainability, and the experiences of air travelers.

They also attempted to investigate the critical effects of the pandemic on cities, with a focus on economic and social environment, mobility patterns, and real estate markets. How transformational changes in public transport services, multimodality, complementary modal choices, hybrid work arrangements, and residential location choices might shape the future cities are the other facets that these chapters attempted to delve into.

Theme 3: Urban Planning and Development offers a comprehensive examination of the critical aspects of urban planning in the post-pandemic world. The chapters in this section discuss the role of municipal planning in driving economic growth and social development, the shortcomings of urban infrastructure, the public transport experiences of the older adults, crowd management in temporary events, and the importance of factors such as social distancing, clean drinking water, and healthy indoor ventilation in pandemic response.

By offering insights from various perspectives and geographical contexts, "Making Sense of Planning and Development for the Post-Pandemic Cities" aims to inform policymakers, urban planners, and academics on the necessary steps to make our cities more resilient, sustainable, and inclusive in the post-pandemic world.

2 Challenges in Urban Spaces During Pandemics

Urban spaces, characterized by their dense populations and diverse socioeconomic structures, are the engines of growth and innovation in modern societies. However, they also present unique challenges during pandemics, as the interconnected nature of urban life can facilitate the rapid spread of infectious diseases and exacerbate the consequences for various aspects of urban living. This section aims to provide a comprehensive understanding of the multifaceted challenges that urban spaces face during pandemics by exploring the following sub-sections:

1. Public Health Concerns: Ensuring the well-being of urban residents during pandemics is of paramount importance, as factors such as high population density and the built environment can contribute to the transmission of infectious diseases. This sub-section will delve into the public health concerns that arise during pandemics in urban spaces, including disease transmission, healthcare infrastructure, vulnerable populations, and the role of the built environment.
2. Accessibility and Mobility Issues: Urban mobility and transportation systems play a crucial role in connecting people to jobs, services, and amenities. However, pandemics can disrupt these systems, necessitating the rethinking of urban transportation to ensure safe, efficient, and accessible mobility options for all residents. This sub-section will explore the challenges and opportunities associated with enhancing mobility and transportation systems during pandemics.
3. Social Distancing and Public Space Usage: The imposition of social distancing measures during pandemics can significantly impact the way public spaces are used in urban environments. This sub-section will discuss the challenges associated with maintaining social distancing in urban spaces and the importance of reimagining the design and function of public spaces to accommodate evolving public health guidelines.
4. Psychological Impact on Residents: The psychological well-being of urban residents can be severely affected during pandemics, as fear of infection, social isolation, and economic hardship can lead to increased stress, anxiety, and depression. This sub-section will explore the mental health challenges faced by urban residents during pandemics and discuss potential strategies for addressing these concerns.

This section aims to provide a comprehensive understanding of the complex challenges faced by urban spaces during pandemics and to contribute to the development of more resilient and adaptable cities in the face of future health crises by examining these sub-sections.

2.1 Public Health Concerns

Public health concerns are of paramount importance when addressing the challenges posed by pandemics in urban spaces. The concentration of people in cities makes them particularly vulnerable to the rapid spread of infectious diseases, and the high population density exacerbates the potential impact of pandemics on various aspects of urban life. In this subsection, we will discuss the most significant public health concerns that arise during pandemics in urban spaces, including disease transmission, healthcare infrastructure, vulnerable populations, inadequate housing and slums, mental health issues, air quality and pollution, food security and nutrition, waste management, health information and communication, and public health preparedness.

2.1.1 Disease Transmission

The transmission of infectious diseases in urban areas is a major concern during pandemics. High population density and the extensive interaction between individuals in public spaces facilitate the rapid spread of infectious diseases, such as the COVID-19 virus [1]. Several factors contribute to the increased risk of disease transmission in urban areas, including crowded living conditions, the use of public transportation, and the presence of large public gatherings [2]. In addition, the built environment, such as the design of buildings and the ventilation systems within them, can also influence the transmission of airborne pathogens [3].

2.1.2 Healthcare Infrastructure

Urban healthcare infrastructure plays a crucial role in responding to pandemics, as it provides essential services for the prevention, diagnosis, treatment, and management of infectious diseases. However, during pandemics, healthcare systems can be quickly overwhelmed due to the rapid surge in demand for medical services and resources [4]. This can lead to the insufficient availability of healthcare facilities, healthcare workers, medical equipment, and supplies (such as personal protective equipment and ventilators), which in turn can compromise the quality and accessibility of healthcare services for the affected population [5].

2.1.3 Vulnerable Populations

Certain groups within urban populations are particularly vulnerable to the adverse effects of pandemics, including the elderly, people with underlying health conditions, low-income individuals, and those living in informal settlements [5]. These vulnerable groups often experience higher rates of morbidity and mortality during pandemics, as they are more susceptible to infection and have limited access to healthcare services [6]. Moreover, the social determinants of health, such as poverty, education, and employment, can exacerbate the health disparities experienced by these vulnerable populations during pandemics [7].

2.1.4 Inadequate Housing and Slums

Inadequate housing and informal settlements (slums) pose significant public health challenges during pandemics. Poor living conditions, such as overcrowding, lack of access to clean water and sanitation, and limited waste management facilities, contribute to the increased risk of infectious disease transmission in these settings [1]. Furthermore, residents of informal settlements often face barriers to accessing healthcare services, including affordability, availability, and geographical accessibility [8]. This can lead to delayed diagnosis and treatment, which in turn can increase the risk of disease transmission and the severity of illness among these populations [1].

2.1.5 Mental Health Issues

Pandemics can have profound effects on the mental health of urban residents, as the fear of infection, social isolation, economic hardship, and the disruption of daily routines can lead to increased levels of stress, anxiety, and depression [9]. Moreover, the closure of schools, workplaces, and recreational facilities, as well as the imposition of lockdowns and social distancing measures, can exacerbate feelings of loneliness and disconnection among urban residents [10]. Mental health services may also be strained during pandemics, as healthcare resources are often diverted to address the immediate needs of managing the outbreak [11]. This can result in reduced access to mental health support and treatment for those in need, further exacerbating mental health challenges in urban spaces [12].

2.1.6 Health Information and Communication

Reliable health information and communication are vital for ensuring effective public health responses during pandemics. Urban residents need timely and accurate information about the risks, symptoms, and preventive measures associated with infectious diseases to make informed decisions about their health and wellbeing [13]. However, the rapid spread of misinformation and disinformation during pandemics can hinder public health efforts and contribute to confusion, fear, and mistrust among urban populations [14]. Ensuring effective communication and countering misinformation are essential for promoting adherence to public health guidelines and mitigating the impact of pandemics on urban spaces [15].

2.1.7 Public Health Preparedness:

Public health preparedness is crucial for managing the challenges posed by pandemics in urban spaces. This includes the development and implementation of comprehensive plans and strategies to prevent, detect, and respond to infectious disease outbreaks [16]. Factors such as insufficient funding, inadequate surveillance and monitoring systems, and limited coordination between different levels of government can undermine public health preparedness in urban areas [17]. Strengthening public health systems and building resilience to future pandemics are essential for protecting the health and wellbeing of urban residents [18].

In conclusion, pandemics pose significant challenges to urban spaces, with public health concerns being at the forefront of these issues. High population density and the concentration of people in cities create environments that are conducive to the rapid spread of infectious diseases, putting immense pressure on healthcare infrastructure and disproportionately affecting vulnerable populations. Additionally, inadequate housing, mental health issues, air quality, food security, waste management, health information, and public health preparedness are all critical aspects that need to be addressed to effectively manage pandemics in urban settings.

Urban planners and policymakers must consider these public health concerns when developing strategies to create pandemic-resilient urban spaces. This includes enhancing healthcare infrastructure, improving living conditions, ensuring access to essential services, addressing mental health issues, mitigating the impacts of air pollution, promoting food security, and implementing effective waste management systems. Moreover, disseminating accurate health information and fostering robust public health preparedness are essential for mitigating the impacts of pandemics on urban residents.

By adopting a comprehensive and interdisciplinary approach to urban planning, cities can build resilience against future pandemics and safeguard the health and wellbeing of their residents. This will not only benefit urban populations during times of crisis but will also contribute to creating healthier, more equitable, and sustainable cities in the long term.

2.2 Accessibility and Mobility Issues

During pandemics, urban spaces face significant challenges related to accessibility and mobility, which can exacerbate existing inequalities and contribute to adverse health, social, and economic outcomes. In this section, we will discuss various aspects of accessibility and mobility issues that arise during pandemics, including reduced public transportation services, impact on non-motorized transport, social distancing and transportation infrastructure, barriers to accessing essential services and employment opportunities, mobility challenges for people with disabilities and the elderly, digital divide and access to information and communication technologies, shifts in travel behavior and demand, and strategies for enhancing mobility and accessibility during pandemics.

Reduced public transportation services are a common challenge during pandemics, as governments and transportation authorities may limit or suspend services to contain the spread of the virus [19]. This can lead to overcrowding and increased wait times for the remaining services, making it difficult for residents to maintain social distancing and increasing the risk of disease transmission [20]. Moreover, reduced transportation services can disproportionately affect low-income and vulnerable populations, who rely more heavily on public transportation and may not have access to alternative modes of transport [21].

The impact on non-motorized transport, such as walking and bicycling, is another important aspect to consider during pandemics. While these modes of transport can offer a safer alternative to public transportation in terms of disease transmission, urban spaces may not be adequately designed to accommodate increased demand and ensure safe travel for pedestrians and cyclists [22]. Moreover, the closure of public spaces and the enforcement of social distancing measures can limit opportunities for walking and cycling as recreational activities, contributing to sedentary lifestyles and adverse health outcomes [23].

Social distancing and transportation infrastructure are closely interrelated during pandemics, as the design and layout of urban spaces can either facilitate or hinder the implementation of effective social distancing measures. Narrow sidewalks, crowded public spaces, and inadequate cycling infrastructure can make it difficult for residents to maintain a safe distance from others and increase the risk of disease transmission [24]. Furthermore, the lack of safe and accessible transportation options can exacerbate existing inequalities and further marginalize vulnerable populations, such as low-income residents, people with disabilities, and the elderly [25].

Barriers to accessing essential services and employment opportunities are another critical challenge during pandemics, as reduced transportation services and social distancing measures can make it more difficult for residents to reach healthcare facilities, grocery stores, and other essential destinations [26]. Moreover, the closure of businesses and the shift to remote work can lead to job losses and reduced income for many residents, particularly those working in the informal sector and in occupations that cannot be performed remotely [27].

Mobility challenges for people with disabilities and the elderly are further exacerbated during pandemics, as these populations may face additional barriers to accessing transportation services and public spaces due to their heightened vulnerability to the virus and potential difficulties in complying with social distancing measures [28]. Moreover, the closure of specialized services and support facilities can further limit their mobility and independence, leading to increased social isolation and adverse mental health outcomes [29].

The digital divide and access to information and communication technologies (ICTs) are also important factors to consider in addressing accessibility and mobility issues during pandemics. With many services and activities shifting online, residents who lack access to digital devices or reliable internet connections may be unable to access essential information, healthcare services, or employment opportunities [30]. This digital divide can disproportionately affect low-income residents, older adults, and those living in underserved areas, further exacerbating existing inequalities and contributing to social exclusion [31].

Shifts in travel behavior and demand during pandemics can have significant implications for urban transportation systems and planning. The fear of disease transmission and the adoption of remote work and online services can lead to a decrease in demand for public transportation and an increase in the use of private vehicles, contributing to traffic congestion and air pollution [32]. Additionally, the increased reliance on e-commerce and home deliveries can lead to a surge in last-mile delivery traffic, further exacerbating congestion and environmental impacts [33].

Strategies for enhancing mobility and accessibility during pandemics should focus on addressing the needs of diverse urban populations and promoting safe, sustainable, and inclusive transportation options. These may include expanding and improving public transportation services, investing in cycling and pedestrian infrastructure, promoting carpooling and ridesharing services, and leveraging digital technologies to facilitate access to information and services [34]. Furthermore, urban planners and policymakers should consider the long-term implications of these shifts in travel behavior and demand, and develop strategies to promote more resilient and adaptable transportation systems in the face of future pandemics and other crises [32].

In conclusion, addressing accessibility and mobility issues during pandemics is essential for ensuring the well-being of urban residents and promoting more resilient and inclusive urban spaces. This requires a comprehensive understanding of the diverse challenges faced by different populations and sectors, as well as the development of innovative and context-sensitive strategies for enhancing mobility and accessibility in the face of uncertainty and change.

2.3 Social Distancing and Public Space Usage

Social distancing and public space usage are critical aspects of urban planning that have been significantly impacted during pandemics. The need to maintain a safe distance from others to curb disease transmission has required a rethinking of how urban spaces are designed and utilized. This section will explore the challenges and implications of social distancing on public space usage in urban settings, drawing on a range of academic literature.

The concept of social distancing has become central to public health strategies during pandemics, such as the COVID-19 crisis [35]. Social distancing refers to maintaining a safe distance between individuals to reduce the risk of disease transmission, with the World Health Organization (WHO) recommending a minimum distance of 1 m (3 feet) between people [36]. This measure has necessitated changes in public space design, usage, and management to accommodate these guidelines and minimize the risk of contagion.

One of the primary challenges of social distancing in urban spaces is the need to balance public health concerns with the desire to maintain social interactions, a vital aspect of urban life [37]. Urban spaces often serve as hubs for social interaction, commerce, and cultural activities. During pandemics, the need to maintain social distancing can disrupt these functions, leading to reduced opportunities for social connection, increased feelings of isolation, and potential negative impacts on mental health [10].

Temporary and adaptive urban design interventions have emerged as a response to these challenges, aiming to create more space for social distancing while preserving the social and economic functions of urban spaces [38]. These interventions have included widening sidewalks, creating temporary pedestrian and bike lanes, and repurposing parking spaces or streets for outdoor dining and recreation [39]. Such measures have been critical in maintaining a sense of community and enabling businesses to continue operating during pandemics.

Urban parks and green spaces have also played a crucial role in providing opportunities for physical activity, relaxation, and social interaction during pandemics [40]. However, the increased demand for these spaces has raised concerns about overcrowding and the potential for disease transmission. Policymakers and urban planners have had to devise strategies for managing park usage, such as implementing capacity limits, timed entry systems, and one-way circulation routes to maintain social distancing and ensure public safety [41].

Supporting local businesses and street life during pandemics has been another significant challenge. With many businesses struggling due to reduced foot traffic and enforced closures, there has been a need for creative solutions to support economic recovery while adhering to social distancing guidelines. Some cities have implemented measures such as expanding outdoor dining, offering financial assistance to small businesses, and promoting local shopping through marketing campaigns [42].

Designing for resilience and adaptability has emerged as a key priority for urban planners in the context of pandemics. This approach involves creating flexible spaces that can be easily reconfigured to accommodate changing needs, such as modular furniture, convertible street spaces, and adaptable building designs [38]. Such strategies can help cities respond more effectively to pandemics and other shocks, ensuring that urban spaces can continue to serve their vital social and economic functions.

Inclusive and participatory decision-making processes have also become increasingly important in addressing the challenges of social distancing and public space usage during pandemics. Engaging diverse stakeholders, including residents, businesses, and community organizations, can help ensure that urban planning decisions reflect the needs and priorities of different groups and promote social equity [43].

In conclusion, the challenges of social distancing and public space usage during pandemics have necessitated a reevaluation of how urban spaces are designed, managed, and utilized. Strategies such as temporary and adaptive urban design interventions, supporting local businesses, and promoting inclusive and participatory decision-making processes have emerged as critical approaches to maintaining social connections and preserving the vitality of urban spaces while protecting public health. By drawing on these insights and engaging in interdisciplinary research and knowledge sharing, urban planners and policymakers can develop more resilient and adaptable urban spaces that are better equipped to cope with pandemics and other future challenges.

2.4 Psychological Impact on Residents

The psychological impact of pandemics on urban residents is a multifaceted issue that extends beyond the immediate health concerns associated with infectious diseases. The COVID-19 pandemic, for instance, has led to widespread feelings of fear, anxiety, stress, and uncertainty, which have been exacerbated by the numerous social, economic, and environmental challenges faced by urban populations during these times [9, 10]. In this sub-section, we will explore the various aspects of the psychological impact of pandemics on urban residents, including mental health consequences, the role of social isolation and loneliness, the effects on children and adolescents, the impacts on frontline workers, and the implications for urban planning and design.

Mental health consequences: The mental health consequences of pandemics have been widely documented, with increased rates of anxiety, depression, post-traumatic stress disorder (PTSD), and substance abuse observed during and after such events [9, 10]. The uncertainty and fear associated with the spread of infectious diseases, coupled with the economic instability, job losses, and other stressors experienced by urban residents during pandemics, contribute to this heightened psychological distress [44]. Additionally, the prolonged duration of pandemics can exacerbate existing mental health issues or lead to the development of new disorders, particularly among vulnerable populations such as the elderly, low-income individuals, and those with pre-existing mental health conditions [9].

Social isolation and loneliness: The implementation of social distancing measures, quarantine protocols, and stay-at-home orders during pandemics has led to increased social isolation and loneliness among urban residents [45]. The loss of social connections, limited access to support networks, and disruptions to daily routines can have profound effects on mental health and well-being, increasing the risk of depression, anxiety, and other mental health disorders [10, 45]. Furthermore, the digital divide in urban areas can exacerbate feelings of isolation for those without access to digital communication tools, creating additional barriers to social interaction and support [46].

Effects on children and adolescents: Pandemics and their associated social and economic disruptions can have significant psychological impacts on children and adolescents [47]. School closures, social isolation, and the loss of extracurricular activities can lead to increased stress, anxiety, and feelings of loneliness among young people [45]. Moreover, children and adolescents from disadvantaged backgrounds may be disproportionately affected by the loss of access to educational and support services during pandemics, further exacerbating existing inequalities [47]. The long-term psychological effects of these disruptions on child and adolescent development are not yet fully understood, but early evidence suggests that they may be substantial [47].

Impacts on frontline workers: Frontline workers, including healthcare professionals, emergency responders, and essential service providers, are at particular risk of experiencing psychological distress during pandemics [48]. The high levels of stress, long working hours, and exposure to traumatic events can contribute to increased rates of burnout, depression, anxiety, and PTSD among these workers [48]. Additionally, the fear of contracting the virus and spreading it to family members may further exacerbate the psychological burden faced by frontline workers during pandemics [48].

Implications for urban planning and design: The psychological impacts of pandemics on urban residents highlight the need for urban planning and design strategies that prioritize mental health and well-being. This includes creating spaces that facilitate social connections while adhering to public health guidelines, ensuring access to essential services and support networks, and promoting equity and inclusiveness in urban environments [49, 50]. For example, the development of green spaces, pedestrian-friendly streets, and mixed-use neighborhoods can encourage social interaction and physical activity, which have been shown to improve mental health outcomes [49, 50]. Additionally, addressing the digital divide in urban areas can help to alleviate feelings of isolation and loneliness by increasing access to digital communication tools and online resources [46].

Urban planners and policymakers should also consider the needs of vulnerable populations, such as children, adolescents, the elderly, and frontline workers, when designing and implementing strategies to mitigate the psychological impacts of pandemics [47, 48]. This might involve developing targeted mental health interventions and support services, as well as ensuring that the urban environment is responsive to the needs and preferences of these groups [47, 48].

In conclusion, the psychological impact of pandemics on urban residents is a complex and multifaceted issue that warrants further attention from urban planners, policymakers, and researchers. By adopting a holistic and inclusive approach to urban planning and design, it is possible to create urban spaces that promote mental health and well-being, foster social connections, and mitigate the negative psychological consequences of pandemics and other public health crises. Urban planning and design strategies should consider the psychological impacts of pandemics on residents by creating spaces that promote mental health, well-being, and social connections while adhering to public health guidelines. This can be achieved by developing green spaces, pedestrian-friendly streets, mixed-use neighborhoods, and ensuring access to essential services and support networks. Addressing the digital divide in urban areas can also help to alleviate feelings of isolation and loneliness by increasing access to digital communication tools and online resources. Furthermore, urban planners and policymakers should consider the needs of vulnerable populations, such as children, adolescents, the elderly, and frontline workers, when designing and implementing strategies to mitigate the psychological impacts of pandemics. This might involve developing targeted mental health interventions and support services, as well as ensuring that the urban environment is responsive to the needs and preferences of these groups.

3 Urban Planning Strategies for Pandemic-Resilient Urban Spaces

Urban spaces, with their high population density and interconnected nature, are particularly susceptible to the impacts of pandemics. The unprecedented challenges brought forth by global health crises, such as the COVID-19 pandemic, have highlighted the need for rethinking urban planning strategies to create more resilient and adaptive urban spaces. This section aims to provide insights into innovative urban planning approaches and strategies that can help mitigate the effects of pandemics and enhance the overall resilience of urban environments. The following sub-sections will be discussed in detail:

1. Flexible and Adaptive Public Spaces: As social distancing and other public health guidelines reshape the way public spaces are used during pandemics, it is essential to consider how urban design can accommodate these changing needs. This sub-section will explore strategies for creating flexible and adaptive public spaces that can support various functions and activities while adhering to evolving health recommendations.
2. Strengthening Urban Green Infrastructure: Green spaces play a crucial role in enhancing urban resilience and promoting public health. This sub-section will delve into the importance of urban green infrastructure during pandemics, discussing the benefits of integrating nature-based solutions in urban planning and management.
3. Enhancing Mobility and Transportation Systems: Efficient and accessible mobility is vital for urban spaces, connecting people to jobs, services, and amenities. This sub-section will discuss the challenges and opportunities associated with enhancing mobility and transportation systems during pandemics, exploring the potential for innovative solutions that prioritize safety, sustainability, and equity.
4. Encouraging Mixed-Use Development and Neighborhood-level Self-sufficiency: Mixed-use development and neighborhood-level self-sufficiency can contribute to more resilient and adaptable urban spaces during pandemics. This sub-section will discuss the benefits of promoting mixed-use development and fostering neighborhood-level self-sufficiency, highlighting the role of urban planning in creating vibrant, sustainable, and pandemic-resilient communities.

By exploring these sub-sections, this section aims to provide a comprehensive understanding of the urban planning strategies that can help create pandemic-resilient urban spaces, ultimately contributing to healthier, more sustainable, and adaptive cities in the face of future health crises.

3.1 Flexible and Adaptive Public Spaces

Flexible and adaptive public spaces play a crucial role in promoting urban resilience and livability during pandemics. The ability to quickly adapt and reconfigure public spaces to meet changing demands can help cities maintain their social and economic vibrancy while protecting public health [51]. This section will discuss key concepts and principles of flexible and adaptive public spaces, and explore a range of innovative urban design interventions and strategies that have emerged in response to the COVID-19 pandemic.

Flexibility and adaptability in public spaces refer to the ability of urban environments to accommodate a wide range of activities, user groups, and changing circumstances [52]. This can involve the use of modular, multi-functional, and reconfigurable design elements, as well as the incorporation of temporary, pop-up, or event-based installations and programming [53]. By creating public spaces that can be easily adapted and repurposed, cities can better respond to the diverse and evolving needs of their residents, while also enhancing their capacity to cope with shocks and stressors, such as pandemics.

Several key principles underpin the design and management of flexible and adaptive public spaces:

- Multi-functionality: Designing spaces that can accommodate a wide variety of activities and user groups, both concurrently and at different times [52].
- Modularity: Incorporating elements and infrastructure that can be easily assembled, disassembled, and rearranged to create different spatial configurations and uses [53].
- Temporality: Embracing temporary, pop-up, and event-based interventions as a means of testing and experimenting with new ideas, fostering innovation, and engaging communities in the process of urban transformation [53].
- Responsiveness: Ensuring that public spaces can quickly adapt and respond to changing needs, preferences, and circumstances, including public health crises and other emergencies [51].

The COVID-19 pandemic has prompted a range of innovative urban design interventions and strategies aimed at creating more flexible and adaptive public spaces. These interventions can be grouped into several broad categories:

- Expanding and repurposing public spaces: Many cities have sought to increase the amount of available public space by repurposing streets, parking lots, and other underutilized areas for pedestrian and recreational use [54]. This has involved the temporary or permanent closure of streets to vehicular traffic, the installation of parklets and pedestrian plazas, and the conversion of parking spaces into outdoor dining areas, among other measures [54].
- Implementing modular and reconfigurable design elements: Cities have also experimented with modular and reconfigurable design elements, such as moveable seating, planters, and barriers, to create flexible and adaptable public spaces that can be easily reconfigured to accommodate different activities and social distancing requirements [55].
- Encouraging temporary and pop-up installations: Temporary and pop-up installations, such as outdoor art exhibits, performances, and markets, have been used to activate public spaces and support local businesses during the pandemic [55]. These interventions can help to foster a sense of community and social cohesion, while also providing opportunities for economic recovery and resilience [55].
- Adopting digital and smart city technologies: Digital and smart city technologies, such as sensors, data analytics, and mobile applications, can be used to monitor and manage the use of public spaces in real-time, helping to ensure that social distancing and other public health guidelines are being followed [56]. For example, cities like Barcelona and Amsterdam have implemented smart crowd

management systems to monitor the flow of pedestrians in public spaces and provide real-time information to residents about the safest and least congested routes [56].

As cities continue to grapple with the challenges posed by pandemics, it is essential to draw upon the lessons learned and best practices that have emerged from the experiences of urban areas around the world. In doing so, policymakers, urban planners, and community stakeholders can work together to create more resilient, inclusive, and sustainable public spaces that are better equipped to withstand future crises.

Several lessons and best practices have emerged from the experiences of cities in implementing flexible and adaptive public spaces during the COVID-19 pandemic:

- Collaboration and partnership: Successful implementation of these interventions often requires close collaboration between various stakeholders, including local governments, businesses, community organizations, and residents [54]. By working together, these stakeholders can pool resources, share expertise, and engage in creative problem-solving to address the complex challenges posed by pandemics [54].
- Experimentation and iteration: The pandemic has highlighted the value of adopting an experimental and iterative approach to urban design and planning, as cities have had to rapidly adapt and innovate in response to the evolving crisis [55]. By embracing temporary, modular, and adaptable interventions, cities can test and refine new ideas, learn from successes and failures, and build a more robust evidence base for future planning and decision-making [55].
- Equity and inclusiveness: It is crucial to ensure that flexible and adaptive public spaces are designed and managed in ways that promote equity and inclusiveness, particularly for vulnerable and marginalized populations who may be disproportionately impacted by pandemics [51]. This may involve prioritizing investments in underserved neighborhoods, conducting inclusive and participatory planning processes, and ensuring that public spaces are accessible and welcoming to people of all ages, abilities, and backgrounds [51].
- Integrating public health considerations into urban design and planning: Public health should be a central concern in the design and management of public spaces, particularly in the context of pandemics [51]. This may involve incorporating features that promote physical distancing, hygiene, and ventilation, as well as ensuring that public spaces can accommodate a range of health-related activities and services, such as testing and vaccination clinics [51].
- Strengthening community engagement and ownership: Engaging local communities in the planning, design, and management of public spaces can help to ensure that these spaces are more responsive to local needs, preferences, and cultural practices [52]. This can involve conducting participatory planning workshops, crowdsourcing ideas and feedback, and involving local residents in the implementation and stewardship of public space projects [52].
- Leveraging partnerships and innovative financing models: The creation and maintenance of flexible and adaptive public spaces often require significant financial

resources, particularly in the context of pandemics when municipal budgets may be strained [54]. By forging partnerships with the private sector, philanthropic organizations, and other stakeholders, cities can tap into new sources of funding and expertise, as well as promote more innovative and cost-effective approaches to public space provision and management [54].

Ultimately, the development and implementation of flexible and adaptive public spaces can serve as a vital strategy for enhancing urban resilience and livability in the face of pandemics and other global challenges. By embracing innovative design interventions, fostering collaboration and partnership, and prioritizing equity and inclusiveness, cities can create public spaces that not only protect public health but also foster social cohesion, economic vitality, and environmental sustainability.

3.2 Strengthening Urban Green Infrastructure

Urban green infrastructure (UGI) refers to an interconnected network of green spaces, including parks, gardens, green roofs, street trees, and natural habitats, which provide a range of environmental, social, and economic benefits to urban residents [57]. In the context of pandemics, UGI can play a crucial role in enhancing urban resilience by promoting physical and mental health, facilitating social distancing, and mitigating the impacts of climate change and other environmental stressors [58]. This section will discuss the importance of UGI in fostering pandemic-resilient urban spaces and provide evidence-based recommendations for strengthening UGI in urban planning and design.

UGI can contribute to pandemic resilience in several ways. First, access to green spaces has been shown to have significant positive effects on physical and mental health, which can be especially important during periods of increased stress and uncertainty [59]. For instance, green spaces can provide opportunities for physical activity, which can help to reduce the risk of chronic diseases, boost immune function, and improve overall well-being [60]. Furthermore, exposure to natural environments can promote psychological restoration and stress reduction, helping to alleviate the mental health impacts of pandemics [59].

Second, UGI can facilitate social distancing by providing additional outdoor spaces for people to gather and engage in activities while maintaining physical distance [58]. During the COVID-19 pandemic, many cities around the world have witnessed a surge in the use of parks and other green spaces, as people sought safe and accessible environments for recreation, socialization, and relaxation [61]. By expanding and diversifying the provision of green spaces, cities can ensure that residents have adequate access to outdoor areas during pandemics, reducing the risk of overcrowding and transmission of infectious diseases [58].

Finally, UGI can help to mitigate the impacts of climate change and other environmental stressors, which can exacerbate the challenges faced by cities during pandemics [57]. For example, green spaces can reduce urban heat island effects, improve air quality, and promote stormwater management, thus helping to create healthier and more resilient urban environments [57]. In this way, UGI can contribute to the overall resilience of cities, supporting their capacity to withstand and recover from various shocks and stressors, including pandemics.

Given the importance of UGI in fostering pandemic-resilient urban spaces, the following strategies can be employed to strengthen UGI in urban planning and design:

Prioritizing equitable access to green spaces: Ensuring that all urban residents have equitable access to green spaces is critical for promoting public health and social cohesion during pandemics [62]. To achieve this, cities can conduct spatial analyses to identify areas with limited access to green spaces, prioritize investments in these areas, and incorporate social equity considerations into their planning and decision-making processes [62].

Creating multifunctional and adaptable green spaces: Designing green spaces that can serve multiple functions and adapt to changing needs and circumstances can help cities better respond to pandemics and other crises [63]. For example, parks can be designed to accommodate a range of activities, such as exercise, relaxation, and socialization, while also providing space for emergency services or temporary health facilities if needed [63]. Incorporating elements such as modular furniture, movable planters, and flexible programming can further enhance the adaptability of green spaces.

Promoting nature-based solutions: Nature-based solutions, such as green roofs, rain gardens, and urban forests, can contribute to the overall resilience of cities by providing ecosystem services and improving environmental quality [64]. By integrating nature-based solutions into urban planning and design, cities can strengthen their green infrastructure networks and enhance their capacity to cope with pandemics and other stressors.

Engaging communities in the planning and management of green spaces: Involving local communities in the planning, design, and management of green spaces can help to ensure that these spaces meet the diverse needs of residents and foster a sense of ownership and stewardship [65]. Community engagement can also facilitate the sharing of local knowledge and the identification of innovative solutions for enhancing the resilience of green spaces during pandemics.

Establishing partnerships and collaboration: Strengthening UGI requires collaboration among various stakeholders, including government agencies, private sector actors, non-governmental organizations, and local communities [66]. Establishing partnerships and fostering cooperation can help to mobilize resources, share knowledge and expertise, and coordinate efforts to enhance the resilience of green spaces during pandemics.

In conclusion, strengthening urban green infrastructure is essential for creating pandemic-resilient urban spaces that support the physical and mental well-being of residents, facilitate social distancing, and mitigate the impacts of climate change and other environmental stressors. By prioritizing equitable access to green spaces, creating multifunctional and adaptable green spaces, promoting nature-based solutions, engaging communities in the planning and management of green spaces, and establishing partnerships and collaboration, cities can enhance their capacity to cope with pandemics and foster more resilient urban environments. Incorporating these strategies into urban planning and design processes will contribute to the development of more sustainable, healthy, and equitable cities in the face of ongoing and future challenges.

3.3 Enhancing Mobility and Transportation Systems

The COVID-19 pandemic has had a profound impact on urban mobility and transportation systems worldwide, with reduced public transportation services, decreased ridership, and increased reliance on private vehicles and alternative modes of transportation, such as walking and cycling [20]. To create pandemic-resilient urban spaces, it is essential to develop strategies that enhance mobility and transportation systems, ensure accessibility, and promote safety and sustainability [21]. This section will discuss various approaches to enhancing mobility and transportation systems, including improving public transportation services, promoting non-motorized transportation, leveraging technology for contactless travel, and developing integrated transportation and land-use planning strategies.

Public transportation plays a critical role in ensuring access to essential services and employment opportunities, particularly for vulnerable populations who may not have access to private vehicles [22]. During pandemics, it is important to maintain and improve public transportation services while ensuring the safety and health of passengers and staff. Strategies to achieve this include increasing service frequency to reduce overcrowding, implementing enhanced cleaning and sanitation protocols, requiring passengers to wear face masks, and promoting physical distancing through measures such as seating arrangements and capacity limits [21]. Additionally, improving communication and real-time information systems can help passengers make informed decisions about their travel plans and avoid crowded vehicles [20].

The pandemic has led to a surge in demand for non-motorized transportation modes, such as walking and cycling, which offer a safe and sustainable alternative to public transportation and private vehicles [22]. Urban planners can promote non-motorized transportation by developing and expanding pedestrian and bicycle infrastructure, such as dedicated bike lanes, widened sidewalks, and traffic-calming measures [67]. Temporary and adaptive urban design interventions, such as "open streets" programs that repurpose street space for pedestrians and cyclists, can also be implemented to support social distancing and encourage active travel during pandemics [32]. Furthermore, providing bike-sharing services and secure bicycle parking facilities can help increase the accessibility and attractiveness of non-motorized transportation options [67].

Technology can play a significant role in enhancing mobility and transportation systems during pandemics by facilitating contactless travel and reducing the risk of disease transmission [21]. Examples of technology-based solutions include contactless fare payment systems, which can minimize physical contact between passengers and staff, and real-time passenger information systems that provide updates on vehicle occupancy levels and waiting times, enabling passengers to avoid crowded vehicles and plan their journeys more effectively [20]. Additionally, digital platforms and mobile applications can be used to support contact tracing efforts and disseminate public health information related to transportation [21].

To create pandemic-resilient urban spaces, it is essential to adopt integrated transportation and land-use planning strategies that promote accessibility, reduce the need for travel, and support mixed-use development and neighborhood-level self-sufficiency [22]. These strategies can include transit-oriented development (TOD), which encourages the development of compact, walkable, and mixed-use neighborhoods around public transportation hubs, and the "15-min city" concept, which aims to ensure that residents can access essential services and amenities within a 15-min walk or bike ride from their homes [68, 69]. By reducing travel distances and promoting active transportation modes, these approaches can help enhance urban resilience and sustainability during pandemics and beyond.

Another essential aspect of integrated transportation and land-use planning strategies is promoting accessibility and equity. This can be achieved by prioritizing investments in public transportation services and non-motorized transportation infrastructure in underserved areas and ensuring that all residents have access to affordable and reliable mobility options [22]. Additionally, incorporating participatory planning processes can help ensure that the needs and preferences of diverse urban populations, including vulnerable groups such as low-income households and people with disabilities, are considered in the development and implementation of transportation and land-use policies [32].

As urban planners and policymakers address the challenges posed by pandemics, they should also consider the long-term implications of their interventions on the overall transportation system. This includes evaluating how investments in public transportation, non-motorized transportation infrastructure, and technology can contribute to improving urban resilience and sustainability in the long run [22].

Furthermore, the COVID-19 pandemic has underscored the need for better data collection, monitoring, and evaluation systems to inform transportation planning and decision-making. This can help ensure that mobility and transportation strategies are evidence-based and adaptive to changing conditions and priorities [21]. For instance, real-time data on public transportation ridership, traffic congestion, air quality, and other indicators can be used to evaluate the performance of different transportation interventions and inform the allocation of resources and investments.

Lastly, the pandemic has highlighted the importance of fostering collaboration and knowledge sharing among various stakeholders, including governments, private sector actors, civil society organizations, and researchers. This can help promote the development and dissemination of innovative mobility solutions, best practices, and lessons learned from different contexts, thereby enhancing the effectiveness of transportation planning and management efforts during pandemics and other crises [32].

In conclusion, enhancing mobility and transportation systems in the face of pandemics requires a comprehensive and forward-looking approach that prioritizes accessibility, equity, and sustainability. By adopting evidence-based strategies, leveraging technology, and fostering collaboration and knowledge sharing, urban planners and policymakers can create pandemic-resilient urban spaces that promote long-term well-being and resilience for all residents.

3.4 Encouraging Mixed-Use Development and Neighborhood-Level Self-Sufficiency

The COVID-19 pandemic has exposed the vulnerabilities of contemporary urban spaces and highlighted the need for more resilient and adaptive urban planning strategies. One such strategy is the promotion of mixed-use development and neighborhood-level self-sufficiency, which can contribute to increased urban resilience during pandemics and other crises [70]. This sub-section will discuss the benefits of mixed-use development and neighborhood-level self-sufficiency, along with potential challenges and policy recommendations for implementing these strategies.

Mixed-use development refers to the integration of various land uses, such as residential, commercial, and recreational spaces, within a single neighborhood or district. This approach to urban planning can provide several benefits during pandemics and other crises, including increased accessibility to essential goods and services, reduced travel demand, and enhanced social cohesion and community resilience [71]. In the context of the COVID-19 pandemic, mixed-use development has been shown to support more flexible and adaptive urban spaces, allowing residents to access essential services, such as grocery stores and healthcare facilities, within walking or biking distance from their homes [72].

Neighborhood-level self-sufficiency, on the other hand, involves the decentralization of urban functions and the promotion of local production and consumption patterns. This can be achieved through various strategies, such as the establishment of community gardens, local food networks, and decentralized energy production systems, which can contribute to increased urban resilience during pandemics and other crises [73]. For instance, during the COVID-19 pandemic, neighborhood-level self-sufficiency initiatives have been shown to support local food security, enhance community resilience, and promote social cohesion, particularly in low-income and vulnerable communities [74].

However, promoting mixed-use development and neighborhood-level self-sufficiency also entails various challenges and potential trade-offs, particularly in the context of existing urban planning policies and practices. For example, conventional zoning regulations often favor single-use development patterns and can hinder the implementation of mixed-use and self-sufficient neighborhoods [70]. Moreover, promoting mixed-use development and neighborhood-level self-sufficiency may require substantial investments in infrastructure, such as public transportation, green spaces, and community facilities, which can be particularly challenging for resource-constrained municipalities and developing countries [72].

To overcome these challenges and promote mixed-use development and neighborhood-level self-sufficiency as part of pandemic-resilient urban spaces, several policy recommendations can be considered:

1. Review and revise zoning regulations and building codes to support mixed-use development and neighborhood-level self-sufficiency, by allowing for the integration of various land uses and the implementation of local production and consumption initiatives [71].
2. Develop and implement incentives and financing mechanisms to encourage private sector investment in mixed-use and self-sufficient neighborhoods, such as tax credits, density bonuses, and public–private partnerships [72].
3. Strengthen urban planning capacity and technical expertise at the local level, to support the design, implementation, and monitoring of mixed-use and self-sufficient neighborhoods [70].
4. Foster community engagement and participatory planning processes, to ensure that mixed-use and self-sufficient neighborhoods respond to local needs and preferences, and promote social cohesion and resilience [73].
5. Encourage interdisciplinary research and knowledge sharing, to advance our understanding of the benefits, challenges, and best practices associated with mixed-use development and neighborhood-level self-sufficiency, particularly in the context of pandemics and other crises [74].
6. Collaborate with international organizations, national governments, and other stakeholders, to develop and disseminate guidelines, toolkits, and case studies on mixed-use development and neighborhood-level self-sufficiency, as part of a comprehensive urban planning framework for pandemic-resilient urban spaces [72].
7. Monitor and evaluate the implementation and impacts of mixed-use and self-sufficient neighborhoods, through the development of appropriate indicators and data collection systems, to inform evidence-based policy making and urban planning practices [70].

In conclusion, promoting mixed-use development and neighborhood-level self-sufficiency can contribute to increased urban resilience during pandemics and other crises, by enhancing accessibility to essential goods and services, reducing travel demand, and fostering social cohesion and community resilience. However, implementing these strategies requires overcoming various challenges and trade-offs, particularly in the context of existing urban planning policies and practices. By adopting a comprehensive and interdisciplinary approach, and by fostering collaboration among various stakeholders, mixed-use development and neighborhood-level self-sufficiency can become key components of pandemic-resilient urban spaces.

4 Structure of the Book

The global COVID-19 pandemic has significantly impacted urban planning and development, leading to a re-evaluation of the ways cities function and adapt to external shocks. This book, "Making Sense of Planning and Development for the Post-Pandemic Cities," aims to delve into the challenges, opportunities, and transformations brought about by the pandemic in different urban contexts. The book is divided into four themes, each covering various aspects of urban planning, infrastructure, transformation, and development in the aftermath of the pandemic.

Theme 1: Urban Infrastructure and Resilience

This theme discusses the role of infrastructure in enhancing urban resilience in the face of pandemics and other external shocks. Chapter 1, by Ewa Korcelli-Olejniczak, focuses on small towns in North-Eastern Poland and examines the relationship between urban size, functions, and resilience to external impacts, especially during the COVID-19 pandemic. The study compares pre- and post-pandemic research to understand the contribution of local governance, urban form, and social practices to resilience.

Chapter 2, by Amena Jahan Urmy, Sadia Binte Rahman and Abue Jawfore Taufique Ahamed Ahade assess the impact of the pandemic on real estate and health infrastructure in urban Bangladesh. It discusses the legal aspects of infrastructure resilience, analyzes current laws and policies, and compares global infrastructure examples to suggest how Bangladesh can overcome its infrastructure shortcomings and become more resilient to future pandemics.

Theme 2: Dynamics of Urban Transformation and Real Estate Market

The second theme delves into the transformations that urban spaces have undergone during the pandemic. As well, it addresses the transformational changes in mobility and the real estate market in post-pandemic cities. Chapter 3, by Sujit Kumar Sikder, Md Moynul Ahsan, Hassan Radoine, Omar Al-Mahadi and Kh Md Nahiduzzaman, explores the changes in urban spatial transformation during the post-COVID era in both developed and developing nations. The chapter highlights the effects of the pandemic on private and public spaces, emphasizing the need for cities to become resilient and adaptable to crises.

In Chapter 4, Wail Bakht introduces the Sustainable Urban Transformation Framework (SUTF) and the Paradoxical Challenges Model (PCM) to address the challenges and complexities of urban sustainability. The author proposes a critical, holistic, and realistic perspective on urban planning and public policy domains.

Chapter 5, by Johannes Bhanye, Mareli Hugo, Abraham Matamanda, and Kgosi Mocwagae, investigates airport travelers' experiences in post-pandemic Africa, proposing practical solutions to improve these experiences. The study reveals negative experiences and high uncertainty among travelers, concluding that enhancing passenger experience should be at the core of airport improvement initiatives in Africa.

Uchendu Eugene Chigbu, in Chapter 6, presents a qualitative urban community evaluation of life in Windhoek, Namibia, post-COVID, framing the pandemic as a natural or human disaster in an urban setting. The chapter prescribes measures for making COVID-stressed cities more resilient to disease disasters.

Chapter 7, by Tiziana Campisi, Talha Tasnim and Kh Md Nahiduzzaman, discusses the effects of the COVID-19 pandemic on cities, specifically the economic and social environment, mobility patterns, and real estate market. The chapter seeks to understand the potential impacts of these changes on contemporary land uses and theorizes how these trends might shape future cities.

Theme 3: Urban Planning and Development

This theme focuses on critical aspects of urban planning and development in the context of the pandemic. Chapter 8, by EM Lentsoane and George Onatu, examines the role of development incentives in South African local government and administration. The chapter investigates the potential of development incentives to attract and retain investment in specific geographic areas and drive economic growth and social development.

In Chapter 9, Arun Baby Wilson and Jayprakash Chadchan compare the pandemic response in urban and rural regions of India, highlighting the shortcomings of urban infrastructure and the need to revisit planning and deployment strategies. The authors emphasize the importance of strengthening the network of medical infrastructure and integrating local/regional concepts in urban planning to better address future pandemic situations.

Chapter 10, by Muhammad Ahmad Al-Rashid, Tiziana Campisi, Muhammad Nadeem, and Bilal Saghir, investigates older adults' experiences of social exclusion in Pakistan during the COVID-19 pandemic. The research identifies restricted access to essential resources, lack of participation, and poor social relations and resources as the main themes. The findings are going to help policymakers and social planners develop socially inclusive aging communities.

Jiyon Shin, Reazul Ahsan, and Soumaya Ezazaa in Chapter 11 examine planning methods for managing crowds at temporary events in limited spaces. The chapter analyzes a crowd crush incident in Seoul, South Korea, and reviews relevant crowd-control measures from other contexts. The aim of the chapter is to provide insights and guidelines for planning safe crowd-raising occasions in constrained spaces.

Finally, Chapter 12 by Kazi Humayun Kabir and Md Ayatullah Khan investigates the relationship between built environment attributes and the COVID-19 pandemic in high-density Asian cities. The chapter argues that reducing urban density is not essential to combat outbreaks and presents evidence from Hong Kong as a prime example of high-density, mixed-use urban planning. It discusses the importance of factors such as social distancing, clean drinking water, and healthy indoor ventilation in pandemic response, and outlines short-term, mid-term, and long-term guidelines for urban planning in the post-pandemic world.

In conclusion, this book brings together a collection of chapters that examine various aspects of urban planning, development, and transformation in the post-pandemic era. By analyzing case studies from different countries and regions, the book aims to provide insights, lessons, and recommendations for policymakers, planners, and researchers to shape resilient, sustainable, and adaptable cities in the face of future pandemics and other external shocks.

References

1. Corburn J, Vlahov D, Mberu B, Riley L, Caiaffa WT, Rashid SF, Ko A, Patel S, Jukur S, Martinez-Herrera E, Jayasinghe S (2020) Slum health: arresting COVID-19 and improving well-being in urban informal settlements. J Urban Health 97(3):348–357. https://doi.org/10.1007/s11524-020-00438-6
2. Rocklöv J, Sjödin H (2020) High population densities catalyze the spread of COVID-19. J Travel Med 27(3):taaa038. https://doi.org/10.1093/jtm/taaa038
3. Morawska L, Tang JW, Bahnfleth W, Bluyssen PM, Boerstra A, Buonanno G, Cao J, Dancer S, Floto A, Franchimon F, Haworth C (2020) How can airborne transmission of COVID-19 indoors be minimised? Environ Int 142:105832. https://doi.org/10.1016/j.envint.2020.105832
4. Chatterjee P, Anand T, Singh KJ, Rasaily R, Singh R, Das S, Singh H, Praharaj I, Gangakhedkar RR, Bhargava B, Panda S (2020) Healthcare workers & SARS-CoV-2 infection in India: a case-control investigation in the time of COVID-19. Indian J Med Res 151(5):459. https://doi.org/10.4103/ijmr.IJMR_2234_20
5. Kumar S, Nayar KR (2020) COVID 19 and its mental health consequences. J Ment Health 29(1):1–2. https://doi.org/10.1080/09638237.2020.1757052
6. Shadmi E, Chen Y, Dourado I, Faran-Perach I, Furler J, Hangoma P, Hanvoravongchai P, Obando C, Willems S, Petrosyan V, Rao KD, Ruano AL (2020) Health equity and COVID-19: global perspectives. Int J Equity Health 19(1):1–16. https://doi.org/10.1186/s12939-020-01218-z
7. Bambra C, Riordan R, Ford J, Matthews F (2020) The COVID-19 pandemic and health inequalities. J Epidemiol Community Health 74(11):964–968. https://doi.org/10.1136/jech-2020-214401
8. Ezeh A, Oyebode O, Satterthwaite D, Chen YF, Ndugwa R, Sartori J, Mberu B, Melendez-Torres GJ, Haregu T, Watson SI, Caiaffa W (2017) The history, geography, and sociology of slums and the health problems of people who live in slums. The Lancet 389(10068):547–558. https://doi.org/10.1016/S0140-6736(16)31650-6
9. Pfefferbaum B, North CS (2020) Mental health and the COVID-19 pandemic. N Engl J Med 383(6):510–512. https://doi.org/10.1056/NEJMp2008017
10. Brooks SK, Webster RK, Smith LE, Woodland L, Wessely S, Greenberg N, Rubin GJ (2020) The psychological impact of quarantine and how to reduce it: rapid review of the evidence. The Lancet 395(10227):912–920. https://doi.org/10.1016/S0140-6736(20)30460-8
11. Torales J, O'Higgins M, Castaldelli-Maia JM, Ventriglio A (2020) The outbreak of COVID-19 coronavirus and its impact on global mental health. Int J Soc Psychiatry 66(4):317–320. https://doi.org/10.1177/0020764020915212
12. Rajkumar RP (2020) COVID-19 and mental health: a review of the existing literature. Asian J Psychiatr 52:102066. https://doi.org/10.1016/j.ajp.2020.102066
13. Okereke M, Ukor NA, Adebisi YA, Ogunkola IO, Favour Iyagbaye E, Adiela Owhor G, Lucero-Prisno III DE (2020) COVID-19 misinformation: mere harmless delusions or much more? A knowledge and attitude cross-sectional study among the general public residing in Nigeria. Pan Afr Med J 37(Suppl 1):53. https://doi.org/10.11604/pamj.supp.2020.37.1.25193
14. Pennycook G, Rand DG (2020) Fighting misinformation on social media using crowdsourced judgments of news source quality. Proc Natl Acad Sci 117(6):2775–2783. https://doi.org/10.1073/pnas.1912444117
15. Lewandowsky S, Cook J, Schmid P, Holford DL, Finn A, Leask J, Thomson A, Lombardi D, Al-Rawi AK, Amazeen MA, Anderson EC, Vraga EK (2020) The COVID-19 vaccine

communication handbook: a practical guide for improving vaccine communication and fighting misinformation. https://sks.to/c19vax
16. Kruk ME, Myers M, Varpilah ST, Dahn BT (2020) What is a resilient health system? Lessons from Ebola. The Lancet 385(9980):1910–1912. https://doi.org/10.1016/S0140-6736(15)60755-3
17. Watson C, Cicero A, Blumenstock J, Fraser M (2020) A national plan to enable comprehensive COVID-19 case finding and contact tracing in the US. Johns Hopkins Bloomberg School of Public Health. https://www.centerforhealthsecurity.org/our-work/pubs_archive/pubs-pdfs/2020/200410-national-plan-to-contact-tracing.pdf
18. Kluge HHP, Wickramasinghe K, Rippin HL, Mendes R, Peters DH, Kontsevaya A, Breda J (2020) Prevention and control of non-communicable diseases in the COVID-19 response. The Lancet 395(10238):1678–1680. https://doi.org/10.1016/S0140-6736(20)31067-9
19. Lai S, Ruktanonchai NW, Zhou L, Prosper O, Luo W, Floyd JR, Wesolowski A, Santillana M, Zhang C, Du X, Yu H (2020) Effect of non-pharmaceutical interventions to contain COVID-19 in China. Nature 585(7825):410–413
20. Tirachini A, Cats O (2020) COVID-19 and public transportation: current assessment, prospects, and research needs. J Public Transp 22(1):1–21
21. Gkiotsalitis K, Cats O (2021) Public transport planning adaption under the COVID-19 pandemic crisis: literature review of research needs and directions. Transp Rev 41(4):492–510
22. De Vos J (2020) The effect of COVID-19 and subsequent social distancing on travel behavior. Transportation Research Interdisciplinary Perspectives 5:100121
23. Knell G, McGill M, Churchwell K, Connelly KA, Spertus JA (2020) American Heart Association Council on Epidemiology and Prevention. Circulation 142(17):e266–e268
24. Bucsky P (2020) Modal share changes due to COVID-19: the case of Budapest. Transportation Research Interdisciplinary Perspectives 8:100141
25. Paez A, Lopez FA, Menezes T (2020) Accessibility, equity, and the journey to work during the COVID-19 pandemic: the case of Toronto. Transp Policy 99:95–104
26. Sy KTL, White L, Nichols M (2020) Population density and basic reproductive number of COVID-19 across United States counties. PLoS ONE 15(10):e0240395
27. Béland D, Marier P, Rocco P, Waddan A (2020) COVID-19 and long-term care policy for older people in Canada. J Aging Soc Policy 32(4–5):358–364
28. World Health Organization (2020) Disability considerations during the COVID-19 outbreak. https://www.who.int/publications/i/item/WHO-2019-nCoV-Disability-2020-1
29. Van Dorn A, Cooney RE, Sabin ML (2020) COVID-19 exacerbating inequalities in the US. The Lancet 395(10232):1243–1244
30. Hilbert M (2020) Digital technologies for social change: the digital divide limits the potential of digital social innovation. Int J Commun 14:3561–3578
31. Niles MT, Bertmann F, Belarmino EH, Wentworth T, Biehl E, Neff R (2020) The early food insecurity impacts of COVID-19. Nutrients 12(7):2096
32. Gössling S, Scott D, Hall CM (2020) Pandemics, tourism, and global change: a rapid assessment of COVID-19. J Sustain Tour 29(1):1–20
33. Wang G, Zhang Y, Zhao J, Zhang J, Jiang F (2020) Mitigate the effects of home confinement on children during the COVID-19 outbreak. The Lancet 395(10228):945–947
34. Cervero R, Kockelman K (1997) Travel demand and the 3Ds: density, diversity, and design. Transp Res Part D: Transp Environ 2(3):199–219
35. Centers for Disease Control and Prevention (CDC) (2020) Social distancing. https://stacks.cdc.gov/view/cdc/90522
36. World Health Organization (WHO) (2020) Coronavirus disease (COVID-19) advice for the public. https://www.who.int/emergencies/diseases/novel-coronavirus-2019/advice-for-public
37. Buckley L (2020) Socially distant cities: the challenge of the COVID-19 pandemic for urban life. City Community 19(2):300–304
38. Sennett R (2020) The pandemic and public space. Project MUSE 48(1):5–10
39. National Association of City Transportation Officials (NACTO) (2020) Streets for Pandemic Response & Recovery. https://nacto.org/publication/streets-for-pandemic-response-recovery/

40. Ugolini F, Massetti L, Calaza-Martínez P, Cariñanos P, Dobbs C, Ostoić SK, Marin AM, Pearlmutter D, Saaroni H, Šaulienė I, Simoneti M (2020) Effects of the COVID-19 pandemic on the use and perceptions of urban green space: an international exploratory study. Urban Forestry & Urban Greening 56:126888
41. Rice WL, Pan B, Satchwell C, Zambrano-Monserrate MA, Arellano CR (2020) Changes in park visitation during the COVID-19 pandemic: a natural experiment in Maryland. Int J Environ Res Public Health 17(22):8450
42. Glaeser EL, Gorback C, Redding SJ (2020) How much does COVID-19 increase with mobility? evidence from New York and four other U.S. cities. NBER Working Paper No. 27519
43. Anguelovski I, Connolly JJT, Masip L, Pearsall H (2020) Expanding the boundaries of justice in urban greening scholarship: toward an emancipatory, antisubordination, intersectional, and relational approach. Ann Am Assoc Geogr 110(6):1766–1784
44. Holmes EA, O'Connor RC, Perry VH, Tracey I, Wessely S, Arseneault L, Ballard C, Christensen H, Silver RC, Everall I, Ford T (2020) Multidisciplinary research priorities for the COVID-19 pandemic: a call for action for mental health science. The Lancet Psychiatry 7(6):547–560
45. Loades ME, Chatburn E, Higson-Sweeney N, Reynolds S, Shafran R, Brigden A, Linney C, McManus MN, Borwick C, Crawley E (2020) Rapid systematic review: the impact of social isolation and loneliness on the mental health of children and adolescents in the context of COVID-19. J Am Acad Child Adolesc Psychiatry 59(11):1218-1239.e3
46. United Nations (2020) Policy brief: education during COVID-19 and beyond. United Nations.
47. Fegert JM, Vitiello B, Plener PL, Clemens V (2020) Challenges and burden of the Coronavirus 2019 (COVID-19) pandemic for child and adolescent mental health: a narrative review to highlight clinical and research needs in the acute phase and the long return to normality. Child Adolesc Psychiatry Ment Health 14(1):1–11
48. Greenberg N, Docherty M, Gnanapragasam S, Wessely S (2020) Managing mental health challenges faced by healthcare workers during COVID-19 pandemic. BMJ 368
49. Galea S, Merchant RM, Lurie N (2020) The mental health consequences of COVID-19 and physical distancing: the need for prevention and early intervention. JAMA Intern Med 180(6):817–818
50. Renalds A, Smith TH, Hale PJ (2010) A systematic review of built environment and health. Fam Community Health 33(1):68–78
51. Meerow S, Newell JP (2016) Urban resilience for whom, what, when, where, and why? Urban Geogr 37(5):309–329
52. Carmona M (2015) Re-theorising contemporary public space: a new narrative and a new normative. Journal of Urbanism: International Research on Placemaking and Urban Sustainability 8(4):373–405
53. Bishop P, Williams L (2012) The temporary city. Routledge
54. Cohen DA, Talarowski M, Awomolo O, Han B (2020) COVID-19 and the early adoption of promising strategies to create and sustain healthy, park-centered communities. Preventive Medicine Reports 20:101240
55. Honey-Rosés J, Anguelovski I, Bohigas J, Chireh V, Daher C, Konijnendijk C, Litt J, Mawani V, McCall M, Orellana A (2020) The impact of COVID-19 on public space: an early review of the emerging questions—design, perceptions and inequities. Cities & Health, 1–17
56. Angelidou M, Karachaliou E, Angelidou T (2020) Smart city planning from a bottom-up approach: local experiences and lessons learnt from the implementation of a smart city pilot project in Amsterdam. Sustainability 12(9):3746
57. Tzoulas K, Korpela K, Venn S, Yli-Pelkonen V, Kaźmierczak A, Niemela J, James P (2007) Promoting ecosystem and human health in urban areas using Green Infrastructure: a literature review. Landsc Urban Plan 81(3):167–178
58. Venter Z, Barton DN, Gundersen V, Figari H, Nowell M (2020) Urban nature in a time of crisis: recreational use of green space increases during the COVID-19 outbreak in Oslo, Norway. Environ Res Lett 15(10):104075

59. Hartig T, Mitchell R, de Vries S, Frumkin H (2014) Nature and health. Annu Rev Public Health 35:207–228
60. Maas J, Verheij RA, Groenewegen PP, de Vries S, Spreeuwenberg P (2009) Green space, urbanity, and health: how strong is the relation? J Epidemiol Community Health 60(7):587–592
61. Rigolon A, Browning M, Jennings V (2021) Inequities in the quality of urban park systems: an environmental justice investigation of cities in the United States. Landsc Urban Plan 205:103952
62. Wolch JR, Byrne J, Newell JP (2014) Urban green space, public health, and environmental justice: the challenge of making cities 'just green enough.' Landsc Urban Plan 125:234–244
63. Rall EL, Bieling C, Zytynska S (2017) Exploring city-wide patterns of cultural ecosystem service perceptions and use. Ecol Ind 77:80–95
64. European Commission (2015) Towards an EU research and innovation policy agenda for nature-based solutions & re-naturing cities. Final Report of the Horizon 2020 Expert Group on Nature-Based Solutions and Re-Naturing Cities. European Commission.
65. Colding J, Barthel S (2013) The potential of 'Urban Green Commons' in the resilience building of cities. Ecol Econ 86:156–166
66. Benedict MA, McMahon ET (2006) Green infrastructure: linking landscapes and communities. Island Press
67. Molloy J, Schatzmann T, Schoeman B, Tchervenkov C (2021) COVID-19, changing travel behaviour and the potential environmental consequences. Travel Behaviour and Society 23:58–71
68. Cervero R, Murakami J (2019) Transit-oriented development: policies, promises, and pitfalls. In: Givoni SI, Banister D (eds) Moving towards low carbon mobility. Edward Elgar Publishing, pp 29–45
69. Moreno-Monroy AI, Orthofer A, Rendon G (2021) The 15-minute city: an opportunity for urban resilience post-COVID-19. Global Development Network Working Paper Series, 116
70. Talen E (2020) Neighborhood: a new framework for thinking about the city. Oxford University Press.
71. Calthorpe P (2017) The next American metropolis: ecology, community, and the American dream. Princeton Architectural Press.
72. Devi S, Saha S, Singh P (2021) The role of mixed-use development in creating sustainable and pandemic resilient cities. In: Saha S, Devi S (eds) Sustainable urbanization and pandemic management. Springer, pp 49–62
73. Hamstead ZA, Kremer P, Larondelle N, McPhearson T, Haase D (2021) Indicators for assessing urban ecosystem services and human well-being in a spatial planning context. Ecol Ind 122:107249
74. Kato Y, Hossain M (2021) Community responses to the COVID-19 pandemic in urban informal settlements in Dhaka. Bangladesh. World Development 141:105405

Seyed Navid Mashhadi Moghaddam is an accomplished urban planner with over thirteen years of experience in the field. He is currently pursuing his second PhD in Human Geography at the University of Ottawa, building on his extensive knowledge and experience in urban planning, land use regulations, public participation, and policy development. Navid has also developed expertise in various disciplines such as EIA, transportation modeling, GIS modeling, DSS techniques, and entrepreneurial city planning.

Urban Infrastructure and Resilience

Poland's Small Towns from Pre- and Post-pandemic Perspective: On Life Quality as a Resilience Measure

Ewa Korcelli-Olejniczak

Abstract The competitiveness of peripheral small towns regarding residential as well as business functions of the supra-local range is discussed in the present paper in the framework of COVID-19 effects. It is found that the towns under study, even though to a varying degree, have revealed a measure of resilience to pandemic-related disturbances. This is traced back to endogenous, territorial capital-related factors. The findings expose the emphasis put by successful socio-economic activities upon the specificity of the market offer, and its linking with tradition, skills, and natural resources. Concerning local government practices, the Covid-19 pandemic has fostered the focus on the quality of life at the community level, while acknowledging the weight of its education and healthcare-related components.

1 Introduction

Small towns form the base, by far the largest in terms of the number of units stratum in the urban hierarchy [1–3], while performing both local, central-place and specialized, long spatial range socioeconomic functions. The role they play in social life and spatial organization receives, however, rather limited attention in scholarly as well as planning and policy literature [4, 5, 6, 7]. These towns situated in predominantly rural, peripheral areas, in Europe and North America especially [8, 9], have over the last decades experienced progressing labor outflow and demographic aging. Yet, under the observed, progressing climate change, their local life quality attributes are of growing importance, by offering some advantage vs. larger cities as a place of residence and workplace location choice.

The COVID-19 pandemic of 2020–2022 has left temporary, but probably also some permanent marks on cities worldwide, their location, heritage and size notwithstanding. This pertains both to the global cities and small towns in peripheral settings. Yet, there may be some relation between urban size and functions on one hand, and

E. Korcelli-Olejniczak (✉)
Institute of Geography and Spatial Organization, Polish Academy of Sciences, Warsaw, Poland
e-mail: ewonia.korcelli@gmail.com; eko@twarda.pan.pl

K. M. Nahiduzzaman (ed.), *Making Sense of Planning and Development for the Post-Pandemic Cities*, City Development: Issues and Best Practices,
https://doi.org/10.1007/978-981-97-5481-6_2

the resilience to external impact observed in social and economic dimensions on the other, related to local governance, urban form, and social practices. In this chapter, some evidence presented, concerning the resilience effects is based on research carried out in 2017–2018, and again in April-June 2022 in selected small towns of North-Eastern Poland—an area, considering its situating in European and national dimensions, here referred to as double-periphery.

It is assumed in the study, that the quality of life comprises an essential factor in the sustenance and development of small towns' place-of-residence functions, whereas the evaluation of its components is to various degree sensitive to the requirements (needs, expectations, living styles) if individual residents and resident groups, and hence the motives standing behind the place of residence choice. On the conceptual side, by referring to the notion of territorial capital, the basic research question posed pertains to sources of subjective life quality, when considered within the small towns' context. It is asked, how the COVID-19 pandemic has affected the perception of quality of life at the local, and community level, and what consequences this may have in terms of small towns' longer-term development prospects.

2 Underlying Concepts

The quality of life, a well-established general concept in social sciences [10, 11], stemming from the perception of living conditions in cities and rural areas [12], has widely been applied in research and public policy, it's both objective and subjective indicators [13] being introduced. Winters and Li [14] refer to subjective well-being, which they identify with an individual's overall life satisfaction when using data on the latter available from the US Behavioral Risk Factor Surveillance System. In their comprehensive, classical study, Andrews and Withey [10], in addition to the life-as-a-whole dimension, distinguish several specific life concerns, including those concerning local government, economic situation, education, services, and facilities, as well as interpersonal relations, the community, and the neighborhood. Alongside the common features of well-being perception, they identify differences among population groups in this respect, depending on age, socioeconomic status, or family life cycle. The community quality of life model by Sirgy [15], based on the personal utility concept, consists of subjective residents' assessments concerning economic and physical conditions, local governmental services, and those focusing on life domain satisfaction. These three dimensions, operationalized by Peters [16], comprise an index averaging the ratings on: the quality of jobs, medical services, public schools, housing, local government services, child care, and senior services. It is generally argued that the subjective, rather than these based on externally defined, objective criteria, such as rents and home values [17] life quality assessments, more often decide on individuals' place of residence choice. The quality of life is interpreted in sociological literature to be strongly interrelated with social capital [18, 19], this link being of particular important for communities that experience demographic, economic, and political change [20–22]. Social capital comprises one of the essential components of territorial capital [23].

Initially, a regional policy-related notion [24, 25], the territorial capital has been conceptualized to integrate local milieu properties that constitute assets in the competition for mobile resources, capital, and skilled labor, between regions, cities, and towns [23, 26, 27]. Such assets encompass both the tangible private, collective, and public goods, including capital stock and infrastructure, alongside landscape and cultural heritage, and the intangible ones—human and social capital, in addition to those of mixed character: relational private services, cooperation networks, connectivity, agglomeration economies, proactive local governance. These factors provide competitive *environmental* tools for economic and social activities. When assessed from a dynamic perspective [28], they should allow for recreating, over time, a given place's competitive advantage. Most of the assets above bear on the perceived, as well as the objective quality of life.

The territorial capital concept shares important common characteristics with the entrepreneurial social infrastructure [29, 30] framework, employed by [22] to identify sources of smart shrinkage in case of small towns, where population decrease over time is found to be accompanied by subjective quality of life improvement. The ESL, which includes the elements of symbolic diversity (depersonalization of politics, constructive controversy in reaching community decisions), resource mobilization (private and collective investments) and the quality of social—horizontal and vertical networks, is considered *a necessary ingredient* [29, p. 48] in successfully linking physical infrastructure with local leadership, and allowing small urban communities to provide for their well-being and development. Both the territorial capital, and the entrepreneurial social infrastructure models focus on networks' quality, together with the quality of public institutions, much relevant to current perspectives on interrelation between the evolving functions and observed demographic change in small towns within rural regions. They also allow for assessing the towns' resilience vs. vulnerability to economic shocks and natural disasters [20, 21, 31], when social resilience [32, p. 347] is defined as: *the ability of groups or communities to cope with external stresses and disturbances as a result of social, political and environmental change.*

3 Data and Methods

North-East Poland, the study area, presents typical attributes of peripheral economic region, including lower than national indicators of population density, urbanization level, technical infrastructure development, as well as GDP per capita. It also features long-term population outmigration and advanced demographic aging. A subset of 19 towns ranging in size from 3 to 10 thousand inhabitants was selected for the purpose of the study. Among several towns excluded from the analysis were suburban settlements and popular summer resorts. Due to the lack of suitable statistical data, local information sources and other, generally unavailable materials collected in the course of fieldwork were used. The main initial database consists of 122 in-depth, semi-structured interviews (six per town, with several exceptions), including three pilot ones, conducted from March to September 2017, and three of updating character held in April and May 2019. A supplementary resource is 55 questionnaire returns, partly open-ended, filled out from May to September 2018, in person-to-person contact with owners or managers of 55 (one to five per town) enterprises in manufacturing and service branches. The interview returns were subjected to content analysis. Both the interview and the questionnaire forms were elaborated for the project. The interview respondents consisted of local stakeholders, including representatives of local government—town mayors or their deputies, school principals and teachers, managers in municipal cultural and medical centers, Catholic and Orthodox priests, local NGO leaders, and local entrepreneurs. The second part of the field study took place in April and June 2022, when three towns selected based on earlier SQL assessments were revisited, to conduct in-depth, open-ended interviews (four per town) with the local stakeholders. The three towns—Ciechanowiec, Reszel, and Myszyniec, represent each of three historical, now administrative regions: Warmia-Mazury, Podlasie, and northern Mazovia, covered in whole or in part by the study area. The towns in question combine the role of local service centers with some specialized, long-range socio-economic functions such as culture, tourism, food processing, and catering services. They rank high, or average (Reszel) on aggregate community quality of life assessments among the 19 small towns under study.

The research approach followed conforms to a post-positivist perspective by assuming that the reality under study comprises an objective fact, one constructed by different subjects, here the interview and the questionnaire respondents. The complete representation is revealed when its various aspects are mutually confronted [33]. The research approach chosen also implies a probability that the findings can be applied to other cases, characterized by an analogous setting and other similar features. In the interview forms, the common block of the study topics, of local quality of life components and their sources, follows the introductory questions adjusted to the respondents' competencies and interests. In an open part, their additional opinions and comments are accommodated. In the contents analysis of interview returns, the quantitative information, and one concerning facts have been separated from opinions held by individual respondents, and from these expressed to represent generalized views of town's residents, or some resident groups [10].

4 Subjective Quality of Life Components

The subjective, local quality of life evaluations, comprise individual opinions of the interviewees as well as, their knowledge concerning opinions on the subject matter that prevail in the respective towns. In at least one interview per town, they also refer to motives standing behind recent town-bound migration moves from rural places, and to those involving the arrival of ex-urbanites, including returning, former town residents. The evaluations refer to the community quality of life dimensions, and additionally to these, concerning local environmental conditions and the respondents' general perception of the towns' future. The relevant ratings were identified by taking into account the proportion of positive and negative opinions, as well as their possible equivocation, when marked by: 'but..', or 'except for..' clause (Table 1). The towns selected for the second part of the study fall in light of the analysis within the first, in case of Ciechanowiec and Myszyniec, and third—Reszel, quartile on the towns' ranking regarding the aggregate quality of life evaluation. In recent research with a focus on small towns, based on longitudinal data [16, 20, 22, 31], the latter is used as the relevant indicator of the towns' resilience *vs.* vulnerability to economic shocks, and population decrease. In the present study, which relies upon the data restricted to a one-time cross-section, the interviews still give some, albeit limited insight into the temporal dimension of the phenomenon, as some of the survey questions explicitly refer to changes observed over the last two decades. The SQL evaluations are based on both the numerical and substantive side of the interview responses. With these limitations in mind, the results are presented below (Table 1).

The high share of positive SQL evaluations among the total, aside from a general appreciation of local environmental conditions, reflects the residents' satisfaction with community public investments, implemented over the last years, numerous with the support of EU funds. Conversely, the local labor markets, featuring mainly negative assessments, still carry the brunt of the systemic transformation of the 1990s, when the small towns' earlier, industrial and agriculture-linked producer service functions were subjected to deep restructuring as a consequence of privatization and

Table 1 Subjective quality of life evaluations

		Number of towns by ratings			
No	SQL components	All positive	Positive prevail	Negative prevail	All negative
1	Local labor market	3	4	7	5
2	Public education	3	7	5	4
3	Health care	3	6	7	3
4	Child and elderly dedicated services	8	10	1	–
5	Cultural activities	9	10	–	–
6	Commercial services	5	9	5	–
7	Technical infrastructure	8	6	5	–
8	Environmental conditions: safety, noise, pollution	12	6	1	–
9	The town's future	4	7	6	2

the opening of national borders to global commodity and investment markets. One of the ensuing effects has been the polarization of small towns' set into these, housing specialized activities of a broad market range, and those, with mostly locally oriented commercial services, and the public sector—schools and local administration units comprising the biggest local employers [34]. Although in certain aspects different, the labor market problems are voiced (with two exceptions) in towns within both small-town subgroups. A common issue, reported by the interviewees, is a poor matching of the demand with the supply of labor. In these towns where specialized, industrial functions play a major part, the scarcity of job offers available to college graduates is emphasized, as in: *This is not a place for educated people.* In towns performing the role of local service centers, where the memory of the former economic activity persists, it is the missing of external investors, preferably in industrial branches. The quote that follows: *Our employment-related problems here will not be solved, unless we gain an establishment with 200–300 hundred jobs* is illustrative in this respect. At the same time, in all urban places surveyed, it is reported that: *The work is available, but not those willing to take it.* The reasons given are wages, often unacceptable when put against social benefits or work commuting to a larger center as an alternative on one hand and decreasing numbers of young labor market entrants on the other.[1]

Small towns occupy an important position in the spatial organization of the public education system, by providing educational services, also for nearby rural communities (even though in part), at both primary and secondary levels. Schools are also foci of a town's social life and are usually paid considerable attention to by the local government. The prevalence, by a narrow margin only of their positive assessments as the local quality of life components, reflects the growing over time educational goals of parents and their offspring, which follow the ongoing economic and demographic change. The critical assessment criterion appears to be, in light of the interviews, the presence and standards at the high school level of the so-called general curriculum, which is the conventional step towards continuing the education at tertiary level. In this regard, the small towns surveyed form three subgroups, or clusters. The first are the ones with all positive assessments, where the local high schools do offer a broad range of subjects to choose from, and have long-proven, good records in preparing for college or university entry. In the second, where either the positive, or the negative assessments prevail, local high schools do operate, but at a more or less limited range, whereas a part of local youth, *those who think about good studies*, choose to attend schools elsewhere, while commuting, or moving to larger urban places. Their share is reported to amount between 20 and 50%, with an increasing tendency. The towns in the last assessment subgroup are those, in which education at the secondary level is now basically restricted to vocational programs.

In the case of health care services, the interview respondents refer first of all to local access to medical specialists, in addition to general practitioners, as well as to emergency aid. In these towns, where the interviewees are satisfied with a public

[1] The source of his apparent contradiction is the rapid transition of national economy from a surplus to a deficit of labor, a compound effect of progressing demographic change, the labor outflow in the years following Poland's EU accession, and economic growth following the 2008–2011 downturn.

offer in this domain, they indicate having a choice between two or more family doctors' centers, also the presence of evening and holiday outpatient services; in a few towns local ambulance and hospital aid are available. In the majority of towns surveyed, however, whereas the basic public healthcare provision is assessed to be rather satisfactory, the necessity is emphasized of seeking specialist advice in a larger town, and (or) acquiring it as a market service (*we are paying for public health insurance, but still buying the medical treatment privately*). In some places, the overall dissatisfaction with local public health care services is illustrated by referring to queues and long waiting times for medical appointments.

Public services dedicated to children and the elderly received in general good opinions from the interviewees. Nurseries and pre-schools, which are found in several towns in the subset, are evaluated as *well functioning*. Aside from, frequently pointed out in the interviews expansion and modernization of playground networks, the provision is documented of other new recreational *cum* educational facilities. Also, children oriented, including therapeutic NGO activities are referred to. Services dedicated to the elderly are shown to receive considerable attention under the already advanced, local population aging. In some among towns, new forms of at-home assistance to those in need are introduced. On the other hand, the so-called active senior citizens, with omnipresent retiree clubs, are indicated to assume important positions in local communities' social, in a few cases also local political life. Children and the elderly appear, in light of the interviews, to be the main recipients of local cultural offers.

The assessments of the latter as subjective community quality of life component need to be looked at for three aspects: the consumption of, participation in, as well as production and reproduction of culture. The first, focusing on entertainment and popular culture is present in events, such as the Town' Days. The second in vocal and dancing group courses, sometimes encounter visiting actors and writers. The third is the sustenance of local traditions—bringing to life and disseminating local folk culture; in one case musical festivals in a renowned composer's birthplace. The expressed in the interviews prevailing satisfaction with local cultural activity reflects over the last years observed improvements in cultural infrastructure, including the renovation and extension of local cultural centers also an approval of the work and initiative of local instructors and organizers. In voices indicating a lack, or scarcity of high culture offers, the limited capability of small towns is recognized in this respect.

In their opinions concerning the quality of local commercial services, the town residents interviewed refer in the first place to changes observed in the structure of the activities and the users' patronage patterns. The entry of supermarkets and other discount chain stores is evaluated according to its impact on locally-owned retail trade. In some towns, *the friendly small shops have retained their regular customers;* more often, however, *the small groceries closed down, and the owners left.* The same happens with specialty stores, selling closing, or house appliances, in sectors that '*gravitate up the settlement hierarchy*' [35]. Concerning other services, the interviewees point out a scarcity of catering businesses open the year around. The availability and variety of personal and technical (building, transportation, etc.) services are assessed positively in general.

Modernization of infrastructure, both social and technical, is a trend noted in interviews held in every town, although its scope and effects observed receive various ratings. Well evaluated are these in particular that bring about an up-grading of public space standards and improve the towns' image, by focusing on the renovation of historical buildings, streets, green and blue infrastructure; also the ones related to sports and culture. In one among the towns they are seen as *filling out a civilization gap;* in another as an indication of *the blurring of living conditions between large and small towns.* In other places, while infrastructure improvements are acknowledged, delays in their provision, and deterioration of some older town quarters are pointed out by the residents. Doubt is also expressed, about whether these are *the investments that can prevent the continuing migration outflow of young people.*

When the interviewees were asked about the main benefits of living in a given small town, their almost universal opinions were *its calmness, security, and slower pace of life.* These basic environmental attributes are, however, subject to gradation. In some towns, the roll is enriched by the traits of *cleanness and greenness*; in a few others, by the town's *friendliness,* with its implicit, both physical and social components. Elsewhere, the town's *walkability,* also: *the larger dwelling space at lower maintenance cost',* are referred to. These attributes' negative aspects: *the town is sleepy; the streets are empty,* though in the minority, are also revealed. In seven of the nineteen towns surveyed, expressed in the interviews prevailing local place of residence perception is in a way institutionalized in the form of membership in the international network. Again, questions can be raised on how, or to what degree it is shared, save for interpersonal level, across different social, in particular demographic groups of towns' residents.

Generalized views on the towns and their future, presenting residents' attitudes, report a split of opinions in the majority of places. At the same time, they show a polarization within the towns' subset regarding future perspectives, with optimist: *this is a town with a future*; *the demography here will come back, even if in a low-seated form; we go for growth* and pessimist: *this little town's future is limited; it will die out sooner or later; the town evolves into a big retirement house* attitude expressed. This division is to some extent congruent with the towns' summary ranking, one showing the contribution of individual subjective quality of life components on a four-point Likert scale [36], where the ratings of 3, 2, 1, and 0 are granted for the positive, positive prevail, negative prevail, and the negative evaluations respectively. Total score is assumed to mark the level of town's resistance to adverse economic and demographic change, as well to the impacts of natural hazards.

5 The Quality of Life Sources

Regarding small towns, the SQL driving forces are identified primarily with social network properties [16, 22]. In this study, the SQL sources are interpreted to include several endogenous, mutually interlocking territorial capital [37] elements. Social

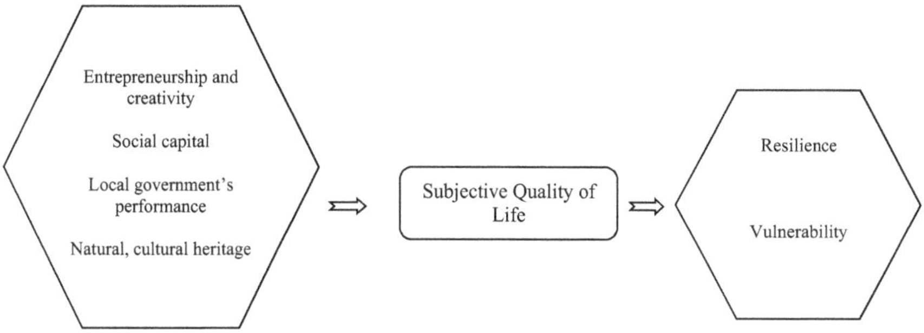

Fig. 1 The SQL sources and manifestations. Author's extension of D.J. Peter's [16] conceptual scheme

capital, local entrepreneurship, local leadership, and assets of the natural and man-made environment (Fig. 1).

Local government's performance, here a proxy for the role of local leadership [38] among factors of urban development, is here interpreted not only via its contribution to the observed standards of community public services [22], but also through efforts aimed at raising the towns' territorial competitiveness [26] and enhancing their general image. As in the case of evaluation of quality of life components, the towns were grouped into four classes according to the presence, and the ascendance of positive or negative opinions expressed. The findings confirm the local leadership's importance among the SQL sources, i.e. its formation factors. In towns of the first subgroup which, except for one, ranks high on the aggregate quality of life ratings, pointed out in the interviews was local governments' multidirectional approach to the modernization of both technical and social infrastructure: *Public investments—a broad spectrum; a lot has been done to improve the living conditions; Our community has gained considerably under the present mayor; It is the mayor's activity that makes this town different.* In three towns, the actions undertaken by the mayors (in one case very successful), were directed towards attracting external industrial investors. In one town, under acute population, including labor supply shrinkage, a social housing financing scheme was designed in an attempt to attract new residents. The small towns, where the acknowledgment of local government's activity was found to be still in the majority (in one, it was praised as complying with *the proper administrative practice rules*) are those, in which, against the mainly positive assessments by the interviewees, some negative, at least indifferent opinions were voiced by entrepreneurs—the firms' survey respondents. In towns in the third and the fourth subgroups, local government's activity in the field of public investments was felt to be delayed, insufficient, or accomplished at the cost of communities' excessive financial debt.

Concerning the role of entrepreneurship and creativity, human capital values, largely determine the quality of labor markets, and their longer development prospects [26]. Hence, the inter-town differences concerning these factors, as they emerge from the interview and the firms' questionnaire survey returns, are relatively close to the ones on the local labor market perception defined as a component of the subjective life quality. This pertains especially to the towns with all positive ratings, where the importance of human capital resources is revealed by reference made to the emergence of local agglomeration economies, the embeddedness of external investments, the role of know-how in the design of custom-made products, and the ability of local firms to adjust to curtailing demand by shifting the market orientation and the production profile. *There is something in the people here that makes them motivated to manage gainful activity; they are open to innovations.* Conversely, in towns with all, or mainly negative local labor market assessments, where specialized economic activities, when present, assume the cost-competitiveness market strategy, *the missing creativity* is pointed out in the interviews. In other towns, including those in the *positive prevail* subgroup, where the based on local entrepreneurship, single innovative and product quality-oriented industrial and service activities are present, they are discerned by the interview respondents as rather *exceptional* and overshadowed by the *passive attitude* and *shallow* labor market characteristics, particularly the missing work offer addressed to skilled labor.

Social capital [18, 19], one of the crucial, intangible territorial capital components [27, 37], is interpreted in the sociological literature on subjective quality of life in the context of the small town [16, 20, 22] as its principal source. These authors also find out that different forms of social networks—the bonding, the bridging, and the linking social ties [39, 40] tend to be in various manners interrelated with individual SQL components. In the present study, some qualitative information on these linkages is derived from the interviewees' responses to questions concerning local civic engagement, community integration, mutual help, as well as the impact of cultural diversity *vs.* homogeneity upon local social relations. Regarding civic engagement, when expressed by the participation in, and the activity level of, the local clubs and NGOs, it is reported to be high, or growing in eleven out of nineteen towns, including three with the lowest summary SQL evaluations. Since it is notably low in two out of three towns' top ranking on labor market quality, as well as on local entrepreneurship, its role may be in a sense of compensatory character, as the high social activity is reported to apply to elderly citizens in particular—the *young retirees,* members of the baby-boom generation of the late 1940s and the early 1950s.

The community social integration [41, 42] receives a bulk of *poor*, or *weakening* opinions in the majority of places, while it shortly revives, as everywhere emphasized, at the time of local charity campaigns devoted to these dwellers affected by acute illness or fire. In three of the towns, it is labeled: *more of the neighborhood, than the community-level society.* This does not, however, imply the presence of strong interpersonal neighborly relations. They are indicated to be sustained and (or) generally positive in four towns, while their lingering over time: *the neighbors used to talk to and meet each other* is observed in two places. In seven towns, rather regularly distributed across the aggregate SQL ratings, the local perception of safety: *I don't need to install any monitoring here; I have got neighbors,* is confronted with a deficit of privacy, referred to in a standard formula: *people sometimes know more about you then you do of yourself.* Against this, reported in all towns are some intra-community linking ties, most often between business activities and local schools, sports clubs, or fire brigades, involving in-kind, less frequently financial assistance, and also the sponsoring of local events. In one town, placed high on the SQL rankings, an open, informal partnership of teachers, administration employees, and entrepreneurs: *acts on the grounds of trust and mutual acceptance to support the local community.* In three towns, external linking ties to regional authorities, and other institutions are pointed out as having allowed for important infrastructural, including culture-related investments. The role of bridging ties, connecting different social groups within local communities, as a factor in the quality of life perception, reveals a specificity of individual towns in this respect. In some, negative effects of persisting local cultural divides, including '*distrust*' and low overall social activity, are identified in the interviews. Elsewhere, the historically shaped cultural diversity is indicated as a town's asset. In a town at the top of summary SQL ratings, the blurring, among the younger generation, of cultural background and place of origin lines, is claimed to be a source of the town's *friendliness.*

Cultural heritage, and landscape assets (i.e. the natural heritage) are listed among impure public goods in R. Camagni's [37] territorial capital taxonomy. Their local perception—the current residents' outlook, may be considered one of the aspects, or dimensions of the subjective quality of life. Their external, i.e. objective evaluation, however, has to be placed among the latter's sources, or formative factors. Following the first approach, concerning the state of the town's built environment, its betterment is emphasized in a majority of interviews. This pertains to the renovation of historical and architecturally valuable buildings and town quarters, but also to the maintenance of housing substance as such: *The town's general appearance has improved; the town center is cared for; the town is charming, it can draw* tourists. At the same time, differences among the towns come to the fore, as illustrated by: *This town is unattractive as a whole.* The natural quality of the towns' rural vicinities is pointed out with few exceptions by the respondents. Their recreational value, however, including potential benefits to the town proper, is seen to be exploited only in a very small part. When seen from the perspective of the future residents' consumption-led preferences, the landscape assets in particular, but the towns' cultural heritage values as well, may be assumed to comprise the priming factors in the place of residence choice.

6 Quality of Life Change in Selected Small Towns—Post-pandemic Effects

Below are summary accounts of in-person interviews conducted with local stakeholders in the town revisited during April and May of 2022. The respondents include those interviewed five years earlier and, in a few cases new ones, such as more recently elected local administration representatives. The interviewees relate to the main local quality of life components and the COVID-19 impact.

Reszel

Respondent 1

Personally—the feeling of security and friendly social interrelations, even if face-to-face contacts, so restricted during the last two years, have not quite revived, and may remain as such. In general—with the local labor market practically missing—an ideal place for the retired. Also for people from Warsaw and Gdańsk, many of whom build their houses or adapt old rural homesteads in the vicinity around. The town as such is also attractive (…). The cultural offer was seriously curtailed during the pandemic and has not quite revived since then. The same concerns the number of tourists. Conversely, the local business has survived. Some of the students, who returned from their academic centers in 2020, are continuing at distance' learning…To sum up: the town has come back in about 70%. It is slower, with no haste. We have seen for the first time that life in a periphery has so many good features.

Respondent 2

No great change. The majority of local service outlets have opened up once the restrictions were lifted. The pandemic was a sort of efficiency test—the well-managed ones have developed; the other disappeared. The returning demand for catering and hairdresser services has been met. Out of our two industrial firms, one is expanding, whereas the other one is affected by decreasing sales… The town in general continues to be a good place for life. Otherwise, the people would not invest in real property here. Some return from a longer stay on the Isles, to retire or work on their own. This trend appears to continue, it is likely to strengthen over time… The children, however, tend increasingly to bypass the local, even though well-equipped vocational school, in favor of high schools in larger towns. Learning, as well as parental control, have suffered during the pandemic, which cannot be quickly improved. At the same time, expectations concerning the future are increasing.

Respondent 3

The situation has come more or less back to normal. Still, the people have learned to appreciate the things they have, once they felt the risk of loss. The COVID has strengthened the feeling of place attachment… The pandemic has changed the functioning of the municipal office for the better. Everyone was prepared to work at a distance. Investments are continued in local roads, though these still call for bigger improvements … some new people lived and worked in the town temporarily. A score of former residents have returned, also from abroad. There is a visible, still rather small stream of newcomers from rural places around, both young and old ones who give up work on the farms.

Respondent 4

More people from out of town build their houses in Reszel. A sort of new neighborhood has emerged over the last couple of years. We are getting phone calls asking about prices and availability of building lots. In rural places near the town, in open space, the new houses are usually such for year-round use, not of the summer cottage kind... Living in a small town is getting better value—for retirement and recreation, but also for those between the ages of 40 to 60. This is not so much in the case of younger ones. Our students, once they leave, do not return for good... The furniture firm here is extending its operations—there seem to be difficulties, however, with finding new workers.

Ciechanowiec

Respondent 1
We are a local center for a rather prosperous, dairy farming area. Our old landmark, the Museum of Agriculture was closed down during the pandemic; now it operates, though at a lower gear—the events which used to bring up to 300 visitors now attract no more than 150. The last years were used for the necessary repair and refurbishing, but it was on the whole a sad time. We felt a lack of tourists... The town's major commercial activity, which is the organization of family, in particular wedding celebrations, somehow did not feel so much the COVID effects. The demand was still present. The firms involved were able to adjust to the restrictions by shipping the food to private homes. Now, it is all normal again.

Respondent 2
Our small town has a certain charm. There are many agro-tourist farms around. In the summer season, the town is full of visitors. It was not quite so in the pandemic time, but even then, the COVID-related social distance rules were perhaps not so strictly enforced as they were in Warsaw. Students returned to their homes in 2020, and participated in the courses via electronic media... At present, the returns of former town residents for retirement are visible. New houses are built in the town and the vicinity. The land lots prices are high. Also, the wedding halls prosper, with as it is heard, notably longer time needed for place reservation.

Respondent 3
Regarding economic activity, the COVID impact can be assessed as minor. The majority of local businesses, including wedding halls, have outlived the pandemic. The closedown period brought about notable progress in the development of information infrastructure and its popular use. Interpersonal relations were seriously affected, and have not quite revived since then. This also applies to cultural events and religious celebrations, as some mental effects have remained... Returns by the town residents, elderly and middle-aged in the majority, including those from abroad, are more numerous than before. Demand for new housing, both private and rental, has increased. Some houses are built, or refurbished, and kept unoccupied for future family use... The municipality's investments focus on tourist and road infrastructure.

Respondent 4
The COVID pandemic period has had visible effects on the town's present-day overall picture. Some are quite positive. In the eyes of visitors, Ciechanowiec has entered a sort of development phase. New public investments have started. The construction branch has gained notably. The local cosmetics services have expanded, attracting

customers even from Warsaw by offering comparable quality for lower prices (….). The more attractive building lots in the town's vicinity are being bought out… There is a shortage of technically skilled workers. Still, some new industrial, as well as commercial investors from out of town express their interest in coming. The duration of this trend is, of course, difficult to assess.

Myszyniec

Respondent 1
Our town's special role as a center of the historical Kurpie region, is the sustenance and promotion of local folk culture. This activity, which includes the organization of workshops and holding fairs devoted to vocal, dancing, and craftsmanship traditions, was seriously hampered during the pandemic period. The flagship annual event—the Honey Prospecting Festival which used to attract around 40 thousand visitors from all over the country and some from abroad, was canceled in 2020, and brought barely half of that number the next year. The COVID affected also other, tourist-related commercial activities…The local labor market is poorly developed. Jobs are available for the less skilled mainly….Our main challenge for the coming years is to upgrade the now very deficient public health services and to invest in local schools. If we cannot offer work here to those with higher occupational aspirations, we need to provide the knowledge allowing them to be competitive elsewhere, and perhaps return in the future to their hometown.

Respondent 2
Local road investments were continued over the pandemic years, though the overall standard is still insufficient; the growing demand for bike lanes needs to be met. New residents come from surrounding rural settlements—the elderly who sell, or pass the farms to their offspring, to get better access to shops, doctors, the church, and young ones who often commute from here to their workplaces. Also former town residents for retirement, including those who migrated years ago to Britain and Ireland; some still work there, while building their houses here. There is a fair number of new houses in the town and the vicinity, but the local society is aging on the whole. The local government is now putting more effort into improving the care dedicated to the old and the disabled.

Respondents 3 and 4
Things have returned to normal more or less. The local firms have survived, though a smaller number of tourists is a problem for some local activities. Several people commute daily to work—in furniture, or food processing plants, within a 20–25 km perimeter…. There were newcomers, among them students and some managers from large cities who stayed and worked from here during the COVID pandemic… Now, we witness more returns of former town residents, from Poland and other countries. Also from surrounding rural places—both the elderly and the young persons. Housing development in the town is a visible effect. This is shown in the number of building decisions approved by the municipality office, the number of which increased by more than 50% from 2019 to 2021.

The three small towns surveyed have, though to varying degrees, revealed some resilience to the COVID-19 pandemic and the ensuing disturbances. Concerning economic activities, the restrictions imposed have had a sort of catalyst effect. The resilience phenomenon, in addition to more general factors, can be attributed to these specific individual towns. In Ciechanowiec, its sources are traced back to the entrepreneurial capacity, the specificity, and the quality of local firms' market offers. In Reszel, the town's cultural, and its vicinity natural assets—territorial capital components, have had a major role in sustaining local life quality, and supporting, both for the retirees and some other amenity migrants, the place of residence function. In Myszyniec, the central place [2] service functions performed for the relatively large and populous agricultural hinterland, one going through a land ownership concentration process, have assured the inflow of new residents and the related housing development. In all three cases, a notable resilience source can also be identified with social network quality [16, 29]. In Myszyniec, as well as Ciechanowiec, these have been external linking ties [43], reaching up to institutions at the regional level, that provided financial cushion allowing the towns' important cultural facilities to go through the closedown period. In Reszel, noted in the interviews good social climate comes out as a resilience factor. Regarding the role of local leadership and local government policies, the pandemic experience appears to have fostered the focus on life quality, while revaluating the contribution of its various components, as the critical condition of small towns' socioeconomic resilience. The concomitance and constellation of these specific components seem not only to strengthen the overall territorial competitiveness of the places but also to make them more crisis resilient.

7 Conclusions with Policy Implications

Presented in this chapter, however partial evidence based on the interview survey, allows one to indicate some more general, both the observed short-term, as well as possible longer consequences of the COVID-19 pandemic within the small towns development policy context.

Considering the sustainability of businesses in towns researched, including the most exposed to crisis commercial services, a majority of firms seem to have survived the lockdown periods. Whereas financial assistance programs at the national level—the anti-crisis shields pertained to all regions and localities, the sustainability factors were rather local, and related to the successful local governance, urban form, and the specificity of social practices. In small towns analyzed, a certain unspoken resistance towards restrictions imposed was observed—the latter related to trust and bonding social capital.

Concerning population movement, aside from the temporary presence of students and professionals from large cities, who were able to study and work remotely from their hometowns, reported an increased, though still rather minor in absolute terms return inflow of ex-residents, including those from abroad, as well as of some other newcomers. In each of the towns surveyed, this is accompanied by a growth of private housing investments, also for future, prospective use. This may signal an improved perception of small-town living in the longer term.

Remote learning and work have imposed actions on local governments, NGOs, and residents, including the extension and up-grading of internet connections, and the dissemination of educational programs for teachers and students, As a result, the small towns have experienced more advanced digital transformation. This progress, catching up with large cities and national core regions, may strengthen the small towns' overall performance and stability, though the effectiveness of this gap narrowing will depend upon the engagement of local actors.

The longer-term pandemic effects remain to be disclosed. As pointed out by Copus [44], concerning technological change and network development, the virus has accelerated changes that were in progress. The same concerns behavioral trends referring to the growing valuation of access to environmental amenities of open space [45]. By the end of the present decade, it should become much clearer whether this, to some extent intergenerational change, along with the stepping-up climate change, will bring about a general tilt of location preferences, also concerning younger-age population groups, in favor of small towns and rural areas.

Acknowledgements This paper is based on a study carried out in the framework of the National Science Centre Poland research project No: 2015/19/B/HS4/00114 on Small Towns in North-Eastern Poland—Patterns and Factors of Functional Change

References

1. Bourne LS, Sinclair R, Dziewoński K (1984) Urbanization and settlement systems: international perspectives. Oxford University Press, Oxford
2. Christaller W (1933) Die Zentralen Orte in Sueddeutschland. Gustav Fischer Verlag, Jena
3. Pumain D (2000) Settlement systems in evolution. Geogr Ann 82B:73–87
4. Bell D, Jayne M (2009) Small cities? Towards a research agenda. Int J Urban Reg Res 33:683–699
5. Korcelli-Olejniczak E (2022) Small towns in settlement systems: a return to the foreground? In: Bański J (ed) The Routledge handbook of small towns. Routledge, New York and Abington, UK
6. Porsche L, Steinfuehrer A, Beetz S, Dehne P, Fina S, Grossmann K, Leibert T, Maass A, Mayer H, Milbert A, Nadler R, Sondermann M (2019) Small town research in Germany—status quo and recommendations. Position Paper of the ARL, vol 114. Akademie fuer Raumforschung und Landesplannung, Hannover
7. Servillo L, Atkinson R, Hamdouch A (2017) Small and medium-sized towns in Europe: conceptual, methodological and policy issues. Tijdschrift voor Ekonomische en Sociale Geografie 108:365–379
8. ESPON 1.4.1. (2008) The role of small and medium-sized towns. The ESPON Monitoring Committee, Luxembourg
9. Liu A, Besser T (2003) Social capital and participation in community improvement activities by elderly residents in small towns and rural communities. Rural Sociol 68:343–365
10. Andrews FM, Withey SB (1976) Social indicators of well-being: perceptions of quality of life. Plenum Press, New York
11. Ferris AL (2004) The quality of life concept in sociology. American Sociologist 35:37–51

12. Berry BJL, Okulicz-Kozaryn A (2011) An urban-rural happiness gradient. Urban Geogr 32(6):871–883
13. Sirgy MJ, Michalos AC, Ferri AL, Easterlin RA, Patrick D, Pavot W (2006) The quality-of-life related movement: past, present and future. Soc Indic Res 76:343–366
14. Winters JV, Li Y (2017) Urbanization, natural amenities and subjective well-being. Evidence from US counties. Urban Studies 47:1956–1973
15. Sirgy MJ (2011) Theoretical perspectives guiding QOL indicator project. Soc Indic Res 103:1–22
16. Peters DJ (2019) Community resiliency in declining small towns: impact of population loss on quality of life over 20 years. Rural Sociology 84:635–668
17. Winters JV (2011) Differences in quality of life estimates using rents and home values. Ann Regnal Sci 51:377–409
18. Coleman JS (1990) Foundations of social theory. Harvard University Press, Cambridge MA
19. Putnam RD (1993) Making democracy work. Princeton University Press, Princeton NJ
20. Besser T (2013) Resilient small rural towns and community shocks. J Rural Community Dev 8:117–134
21. Haase A, Rink D, Grossmann K, Bernt M, Mykhnenko V (2014) Conceptualizing urban shrinkage. Environ Plan A 46:1519–1534
22. Peters DJ, Hamideh S, Zarecor KE, Ghandour M (2018) Using entrepreneurial social infrastructure to understand smart shrinkage in small towns. J Rural Stud 64:39–49
23. Camagni R (2008) Regional competitiveness: towards a concept of territorial capital. In: Capello R, Camagni R, Chissolini B, Fratesi U (eds) Modelling regional scenarios for the enlarged Europe. Springer, Berlin, pp 33–47
24. Organization for Economic Co-operation and Development (2001) OECD territorial outlook. OECD, Paris
25. European Commission (2005) Territorial state and perspectives of the European Union. European Commission, Brussels
26. Camagni R (2002) On the concept of territorial competitiveness: sound or misleading? Urban Studies 39:2395–2411
27. Camagni R, Capello R (2013) Regional competitiveness and territorial capital: a conceptual approach and empirical evidence from the European Union. Reg Stud 49:1383–1402
28. Porter M (1998) Competitive advantage, agglomeration economies, and regional policy. Int Reg Sci Rev 19:85–90
29. Flora C, Flora J (1993) Entrepreneurial social infrastructure: a necessary ingredient. Ann Am Acad Polit Soc Sci 529:48–58
30. Sharp J, Agnitsch K, Ryan V, Flora J (2002) Social infrastructure and community economic development strategies: the case of self-development and industrial recruitment in rural Iowa. J Rural Stud 18:405–417
31. Besser TL, Recker N, Agnitsch K (2008) The impact of economic shocks on quality of life and social capital in small towns. Rural Sociol 73:580–604
32. Adger WN (2000) Social and ecological resilience: are they related? Prog Hum Geogr 24:347–364
33. Guba EG, Lincoln YS (1994) Competing paradigms in qualitative research. In: Denzin NK, Lincoln YS (eds) Handbook of qualitative research. Sage, Thousand Oaks, pp 105–117
34. Korcelli-Olejniczak E, Piotrowski F, Melnyk M (2022) Functional polarization of Poland's North-Eastern small towns in the light of R. Camagni's concept of territorial Competitiveness. Geographia Polonica 95(4):327–346
35. Powe NA, Shaw T (2004) Exploring the current and future role of market towns in servicing their hinterland: a case study of Alnwick in the north east of England. J Rural Stud 20:405–418
36. Skobba K, Osinubi A, Tinsley KL (2020) What about rural blight? Housing and neighborhood conditions in Southeastern small towns. Rural Sociol 85:85–110
37. Camagni R (2009) Territorial capital and regional development. In: Capello R, Nijkamp P (eds) Handbook of regional growth and development theories. Edward Elgar, Cheltenham
38. Bell D, Jayne (2006) Small cities. Experience beyond the metropolis. Routledge, Abingdon

39. Granovetter MS (1973) The strength of weak ties. Am J Sociol 58:182–196
40. Paxton P (1999) Is social capital declining in the United States? A multiple indicator assessment. Am J Sociol 105:88–127
41. Agnitsch K, Flora JL, Ryan V (2006) Bonding and bridging social capital: the interactive effects on community action. J Community Dev Soc 37:36–51
42. Sunblad DR, Sapp SG (2011) The persistence of neighboring as a determinant of community attachment: a community field perspective. Rural Sociol 76:511–534
43. Meijer M, Syssner J (2017) Getting ahead in depopulating areas: how linking social capital is used for informal planning practices in Sweden and the Netherlands. J Rural Stud 55:59–70
44. Copus A (2022) Urban growth engines or relational proximity? What can we learn from enterprise population and business demography indicators in the context of rural and small-town Scotland? In: Bański J (ed) The Routledge handbook of small towns. Routledge, New York and Abington, UK
45. Geyer HS (1996) Expanding theoretical foundation of differential urbanization. Tijdschr Econ Soc Geogr 87:44–59

Ewa Korcelli-Olejniczak holds a PhD in Human Geography (IGSO PAS) and since her Habilitation in 2013 has been working as an Associate Professor at the Institute of Geography and Spatial Organization, Polish Academy of Sciences. She has also experience as a lecturer (Pedagogical Department, University of Warsaw), as well as a local coordinator or participant in numerous research projects, including 7 EU FP DIVERCITIES and ESPON 'Best Metropolises'. At present, she is coordinating a Living Lab in the HORIZON EUROPE 'RUSTIK' project that focuses on the transition challenges of rural areas. Her achievements include authorship or co-authorship of over 80 publications, including five books. She represents an interdisciplinary approach that encompasses a range of urban and regional studies, sustaining constant interest in socio-economic development and community. She also focuses on the practical dimension of science and participates in public and social initiatives related to urban development.

Infrastructure Resilience in the Post Pandemic Urban Areas of Bangladesh

Amena Jahan Urmy, Sadia Binte Rahman, and Abue Jawfore Taufique Ahamed Ahade

Abstract Being a highly populated country, Bangladesh with immense population density has always faced various challenges. The COVID-19 pandemic has shaken the overall real estate and health infrastructure of it. It has been noted that the intensity of the pandemic and the number of cases and deaths are statistically higher in Dhaka metropolis and its surrounding regions, such as Narayanganj and Gazipur; identifying the urban areas as the hotspot of outbreak. Experiencing this conundrum, this country realized the necessity of having a comprehensive planning and development overhaul in these urban areas which make them resilient to any similar pandemic situation. This chapter has discussed this pandemic resiliency shift of Dhaka and other urban areas in Bangladesh. Visiting the journey from the perspective of laws and highlighting how this infrastructure resilience can be achieved through legal implementations is the key discussion of this study. Along with analyzing the current laws and policies, what new attempts and initiatives have been taken and how this works towards combatting similar crises are also discussed here. Adding the current scenario where the lack of cautiousness concerning the given factor and the consequences are explored accordingly. To get an optimum latitude of such resilience profile, the global infrastructure demo has also been compared. In an assertive tone, it can be expected that adhering to the existing laws and undertaken policies; and generating new planning and development ideas from the lesson of other states, Bangladesh can overcome the infrastructure shortcomings it had and become resilience to the future pandemic. A proper framework that discusses the overall placement of the fact, the results and the possible determination that might accelerate the footing in a better position.

A. J. Urmy (✉) · S. B. Rahman
Department of Law, Bangladesh University of Professionals, Dhaka, Bangladesh
e-mail: amena.urmy@bup.edu.bd

S. B. Rahman
e-mail: sadia.rahman@bup.edu.bd

A. J. T. A. Ahade
Department of Maritime Law and Policy, Bangabandhu Sheikh Mujibur Rahman Maritime University, Dhaka, Bangladesh
e-mail: taufique.mlp@bsmrmu.edu.bd

K. M. Nahiduzzaman (ed.), *Making Sense of Planning and Development for the Post-Pandemic Cities*, City Development: Issues and Best Practices,
https://doi.org/10.1007/978-981-97-5481-6_3

Keywords Infrastructure resilience · Development · Planning · Pandemic · Urban area

1 Introduction

Bangladesh is a highly populated country containing a huge population of 171,258,374 within the limited surface area of 147,570 square kilometer. It has the tenth-highest population density in the world with 1115.62 people per square kilometer. The population is mostly dense in the urban areas and surroundings of such cities. Dhaka, the country's capital and largest city holds 14.4 million residents and a population density of 19,447 people per square mile (50,368/square mile) [1]. Repercussions of having unplanned densely populated cities are explicitly witnessed in recent events, especially from the response of the urban areas of Bangladesh in the pandemic situations like Covid-19. In 2020, Covid-19 pandemic has struck upon the world and Bangladesh also gone through the same drastic phase. It has been noted that the intensity of the pandemic and the number of cases and deaths are statistically higher in Dhaka metropolis and its surrounding regions, such as Narayanganj and Gazipur. These regions have been recognized as the main hotspots of the COVID-19 pandemic in Bangladesh in a variety of gray literatures, scientific publications, and local and international newspaper reporting. Numerous studies conclusively show that one of the main drivers of COVID-19's fast development is dense population [2]. Experiencing this conundrum, this country realized the necessity of having a comprehensive planning and development overhaul in these urban areas which make them resilient to any similar pandemic situation.

This chapter has depicted the concrete resiliency level Dhaka city was ready with before the pandemic situation- which was, at least from the perspective of the authors, not adequate to combat the pandemic situation. The readers will also be able to visit the resiliency transformation of the urban cities of Bangladesh through a legal perspective. Albeit, it cannot be denied that the existing laws and policies of Bangladesh, if implemented carefully, such retrogressive, dreadful situation would not take place. Nonchalant treatment of the concerned authority and leniency of the people regarding their rights can be identified as factors behind such failure. Accepting that limitation, the chapter will try to find out the loopholes of the existing legal framework of the country which refrained her to achieve infrastructural resilience. On the same note, it will focus how to overhaul the existing system based on the newer innovative changes taken places in other countries through analyzing in a comparative approach. Apart from that, the multi-strata initiatives of the Government of Bangladesh which also includes framing new policies to combat such pandemic akin conditions in future were scrutinized by the authors in order to discover their efficiency in the long run and to make an infrastructurally sound and resilient urban areas in the post-pandemic Bangladesh.

2 Literature Review

As the Covid-19 pandemic situation comparatively new, the literatures regarding it are ongoing. Though some can be found on the pandemic phase or pre-pandemic phase, the numbers are few on the post-pandemic works and gets fewer when it is about the infrastructural resiliency, particularly on urban areas of Bangladesh. 'COVID-19 Pandemic: Rethinking Strategies for Resilient Urban Design, Perceptions, and Planning' is one of such pieces which has focused on the major lessons that can be adopted for post-pandemic urban resilient planning related to disaster management and climate change adaptation, preventing extensive challenges of sustainability apart [3]. Financing, policy making more and making the impact of any future pandemic less effective on our infrastructural sectors is important. The G20 plan where it tends increasing the resilience of infrastructure against risks covering from pandemics. Communication infrastructure has been increased, the businesses are following the digital platforms to run their daylong works, but the accessibility is differing from region to region. Overall, infrastructure resilience depends on the physical aspects, operations, the proper governing for risk allocation then the adequate financing and afterwards the initiated service accessibility and their continuance process. The strategies made should carry the protection and prevention atmosphere, be flexible-easily adaptable and ensure recovery and responses [4]. New infrastructural projects should meet up the economic and social resilience, the concept of sustainable infrastructure promoting social inclusion and addresses different dimensions of inequality. This initiative should be finalized this way that comes across solving the infrastructural issues in the post pandemic state [5]. Financial backups, productivity and environmental factors are the turning points in working on the mechanisms on the post- pandemic infrastructural issues. The Malaysian government has taken initiative to work on this current pattern [6]. Infrastructural projects which are under procurement and under construction should be reassessed. The necessary undertakings should follow the short, mid-term, long term durability. A robust performance and risk framework with adequate fine-tuning and stress-testing is required for the ultimate COVID-19 response and resilience-building strategy. Also, a globally connected Project Management Office (PMO) can help in furtherance [7]. The planning, financing, durability, sustainability, and recovery-responses should be the guiding factor in the case of designing the infrastructural concept. In the post COVID-19 pandemic, it has become more significant that the pertinent measures must be brought about.

3 Methodology

Authors have followed the thematic method in this book chapter to address the key problem and results therefrom. This method requires a qualitative database. The study obtained the results through secondary data-based research including news, statistics, and literature. The review study process accumulates the relevant literature, articles, journals of different authors, analyzing them with a new set of objectives and applying the result in the work having pertinent proportions. The database that has been used here is based on legal databases which are used in the field of law, computer assisted legal research is another format here.

3.1 Data Collection

The conceptions, data derived from the secondary sources and here the data is used here are from various sources, such as articles, journals, news etc. there, the thematic method has been applied to analyze the qualitative data. The case study that has been incorporated here is based on a secondary derivation. The collected data were reviewed to connect with the given problem in this paper. Though a methodical interview could not be taken, the paper contains all the necessary information, analysis that leads to the proper understanding of the given issue here through the thematic process.

The inclusion and exclusion part concerning the methodology is, adding the pertinent that supports the infrastructural resilience in urban places mentioned in numerous writings, the impact of COVID-19 on this regard and the necessary implementations while not incorporating the rural area factors with a view the pandemic impact and the matter of resilience.

3.2 Data Sources

Primary source	Relevant laws concerning the issue
Secondary source	Necessary statistical records from numerous articles, journals, literatures based on the given theme, data from newspapers etc.

3.3 Data Analysis

Through secondary information, the thematic approach is given as an analyzing tool for qualitative data. The literatures, perceptions are here have been analyzed with the key concept of the paper. The resilience of the infrastructural issue has been addressed in many of the places which are a major site of this paper and they have been explained, explored with the theme parameter. The keywords are used here, focusing on infrastructural issues, development, the impact of COVID-19, resilience and effect on the urban areas are utilized here throughout the study with a proper explanation and necessary modifications.

A proper methodical discussion or interview was not feasible in this study as it takes into consideration the while urban areas in Bangladesh. Case study research, likely, does not always represent a sample and the researcher's goal is not to enumerate frequencies (as, for instance, in statistical generalization) but to expand and generalize existing theories. However, this chapter primarily has focused on the existing urban infrastructure planning in Bangladesh, how it was running throughout these years and after the new-normal phenomenon of COVID-19 pandemic. The legal and policy work on this issue is another crucial discussion point and afterwards the Government initiatives have been discussed focusing on the urban sector to find out what particular changes can be brought.

3.4 Limitations of the Methodology

The one limitation that the authors have faced is that a proper interview could not be arranged for this study, which would have met the current requirement of the study and furnish better result. Another limitation was the secondary data based that was dealt cannot be said adequate to balance the paper's entire theme.

4 Analyzing Infrastructure Resilience in the Post Pandemic Urban Areas of Bangladesh

4.1 How the COVID-19 Pandemic Has Affected the Urban Areas in Bangladesh

Urban growth is still "big cities-centric" nowadays. Around 38% of Bangladesh's population lives in urban areas, mostly in and around the cities of Dhaka and Chattogram. Each district and divisional headquarters in Bangladesh have at least one secondary town, another significant town, including a small township at each upazila. Although cities account for between 70 and 80% of the population in high-income countries, historically no nation has been able to rise to this level. By 2031, the nation wants to achieve the lofty goal of being an upper-middle-income nation [8]. Bangladesh regards the five-year plan as the country's overall development strategy. The government has a number of tools at its disposal to address the housing requirements of all income populations, and these plans are just one of them. Capital Development Authority—*Rajdhani Unnayan Kartripakkha* (RAJUK) began work on the Dhaka Metropolitan Development Plan (DMDP) in 1995 using a strategic planning methodology for the city of Dhaka.

We all thought the existing policies, legislations and planned infrastructure are capable to combat any situation including natural disaster or pandemic. Then there came COVID-19 pandemic, a long-term state where our usual life experienced a significant change. It has affected basically every area around the world where normal life was stuck in a new normal position and Bangladesh was not an exception as well. Hence the urban planning was not so well planned for about a long time, there this pandemic has proved it to be failed viciously. In a developing country like ours, we have faced tremendous hardship with such impromptu poor urban planning and now in the post pandemic era we are still working on placing the things [8].

The infrastructure of health sector was the worst effected ones in this Covid-19 pandemic. The first phase was attacked into the urban informal settlements (slums) where we do have concerns relating to the overcrowding, unhygienic waste management and pollution issues leading towards severe health impact from COVID-19 [9]. The public and private health facilities were affected as the lack of adequate healthcare providers to treat general patients, service hours were shortened. The public test centers were affordable but overcrowded with patients on the other hand the private centers were easier to access but unaffordable for the urban people, mostly the slum dwellers [9]. The practice of that period is still ongoing in some of the places in the post pandemic era. Restructuring the health sector comprehensively seems to be a dire necessity for protecting the present as well as the future conundrum.

Public health initiative is one of the most cited topics where it includes the collective social contract, hospitals, urban trees, playfields, footpaths, clean rivers and kacha bazars—can all be considered here. Biking, cycling is always an option that promotes proper urban planning as it does not emit carbon, but in Bangladesh we hardly get places where cycling is feasible. The roads are not safe, the footpaths are occupied by informal markets here [10].

On the other hand, the urban dwellers were supposed to live within a safety net, but a recent study shows that only 17.84 areas are covered by this [11]. The urban planning after the pandemic should be reconstructed but there are certain factors that need to be checked. A proper balance between densification and disaggregation is a key question here. This pandemic has shown us our urban planning was not firmly organized and that it creates hassles after the pandemic [8].

The urban structure due to the health service deuteriation, the economic hazard, is not properly maintained. The amount of non-resolved housing arose in our country as the urban areas for living were not adequately planned, there are overcrowded places, unhygienic-sanitation issues which lead to the increasing number of affected people. We might face such a pandemic condition later, so we should be focused on planning the urban areas. The present context in Bangladesh after the post pandemic era, focusing on urban structure that clearly defines us to work diligently on those factors, otherwise the next pandemic might make the state even more vulnerable. An environment friendly atmosphere is consistently a required entity, here suitable urban organization is needed.

4.2 Resiliency Under the Existing Legal Framework of Bangladesh

Bangladesh has already framed significant number of legislations and policies which, to some extent, can make its' urban cities resilient to any future akin pandemic. The very motivation behind framing various national framework was basically two- firstly, considering the quizzically increasing real estate and housing industry in the urban and suburban areas; secondly and most importantly, to fulfill the ultimate obligation under the international instruments to which Bangladesh is incumbent.

For an example, Article 25 of the Universal Declaration of Human Rights (UDHR) and Article 11 of the International Covenant on Economic, Social and Cultural Rights (ICESCR) unequivocally recognizes the right to an "adequate standard of living" of every human being and one of the denominating factors of such standard of living is ensuring housing. This recognition implies that the right to housing needs to be ensured by the state in line of ensuring adequate standard of living [8]. In line of such notion, to express the obligation derives to Bangladesh from signing and ratifying several international instruments, Article 15(a) of the Constitution of Bangladesh inserted, *"It shall be a fundamental responsibility of the State to attain, through planned economic growth, a constant increase of productive forces and a steady improvement in the material and cultural standard of living of the people, with a view to securing to its citizens- a) the provision of the basic necessities of life, including food, clothing, shelter..."* The right mandates a range of protections including availability of services, facilities, and infrastructure; accessibility and habitability. Enforcement of the housing right is also seemed to be enforced by the Apex Court of Bangladesh in Jhilpara Slum Eviction Case (2008). In order to protect this right for all the people of Bangladesh, including the toiling masses, the Bangladesh Legal Aid and Service Trust (BLAST), the Centre for Laws and Litigations (also known as Ain O Shalish Kendra) and others has filed public interest litigation/ PIL to the High Court Division of the Bangladesh Supreme Court [8].

Similar assurance is repeatedly reflected in the national legislations and policies of Bangladesh, fulfillment of which, would partially make the road easier for the country in planning and developing a resilient city in post pandemic. The Real Estate (Development and Management) Act, 2010 was enacted with a view to establishing regulation and promotion of the real estate sector. This Act pertinently ensures civic benefits to be, as far as possible, be ensured according to the Land Development for Private Residential Project Rules 2004 (Section 7). Section 11 of this Act stipulates the mandatory facilities of the real estate housing including sufficient air circulation and all types of utility services, open space and most importantly maintain standard distance in between the buildings- which are proved to be helpful to prevent any contagious diseases or similar natured pandemic. On the other hand, it imposes penalties for unauthorized real estate projects (Section 20), change of approved design or facilities (Sections 21 and 26) that implies the constructions which are not build or designed as per the rule of this Act will be subjected to necessary actions.

Before enacting the Real Estate Act 2010, Building Construction Act 1952, was promulgated in order to control haphazard and unplanned development of buildings, excavation of tank and cutting of hills by putting restriction (Section 3), demolishing construction if necessary (Section 3B). It is to be noted that this Act particularly puts circumscription on improper use of land; which is determined by the distribution and classification of lands based on the purpose of use, such as residential/commercial/industrial areas. Now, this Act of 1952, is not only environment friendly and green law in nature, but also can be considered as an effective legislature to build resiliency in the urban areas.

Harmoniously, Bangladesh National Building Code 2015 is the one comprehensive code regarding housing which tries to engage with every possible matter regarding planning, design and construction of buildings and associated service facilities including electrical, mechanical, sanitary etc. It is being followed to ensure safety, minimization of wastage in construction and optimum return for the user. It particularly puts emphasis on sustainable development such as Green Building Technology, Renewable energy, Waste Management, Climate change and so on.

Apart from all these laws, there are certain legislations which are concerned with the environment and sustainable development and planning of housing such as Playgrounds, open spaces, parks and natural reservoirs conservation law in all municipal areas of the country including metropolitan, divisional cities and municipalities of district cities, 2000; The Brick Manufacturing and Brick Kilns Establishment (Control) Act 2013; Building Construction Rules 2008; Water Body Conservation Act, 2000 etc.

When it comes to specifically planning and development of urban areas, Bangladesh has its own Town Improvement Act (1953) which contains provisions regarding the development, improvement and expansion of the Capital of the Republic and its' peripherals by *"opening up congested areas, laying out or altering streets, providing open spaces for purposes of ventilation or recreation, demolishing or constructing buildings, acquiring land for the said purposes and for the re-housing of persons displaced by the execution of improvement schemes*". RAJUK is working to create Dhaka, the capital town, a sustainable, habitable, and environmentally responsible metropolis. It has taken the initiative to create a livable city in order to achieve the planned city's objective by addressing housing, transportation, and large-scale water-based public space/open space issues. The organization's primary tasks included the creation of development plans, road expansion to relieve traffic congestion, the design of better traffic flow plans, the establishment of open spaces for relaxation, the dismantling or construction of buildings, and the building of roads, bridges, and culverts. It also undertook an Urban Resilience Project (URP), whose main goal is to increase the capacity of related agencies to respond quickly and effectively to emergencies within Dhaka and Sylhet. A secondary goal of the URP is to raise the level of quality of new buildings in the two cities, with completion scheduled for around 2022 [12].

Apart from the infrastructural legal instruments, the post pandemic cities can be made resilient through efficacious decentralization which is already embodied in the Article 59 and 60 of the Constitution of Bangladesh which guides and classifies the republic's administrative units in local government bodies. Complying with the constitution, the Local Government Act of 1997, which was followed by the Upazila and Zilla Parishad Acts of 1998 and 2000, established a powerful four-tier local government commission with the goal of ensuring the decentralization of power in every stage and emphasizing public engagement. In line with the system of administrative decentralization, the government of Bangladesh has undertaken measures to enhance the economy by shifting authority away from the core to the margin [13].

These provisions were supposed to be sufficient to make the urban areas of Bangladesh resilient to any disaster, climate change or similar issues. But if we singularly think about the post pandemic resilience which would combat the similar pandemic future, then legal frameworks relating to planning of health sectors' infrastructure is something that needs to be taken into account. That is because, though we designed shelter home for facing natural disasters or emergency situation, but it did not foresee any pandemic situations like COVID-19 from where isolation, social distance etc. concepts might arise.

It needs to be realized at the same time, the real estate industry does not entail solely to the right to shelter, also involves other rights such as right to health, right to privacy, right to safe and healthy environment etc. This pandemic resiliency requires to highlight upon right to health in particular.

The right to health has been recognized in several international instruments including ICESCR by providing "*the highest attainable standard of physical and mental health.*" Governments are obligated to take effective steps for the "*prevention, treatment and control of epidemic, endemic, occupational and other diseases.*" Article 12(1) provides a definition of the right to health, while article 12(2) enumerates illustrative, non-exhaustive examples of States parties' obligations. While the Covenant provides for progressive realization and acknowledges the constraints due to the limits of available resources, it also imposes on States parties' various obligations which are of immediate effect. States parties have immediate obligations in relation to the right to health, such as the guarantee that the right will be exercised without discrimination of any kind (Art. 2.2) and the obligation to take steps (Art. 2.1) towards the full realization of Article 12. Such steps must be deliberate, concrete and targeted towards the full realization of the right to health [14].

When the question of the obligation of the state arises, the first legal instrument that is scrutinized to get the answer is the Constitution of that state. Article 15 of the Constitution of Bangladesh states that it shall be a fundamental responsibility of the State to secure for its citizens the provision of necessities of life, one of those is the medical care. Additionally, Article 18 states that the State shall regard the raising of the level of nutrition and the improvement of public health as among its primary duties. These are fundamental to the interpretation of our Constitution and must be applied by the State in formulating laws. Though rights enshrined in Part II of the constitution is not judicially enforceable, justiciability is not the sole criterion to assess the magnitude of a constitutional obligation. If we consider Fundamental Rights of Part III of the constitution, Article 32 of the Constitution provides for every citizen the right to life, which has been widely interpreted in cases to include the right to health [15].

Practical implementation of these rights is overlooked when the people of Bangladesh, especially the citizens of cities- where the covid patients were comparatively higher, were deprived of any basic medical care or health facilities. This indicates how the country failed to design a sustainable health infrastructure for its residents and how an overhaul is required in this sector [16].

Analyzing the current laws we have in Bangladesh right now it is clear, these can be helpful to combat any pandemic situation like Covid-19, but as said before, not adequate to build apt urban resiliency in Bangladesh. That's why the Government of Bangladesh has taken several measures so that Bangladesh can be resilient towards similar situation, focusing the urban areas as they were affected more comparing the rural areas. In the next segment, those initiatives have been scrutinized thoroughly to evaluate the resiliency level.

4.3 Analyzing the Trajectory of Government in Planning and Development for Post Pandemic Cities

Bangladesh government and the World Bank finalized a \$300 million financing agreement to support the state's urban local administrations in responding to and recovering from the COVID-19 outbreak and enhancing readiness for forthcoming pandemic. 39.9 million urban people across all 8 divisions will benefit from the 'Local Government COVID-19 Response and Recovery Project'. As they have already recovered from the pandemic and are preparing themselves for upcoming shocks like climate change, natural catastrophes, and disease outbreaks, it will help its cities and towns to rebuild more effectively. Additionally, the initiative will provide funding twice a year to 329 towns and 10 city corporations to upgrade vital urban services and infrastructures in order to prepare for and respond to the effects of climate change, natural catastrophes, and upcoming disease epidemics [10].

This project will assist local governmental entities in adopting the necessary steps to advance climate-smart urbanization and get ready for upcoming shock vulnerability. It will execute labor-intensive public works that will both ensure water supply, sanitation, drainage, and other essential services which would be useful to low-income areas, shantytowns, and zones with a high disease outbreak and disaster risks, while also increasing employment for the underprivileged urban population. It will generate 10,000 jobs and 1.5 million days of temporary employment for women underneath this public works program. Every piece of infrastructure will have energy-saving features, such solar panels and cool roofs in the public sectors. The initiative would enhance sanitation in public areas by installing hand-washing stations, bathrooms, and separate restrooms for women.

Moreover, RAJUK proposed and subsequently framed a *Detailed Area Plan (DAP)* in December 2021 and its' summary was finally approved by the Government of Bangladesh on Sixth July 2022, followed by the final approval on 23 August 2022 [17, 18]. Mentioning the plan in this chapter has a definite connection with the resiliency model this chapter is seeking for. This plan has been formulated from the very fresh experience of combatting Covid-19 pandemic and in many segments of it reflects beefing up the overall planning and development of Dhaka city for any facing any situation concerning environmental, disaster or erupting any similar pandemic, though precariously. This 20-year master plan is visioned to convert Dhaka city into a livable and modern mega city by the year 2035 [19]. Replacing the traditional 'plot-based development', it came up with the idea of 'block-based development system' which would the Government believe will be able to fulfill the housing needs of Dhaka and its peripherals stretching across 1582 sqkm. by means of easing pollution, traffic congestion, waterlogging and other issues through effective use of land [17, 18].

One of the salient features of DAP is it will permit construction of high-rise buildings and merge the small units of houses. There will be two folded repercussions of such step, firstly it would offer a lot of open space and limitations of land will be solved- which will be useful for healthy environment and adequate air ventilation; and secondly, it will also decrease haphazard expansion of the urban area and protect low-lying urban and agricultural lands.

During the prolonged lockdown period, heavy migration of the lower and lower middle-class people from urban to rural areas was observed. Nearly 50,000 low-income individuals have fled the city for their native lands after losing their employment in the pandemic, leaving them with uncertainty. They became drastically helpless and homeless which hits hard to the reality that these categories of people cannot afford any housing facilities in the cities [20]. Realizing this discriminatory planning of urban areas, the new DAP has planned a housing project in small size flats within 650–700 sq ft for lower and lower middle-class segment of the society in the 57 locations. Thus, if any pandemic occurs in future, the economic shock at least will not lead to make million people homeless and unsafe situations. On the other hand, the Real Estate and Housing Association of Bangladesh (REHAB) claims that implementation of this plan would cause reduction of the overall flat sizes to 33–35% as the open spaces aptly upsurges, but at the same time it would cause price hike of the real estate to 50%. Such an uplifted price will go beyond the capability of the lower-middle- and lower-class people to purchase any shelter for them [21].

Decentralizing the urban population and reducing population density and at the same time ensuring civic immunities are the concern of DAP; which creates resiliency to the future pandemic ensuring social distancing and offering better medical infrastructure. If the plan can be executed, it aspires to give such housing facilities at least 2.6 crore people. Under this mega plan, the Dhaka and its designated surrounded areas has been divided into 6 discrete regions and seventy-five sub-regions; these divisions and sub-divisions are further classified under different category as per its' purpose of utility such as residential, commercial, industrial, agricultural, institutional, waterbodies, forest, open space, and flood prone areas.

One of the key requirements of decentralization of the urban cities is that there should be an accessible rapid communication system among the urban cities and other areas. According to the DAP, currently, there are 228 km inter-regional connecting roads in Dhaka which would be increased to 291 km. It also plans to extend the collector road by 1,200 km. There is also a plan for creating a 202-km cycle lane and 574 km waterways in the city. Moreover, the DAP proposes construction of six Metro rail lines, two bus rapid transits, six elevated expressways, two main roads parallel to Dhaka-Mymensingh Road, radial roads connected to two ring roads and circular waterways. It also proposes to construct 13 inter-district bus terminals and two truck terminals around Dhaka. These long-term communication projects will expedite the transportation and hence, capable the populace to footloose all over Bangladesh. Besides 627 schools, 285 colleges are to be built, based on which school zone will be determined to restrict mobility and traffic congestion [20].

To combat with the environmental issues including global warming and climate change; disaster management as well outbreak of any ailment- DAP aims to provide an environment friendly urban area by constructing five large regional parks, 55 water parks, 13 large eco-parks (including Bhawal Forest), 13 other parks and playgrounds, a cultural zone adjacent to the Buriganga River and a park with an area of about 425 acres in Keraniganj.

Visiting the previous experience where the infection and death rate due to pandemic were higher than any other regions, Bangladesh realized its' subpar, piece-meal health infrastructure was meagre to combat such situation, a large infrastructural shaft has been opted in this plan and proposed to build 287 hospitals in the zones with a view to ensuring quality healthcare. There is no alternative to implementing such a mega plan, if this country wants in real life to create resiliency in the post pandemic cities.

4.4 What Bangladesh Can Learn from Other Countries' Urban Planning

As we know urban planning needs to be developed due to the post pandemic state, there we can refer to some of the measures taken by some other countries to have an appropriate understanding. The actions taken by the countries are substantially based on rebuilding their urban structure including the marketplace arrangements, hospitals, educational institutions, living places, environmental protection as well. The new urban design should be long term, having some of the equitable benefits and preventive measures.

4.4.1 Mobility and Cycling Initiatives in Other Countries

Easing the mobility can be one crucial point here, street space, carbon emission controlling can be under this phase. Here Berlin has introduced the cycling service having cycle lanes to save time and thus the emission can be controlled as well. In Paris, they have lined up with the cycling service to '15 min city' to move around on time. London is following the bike service as an easy medium of transport. In our country, we can follow the similar strategy of having cycle lanes for smooth communication processes which would be environmentally friendly as well. So, this smart city pathway can be one of the undertakings towards the urban development goal.

4.4.2 Digital Infrastructure and Housing Reformation

The authority should also concentrate on the digital infrastructure system that might lead to a sustainable urban development providing affordable and adequate shelter places for all. The motto of redesigning the housing plans may assist us with our future accommodation needs [8]. In the past, during the Industrial Revolution, unsanitary living was an issue and that led to several discomforts among the people. Thus, in London, Paris and New York they worked towards housing reform and modernized their sewage infrastructures. So, reformation is not a new concept, urban planning has always been a concerning issue.

Planning for a proper pedestrian culture is required in Bangladesh as that is practiced in European countries. The cooperation of the urban administrators, planners, architects, healthcare professionals and social workers are needed to re-conceptualize the footpath as a public health infrastructure assisting to reduce obesity, resist diabetes, and encourage people to better experience their city [10].

4.4.3 Technological and Constructional Development

Talking about the housing scheme, here Singapore has worked on the land development that connects with the housing. The other initiatives taken by their government for mangrove restoration as a coastal defense program which is funded partly through the proceeds of its urban development. In Bangladesh, we may work on these factors. In Nigeria, their efforts towards infrastructure and construction on urban land development encompasses green industrial activities, released for housing development. In the UK, the North of England has expertise in renewable energy generation; the West Midlands is working on vehicle components all gathered in the process of public transport. Manilla, Rio de Janeiro have upgraded the public transport system so that it may reach each urban community having less risk and less pollution [22]. Tamil Nadu has worked on their housing project that refers to climate resilient housing for about 6,000 urban households, Delhi is having a long-term master plan with a well oriented railway system. China has planned towards water resource management for urban and rural sections after the pandemic. Fiji has implemented a blockchain based land management system that helps in land tax related function. The space-based technologies have consistently accommodated Indonesia for proper location and design-based infrastructure having the environment sustainability [23]. Living in a digital arena, the application of technologies may reduce the work pressure concerning the land management-housing and urban planning factor.

In Bangladesh, we do have several scopes to implement such features that have been developed or introduced in other countries. As we know, after the pandemic it has been quite strenuous to place things as before; here the urban planning is one of the most crucial parts. The planning taken from the other countries can be a learning point for Bangladesh to deal with these businesses in the post pandemic state. Mercy Tembon, World Bank Country Director for Bangladesh and Bhutan has expressed his hope stating- "*around 36 percent of the population living in urban areas the city corporations and the municipalities can play a critical role in helping the urban poor recover from the pandemic as well as prepare to handle future shocks*" [10].

5 What More Right Actions Can Be Taken?

Some essential measures are needed to be taken, to be emphasized in order to betterment of the entire state. Firstly, restructuring the health sector, building more hospitals and arrangement like quarantine centers to combat the unforeseeable pandemics and emergencies. In Urban area, the proper maintaining of hygiene is needed more as the industrial growth is higher in these areas, so practically there is high chance of any disease spreading faster, just as we have witnessed during Covid-19. So, hospital, or any health care institution should be cleaned properly with authorized persons who have received training of hygiene.

Secondly, there are changes need to be adopted in the architectural facilities in the post pandemic era. In residential place the proper use of open spaces should be ensured as that was regarded a potential hotspot of spreading the virus. Maintaining the lanes in front of the buildings and the standard spaces among the buildings, bike friendly spaces- are imperative where in some of the areas it has been considered but most of them are lagging behind [24]. Designing smart home, with technological advantages is also crucial these days for maintaining a sustained area to receive facilities, for example, having emergency evacuation plans and quarantine area in every such building can bring a big change.

Thirdly, the post pandemic phase requires some of the structural changes in residential areas, as the condition made us stay indoors so the big windows, cross ventilation process, a balcony or terrace, a cornered home office are new requirements. This diversity is not confine to housing places, the restaurants, academic institutions, open spaces are also demanded to be planned in a new way [25] alongside, warranting safety net for the urban dweller.

Fourthly, the long-planned decentralization of Dhaka city and other mega cities should be implemented as soon as possible, though several projects have been taken by Government for the same purpose, but the ultimate fascination to stay at the heart of cities are cutting those initiatives. Unless the similar privileges and opportunities cannot be offered in the peripheral areas of the urban cities, spontaneous decentralization will be not made possible.

Then, housing places whether residential, commercial or industrial places should be constructed accordingly so that we can be prepared for any other future pandemic issues. The public health initiative refers to many aspects where the construction of hospitals should be placed properly, with safe and wide spacious road, markets, malls and etc., all should be build following the given infrastructure policies with environment friendly process.

Some steps that have already be implemented in other countries such as the cycle lanes, digital infrastructure system can be introduced in this regard. The overall development of mangrove coastal areas is another vital point here as Bangladesh is having a large area in this aspect. Urban land developments should emphasize on green industrial activism. A proper transportation allocation along with well oriented railway system can be introduced. The planning concerning the infrastructure and land management prices should follow the concept of environmental sustainability.

Finally, making sure of proper implementation of the current laws can mitigate the pandemic situation to a greater scale. Also new policies and laws can be framed to prepare the urban areas with pandemic resiliency which can help us to combat the contiguous virus or endemics or pandemics.

6 Conclusions

Our current standing, and process of forwarding one more step has been clearly discussed above. The post pandemic has given us a new reality to settle things with an advanced approach. The urban planning must be resolved so that further, we do not need to reshape it due to any future epidemic. The post pandemic state indicated that where we are, the existing legal framework referred to what we had done and forthwith what needs to be adopted. The planning of the Government for the future resilience is a clear sign that the measures framed under the legal regime have been accumulated under their action, also here the concern for climate change and environmental issues is vital. Learning from different countries is another way towards expanding our contemplate for resilience; their redesigning might assist us to shape our infrastructural modification. The learnings and other knowledge would have a significant impact if applied with due care.

A part of our infrastructural establishments is the land management system. From housing, schools, hospitals, roads, shops, everything should be brought under one conventional design in urban places. Here the conversion of planning in the urban slums cannot be neglected. Urban services, construction and infrastructures can be based on technology, to have the digital solution. From water supplies to other service providers, everywhere the implementation is needed. Strengthening financial sustainability and building governance capacity is essential as due to pandemic the people have suffered through acute financial challenges. The City authorities should produce a systematic way about safety and resilience and put them into urban governance [23]. Environmental sustainability should be a rising factor in case of any urban structural groundwork. The land management and infrastructure must follow the environmental conservation rules and procedure. Overall, the infrastructure resilience process in the urban areas in the post COVID-19 pandemic should be planned and implemented in a progressive way, making it a worthwhile standard. Existing cases with relevant policy accomplishment and a proper monitoring mechanism is all we expect further.

References

1. World Population Prospects The 2002 Revision (2004) Department of Economic and Social Affairs Population Division, United Nations. https://www.un.org/development/desa/pd/sites/www.un.org.development.desa.pd/files/files/documents/2020/Jan/un_2002_world_population_prospects-2002_revision_volume-iii.pdf
2. Das A, Roy S, Parvez M, Hasan M (2021) Decentralized activity centers in rural Bangladesh: a step towards effective emergency situation management in the post-pandemic cities. J Urban Manag 10:242–254. https://doi.org/10.1016/j.jum.2021.07.002
3. Afrin S, Chowdhury F, Rahman M (2021) COVID-19 Pandemic: rethinking strategies for resilient urban design, perceptions, and planning. Front Sustain Cities 3:1–3. https://doi.org/10.3389/frsc.2021.668263
4. COVID-19 and a new resilient infrastructure landscape (2021) OECD. https://www.oecd.org/coronavirus/policy-responses/covid-19-and-a-new-resilient-infrastructure-landscape-d40a19e3/. Accessed 27 Apr 2023
5. Puliti R (2022) The infrastructure of recovery. World Bank Blogs. https://blogs.worldbank.org/voices/infrastructure-recovery. Accessed 27 Apr 2023
6. King et al (2021) Mechanisms for addressing the impact of COVID-19 on infrastructure projects. IOP Science. https://doi.org/10.1088/1755-1315/682/1/012047. Accessed 27 Apr 2023
7. Infrastructure projects: Responding to the impact of the COVID-19 global crisis. PWC. https://www.pwc.com/sg/en/publications/a-resilient-tomorrow-covid-19-response-and-transformation/covid-proofing-infrastructure-projects.html. Accessed 28 Apr 2023
8. Rahman (2020) Reimagining post-pandemic cities and towns. The Daily Star, 13 Sept 2020
9. Mahmood SS, Hasan MZ, Hasan AMR, et al (2022) Health system impact of COVID-19 on urban slum population of Bangladesh: a mixed-method rapid assessment study. BMJ Open 12:2. https://doi.org/10.1136/bmjopen-2021-057402
10. Morshed (2020) How can Dhaka becom
11. Samin (2020) Low-income people leaving Dhaka. Business Standard, 23 Jun 2020
12. Urban Resilience Project (2015) RAJUK. https://urp.gov.bd/rajuk/#:~:text=The%20overall%20objective%20is%3A,in%20Dhaka%20and%20Sylhet%20City. Accessed 10 Sept 2022
13. Shawkat (2004, Mar 14) Constitutional amendment on local government: what is the rationale? Daily Star
14. The Right to the Highest Attainable Standard of Health (Art. 12) (2000) CESCR General Comment No. 14. Twenty Seventh Session of the Committee on Economic, Social and Cultural Rights
15. Ekramul H (2011) The Bangladesh Constitutional Framework and Human Rights. Dhaka Univ Law J 22:78
16. Government of the People's Republic of Bangladesh, National Preparedness and Response Plan for COVID-19 (2022) Bangladesh Directorate General of Health Services, Health Service Division, Ministry of Health and, Family Welfare. https://reliefweb.int/report/bangladesh/national-preparedness-and-response-plan-covid-19-bangladesh. Accessed 17 Apr 2023
17. Staff reporter (2022) Govt approves new DAP to keep Dhaka habitable. Financial Express, 24 Aug 2022
18. Staff reporter (2022) Expectations from the new DAP. Financial Express, 30 Aug 2022
19. RAJUK at a glance (2022) RAJUK. http://www.rajuk.gov.bd/site/page/9f3cea70-ae93-4f53-975f-fe391a8b9d5e/. Accessed 13 Sept 2022

20. Jahidul (2022) No more plot housing scheme for Dhaka as gazette issued finalising DAP. The Business Standard, 23 Aug 2022
21. Detailed Area Plan (DAP): Urban Decentralization (2022) Investment Decision. https://btibd.com/investment-decision/detailed-area-plan-dap-urban-decentralization/. Accessed 10 Sept 2022
22. Vianello M (2022) Urban Planning after COVID-19; Supporting a global sustainable recovery. RTPI. https://www.rtpi.org.uk/research/2021/october/urban-planning-after-covid-19/. Accessed 13 Sept 2022
23. Suchantono B (2021) How Asia's Cities Can Become More Livable After the Pandemic. ADB. https://blogs.adb.org/blog/how-asia-s-cities-can-become-more-livable-after-pandemic. Accessed 14 Sept 2022
24. Jeba (2020) Architecture for a post-pandemic world. The Business Standard, 14 Sept 2022
25. Ahmed and Marzan (2020) What will the post-Covid-19 city look like? The Business Standard, 10 Sept 2020

Amena Jahan Urmy is a Lecturer in Department of Law, Bangladesh University of Professionals (BUP) where she also serves as the Moderator of the BUP Law and Moot Court Club (BUP LMCC). She has passed her SSC and HSC from Viqarunnisa Noon School and College and then completed her LLB and LLM from Bangladesh University of Professionals with highest distinction in both. She has attained Vice Chancellor Scholarship of BUP for her academic excellence. Ms. Urmy has also led one of the premier mooting societies in Bangladesh as Vice President of BUP Law and Moot Court Club, became Champion of BUP Moot Court Competition 2019 and ranked globally as 25th Best Advocate in FDI International Arbitration Moot Court Competition 2020. She has publications on Constitutional Law published in PEN ACCLAIMS Journal in 2020 and on Human Rights in the disputed territories. Her field of interest includes Constitutional Law, Human Rights Law, Intellectual Property Law and Criminal Law. She is interested in a wide range of opportunities, ranging from academia, legal and political writing and willing to contribute her knowledge, creativity and innovative thinking in social welfares.

Sadia Binte Rahman is currently appointed as a Lecturer in Department of Law at Bangladesh University of Professionals (BUP). She has pursued her LL.B. (Bachelor of Laws) and LL.M. (Master of Laws) degrees from Bangladesh University of Professionals. She had participated in moot court competition as a researcher, later worked as an adjudicator and coach in there. She is passionate about writing on several legal matter. Her research interest covers the Civil law added to the procedural and substantive aspects, Environmental Law and Constitutional Law. Following her academic research work in undergraduate and post-graduate level, she is interested in some other areas such as; Labour and Industrial Law alongside Human Rights Law.

Abue Jawfore Taufique Ahamed Ahade is a Lecturer in the Department of Maritime Law and Policy at Bangabandhu Sheikh Mujibur Rahman Maritime University, Bangladesh (BSMRMU). He is one of the coordinators of BSMRMU Law and Moot Court Society. He earned his LL.B. and LL.M. degree from Bangladesh University of Professionals (BUP). Mr. Taufique is an enthusiastic mooter who has participated, mentored, coached, and adjudicated various moot court competitions. He possesses interest in doing research on the topic of Constitutional law, Criminal Law and Family law.

Dynamics of Urban Transformation and Real Estate Market

The Variability of Dynamics in Urban Spatial Transformation During the Post-COVID Era: Perspectives of the Global South and North

Sujit Kumar Sikder, Md Moynul Ahsan, Hassan Radoine, Omar Al-Mahadi, and Kh Md Nahiduzzaman

Abstract We explore the urban space transformation dimension in response to COVID-19, emphasizing adaptability of cities across global south and global north. It sheds light on how pandemic responses reshape urban space utilization. It is evident that the global south nations are requiring a delicate balance between affordability and effective transformation due to the financial constraints hinder comprehensive changes. Densely populated areas struggle to create open spaces, complicating adherence to social distancing, especially in informal settlements areas and in public transports modes. Global north nations, despite greater financial capacities, face challenges in reorganizing existing structures and businesses. The shift to smaller office spaces and home-based work challenges the collaborative nature of urban spaces. Above all, the innovative planning strategies are necessary a context-sensitive modeling approach, generating scenarios to guide the (re-)organization and (re-) design of urban spaces in consideration of mobility behavior changes, model shifts

---

S. K. Sikder (✉)
Leibniz Institute of Ecological Urban and Regional Development, Weberplatz 1, 01217 Dresden, Germany
e-mail: s.sikder@ioer.de

M. M. Ahsan
Department of Real Estate Development and Management, Ankara University, Ankara, Turkey
e-mail: moynulurp01@gmail.com

H. Radoine
School of Architecture, Planning and Design (SAP+D), Mohammed VI Polytechnic University (UM6P), Ben Guerir, Morocco
e-mail: Hassan.Radoine@um6p.ma

O. Al-Mahadi
Architecture and City Design Department, King Fahd University of Petroleum & Minerals, Dhahran 31261, Saudi Arabia

K. M. Nahiduzzaman
Faculty of Applied Science, The University of British Columbia (UBC) Okanagan, 1137 Alumni Ave, Kelowna, BC V1V 1V7, Canada
e-mail: Kh.Nahiduzzaman@ubc.ca; KhMd.Nahiduzzaman@um6p.ma

K. M. Nahiduzzaman (ed.), *Making Sense of Planning and Development for the Post-Pandemic Cities*, City Development: Issues and Best Practices,
https://doi.org/10.1007/978-981-97-5481-6_4

and location dynamics. This should be seen as an opportunity to achieve desired goal of broader sustainability. Therefore, the dynamic decision models such as stochastic approach can be a candidate to foster resilient, inclusive, and responsive urban built-environment. The study emphasizes the need for deeper cooperation between global south and global north countries for overcoming financial constraints and knowledge gaps. Collaborative culture in data sharing and technology are crucial for maintaining coherent transformation, especially as virtual jobs peak, making inter-country cooperation essential for evolving urban spaces and rethinking planning.

1 Introduction—Setting the Scene

Contemporary global challenges, exemplified by the frequent occurrence and escalating intensity of natural hazards, are exacerbated by our insensitive behavior, marked by excessive and inconsiderate consumption of both renewable and non-renewable resources. This compounding magnitude of impacts is contributing to the deterioration of existing infrastructure worldwide, leading to simultaneous challenges in the global north and south. The repercussions extend to the loss of biodiversity, the emergence of new forms of inequality, and environmental injustice. In response to these issues, scholars and societal actors are urging a reimagining of paradigms, emphasizing concepts such as de-growth, environmental justice, co-creation, and broader social changes to foster a resilient transformation. It is within this context that the COVID-19 pandemic emerged as a global crisis, introducing an additional layer of challenges to the pre-existing global predicament.

The reassessment of urban design and city responses to pandemics triggered by the COVID-19 pandemic aligns with historical patterns influencing cities' reactions to such crises. Evidence suggests an inevitable second wave of COVID-19, emphasizing the importance of applying proven guidelines from past pandemics like cholera, the Black Death, the Spanish flu, and the Swine flu. These guidelines advocate for the prompt implementation of physical distancing protocols and the restricted, time-bound reopening of urban facilities [1].

Cities that adhered to such regulations during historical pandemics, like the Spanish flu, successfully flattened the curve, although some faced subsequent waves with increased infections and deaths. The timing and duration of social distancing measures emerged as key factors in effectively combating pandemics, as observed in the experiences of the UK during the second wave of the Swine flu and the Spanish flu [2]. Ahsan [3] illustrates Turkey's historical response, citing examples from the Byzantine and Ottoman periods, highlighting how plagues led to technical transformations in urban life and municipal service delivery. It is critical that the ability to implement and sustain effective measures varies between rich and poor countries, shaping the overall success in combating pandemics.

Achieving sustainable urban transformation has always been a significant challenge and associated with complexities, especially about identifying the factors causing transformation and their variable dynamic impacts on physical and virtual domains in both public and private spaces. While global climate changes are the key cause of challenges including frequent and increasing intensity of foods, wildfires, cyclones, heat waves, earthquakes, etc., other disruptive forces including online based services e.g., e-commerce, virtual and hybrid mode of employment, online schooling, etc. are understood to be causes for spatial transformation in the urban, rural and remote regions [4, 1]. The dependency of online base services was acute during the COVID-19 pandemic when people's in-person mobility was severely restricted. During that time, the dynamics and momentum of physical transformation reflected in daily mobility, use of public transit, public space, including marketplaces, playgrounds, parks, and other services were increasingly visible. The continuum of urban transformation dynamics got new momentum and the dynamics continue "during" and "post" pandemic that are understood to be a "new" normal where virtuality has been immersed with contemporary "physicality", while their marital bonding is almost inseparable because of the enormity of mutual benefits that it offered to the people across the cities and remote regions. Under this dynasty of "new" normalcy, it becomes a critical area of convention about the nature of transition and its impacts of remote and hybrid (online and in-person) work, remote living, neighborhood accessibility (e.g., 15 min city), multi-modal mobility, quality of urban green and the rise of social inequality on the spatial features of the cities e.g., locational choice of residences, conventional physical presence of the commercial office spaces, retail stores, restaurants, convention halls, etc. on defining people's movement within the cities and across the regions [4, 5]. During the pandemic, new virtual-induced-physical urban living pushed citizens to adopt to a "new" and diverse working habits, characterized by virtual induced in-person activities. That indulged into creating both a digital-in person "unity" as well as "division" [6, 7]. This paper attempts to unearth the nature of such division that many countries across the globe experienced in responding to the restrictions imposed to contain the spread of COVID-19.

The ways countries responded to the COVID-19 restrictions, and adopted a rather new lifestyle emanated from their respective social, economic, and governing systems. Often countries were severely hit by the tele-couple effect of the globally networked economic and logistic (supply-chain) system. Local conflicts and unrest with anger also have been observed due to hard lockdown as lower and middle-income people needed to go out to their formal and informal occupations e.g., desk job, retail sales in the open market, street vending, begging, domestic house services, etc. to earn money for (daily) living. Global south nations with high population and building density had challenges to main physical distancing, access to shared basic services e.g., water and infrastructure, limited urban public spaces, and lacked public transport services that barely complied with the COVID protocols [8, 9] . People living in the shantytowns e.g., slum and squatter settlements needed to choose between (1) hunger and death without food while fighting against the COVID-19, and (2) risking themselves to continue to work where maintaining the COVID protocols is nothing but a luxury. The latter was the predominant scenario of the people within the low income bracket across the global south nations. On such a thrive for survival, "during-" and "post-" COVID urban transformation in the global south countries arguably remained unchanged. In this prelude, this study aims to unearth such "differential" dynamics of urban transformation that countries in the global north and south have witnessed during as well as the post-pandemic era.

During Covid-19, private space became scarcer in the home environment, and hard lockdowns also restricted citizens' access to urban greens including public parks, playgrounds and other open spaces. This has been evidently reflected in studies, for instances, high covid risk in less nature-accessible neighborhoods [10], green space use diversity associated with COVID restriction [11], spatial association of urban green space-related to COVID [12], spatial causality of urban sociodemographic factors on COVID incidence [13]. Therefore, recommendations were pouring in for the cities to take rightful measures to make them resilient based on the lessons learned from the pandemic [3], and that might equally contribute to mainstreaming sustainability transformation. However, such evidence based recommendations and scholarships were overly based on the experiences that the rich and middle-income counties experienced. They don't necessarily reflect the status quo that most of the global south nations went through. In this backdrop, this study took an attempt to unearth the differential experience that countries in the global north went through during the pandemic, what that means for the post-pandemic urban transformation, and how different they are from the rest of the world. A total count of such rather contrast scenarios is critical to define the planning and what that means for the future transformation of the global north and global south nations.

2 A Contextual yet Differentiated Dynamics of Urban Spatial Transformation

The urban transformation has been largely differentiated across the contexts of the global cities. Below we attempt to shed light on the variability of the response to the use, plan and development of urban spaces.

2.1 Public Spaces

Amid the COVID crisis, the question of "who has the right to the city" emerged as a crucial factor in regulating urban public space. Global north cities globally approached their spaces through the lens of values, culture, and growth within a governance perspective. In contrast, the concept of "the right to the city" in global south nations often reflects challenges stemming from inadequate governance practices [14].

Cities in global south economies have sought innovative uses for public spaces, such as repurposing school playgrounds for daily marketplaces. However, this adaptation has introduced new challenges, including issues of solid waste management and administrative complexities [15]. In advance economies, global cities swiftly implemented changes like creating bicycle lanes along major urban highways overnight, demonstrating a level of agility that was previously deemed unthinkable or complex before the COVID era.

2.2 Public Transit

The appeal of public transit services has waned due to concerns about close contact, leading to reduced demand for mobility. Accessibility equity in public transit services has become a prominent topic, with studies noting a surge in individual car-based mobility. In global-south countries, the already low quality of public transit or informal mobility options became costlier. To address the shift in behavior, transit authorities need to rethink and offer more attractive options compared to individual mobility, integrating multi-modal and shared services.

Cities in the global-north countries, the pandemic prompted urban planners to address unforeseen challenges, adapting to shifts in spatial dynamics and energy consumption as residential areas gained prominence over office districts. The stress of urban densification emerged with a surge in housing demand in urban fringe or countryside areas in the new normal [16]. Planners in those countries could leverage the benefits of new digital datasets gathered during the pandemic to inform public policy. Accessibility and utilization of these datasets were key considerations. Conversely, cities in global-south nations faced limitations. Spatial and energy shifts were less manageable due to resource constraints. The challenges of urban sprawl and increased housing demand were exacerbated, often surpassing the capacity for effective planning. Access to and utilization of digital datasets were hindered by issues of infrastructure, technology, and governance, limiting their potential for informing public policy. Managing service infrastructure, especially in the realms of digital and local supply services, presented formidable challenges, surpassing routine maintenance capacities in cities with constrained resources [17].

2.3 Building Space

The real estate market has suffered with rapid uncertainty due to shifts in the demand factors, new working culture, more quality housing, and locational restrictions. The dynamics of real estate markets show that real estate in urban fringes, newly global north residential areas and the rural areas has increased and the real estate characteristics have changed to overcome the risks of pandemics [16]. The real estate sector urged an urgent need for the transformation of offices in big cities to more especially appropriate/other uses and radical changes in the use of office spaces. An increase in demand for smaller office buildings and home-office spaces instead of building large towers, headquarters and regional directorate buildings has been observed [18]. As these transformative ideas shape the real estate market, global-south nations faced significant challenges in providing spaces that align with COVID restrictions in both residential and commercial structures [18]. In multi-storied apartment buildings, adhering to recommended distances became a luxury due to the inherently small spaces, making it difficult to ensure sanitized and safe interactions in common facilities like elevators, parking lots, and stairs. The same difficulties extended to commercial buildings. Figure 1 illustrates the combination of high building and population density, highlighting the immense challenges in maintaining COVID protocols for a safer environment.

2.4 E-commerce and Logistics

E-commerce and the logistics system flourished globally, with local e-commerce businesses emerging as a response to both demand and the search for new income opportunities following job losses due to COVID restrictions. Digital literacy played a crucial role in adapting to new forms of business, although the demographic structure of countries or cities became a determining factor.

With the current effects of the global pandemic COVID-19, urban spaces are experiencing a rapid transformation besides the "private", and "community" spaces. Such a dynamic could be interpreted as an immediate response to the unprecedented situation while others could be understood as an indicative pathway where the urban space could transform into the future [1, 26–29]. To put this into context, a home-based office as a private space is seen to become a salient feature in the global pandemic. This has turned the home into a mixed-use "semi-private" space with homeschooling provisions for the kids [2]. This has been a manifestation of the rapid transformational use of space, aided by smart technologies and ICT infrastructure. Such a transformational response could be understood as an "immediate resilience" to respond as well as "adapt" to the ongoing pandemic. In fact, the new home-based "semi-public" spaces (balconies, back and front yards) have to be used for daily amenities and community interaction between the immediate neighbors [15, 30]. Yet, these attributes are not consistently present in high-density apartment buildings. Moreover, it becomes extremely challenging to avoid physical interaction and maintain the required physical distance in public spaces, such as neighborhood parks, within a high-density built environment. On a broader scale, urban public spaces, especially regional parks, recreation facilities, and playgrounds, face an intense challenge. The foundational assumptions of (i) enhanced physical cohesion and (ii) increased interaction confront significant hurdles, given that COVID-19 necessitates greater physical distancing, moving from merely distancing to a complete absence of any physical interaction. This profound shift puts the objective of cohesive and highly interactive utilization of urban spaces in a state of severe challenge. Particularly challenging is the fact that while conventional wisdom on resilient and inclusive communities advocates for dense and highly interactive mixed-use development, the present circumstances of COVID-19 demand somewhat the opposite [15, 31].

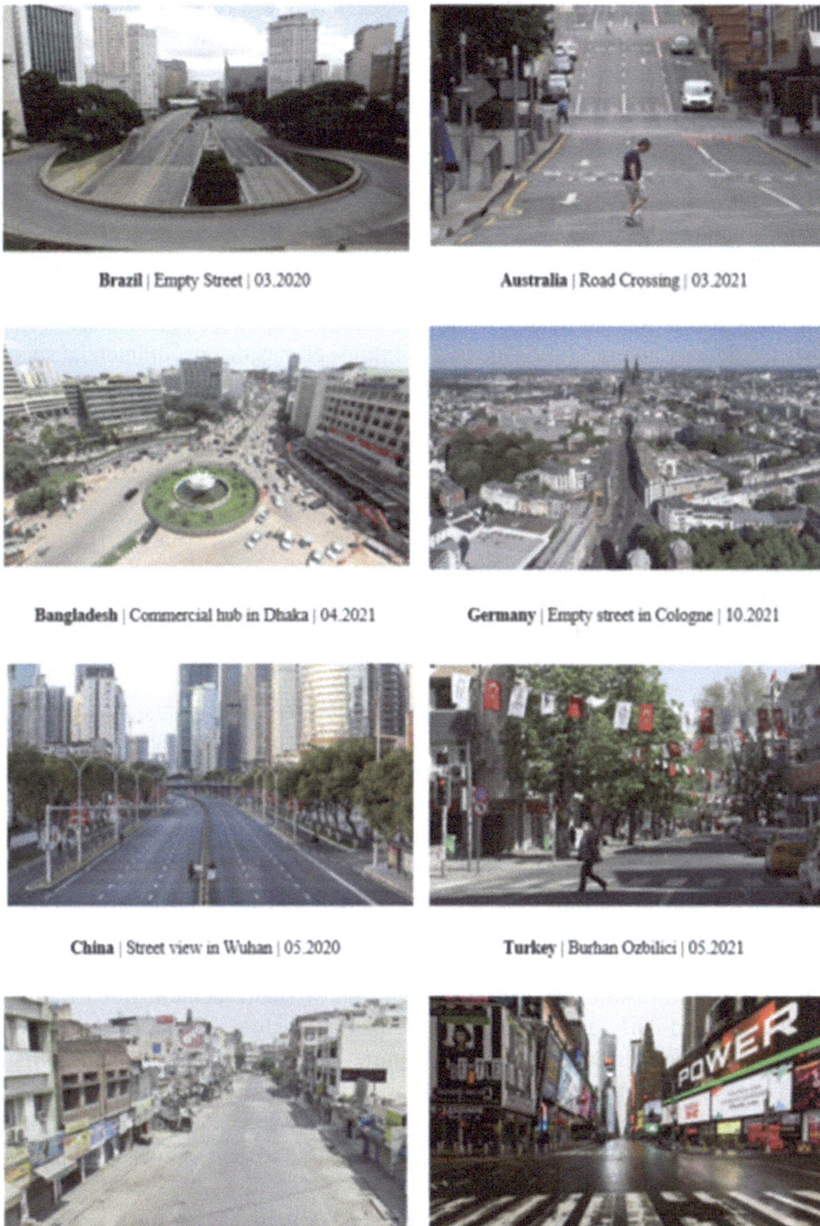

Brazil | Empty Street | 03.2020

Australia | Road Crossing | 03.2021

Bangladesh | Commercial hub in Dhaka | 04.2021

Germany | Empty street in Cologne | 10.2021

China | Street view in Wuhan | 05.2020

Turkey | Burhan Ozbilici | 05.2021

India | Empty street at Amritsar | 05.2021

USA | Empty Time Square in New York | 06.2020

Fig. 1 Desert urban space around the world that reported in electronic media; *Photo Source* Brazil [19], Australia [20], Bangladesh [21], Germany [22], China [23], Turkey [24], India [25], USA [24]

However, in global south countries, the real estate market and e-commerce sector faced distinct challenges. The transformation of urban spaces, especially in terms of home-based offices, was hindered by constraints in infrastructure, technology, and societal factors. The adoption of smart technologies and ICT infrastructure for rapid spatial transformation was less prevalent due to limitations in resources. High-density apartment buildings lacked the features conducive to the new normal, such as home-based semi-public spaces, hindering daily amenities and community interactions among neighbors. Physical distancing requirements in public spaces, particularly in densely built environments, presented challenges for cohesive and interactive urban usage, contrary to the consensual knowledge of resilient and inclusive communities advocating for dense and interactive mixed-use development. For instance, Fig. 1 shows the density and type of residential buildings alongside the commercial spaces which literally make them inseparable from each other. Although the number of traffic looked to be relatively lower on the major thoroughfare, the congestion of people and traffic almost remained the same on the local road connecting high density to the office and residential buildings.

3 Critical Link of Pandemic and Planning, Development and Management of Urban Space

The current pandemic once again critically questions the fundamental principles and approaches to planning, development and management of urban spaces. Major trends and practices, such as sustainability and resilience cannot be mastered with current methods for planning and management [32]. As global climate change continues to exert a profound transformative influence, the advent of COVID-19 has introduced challenges in the management of urban spaces. Consequently, there has never been a more crucial time to contemplate potential pathways requiring a nuanced shift in contemporary planning and design principles for urban spaces. Key aspects such as land use regulations, inclusive community planning and amenities, provisions for basic infrastructure, and the integration of smart technologies for digital socialization and daily activities should be meticulously re-evaluated through the lens of "resilience" in anticipation of a future characterized by a "new normal" [15]. These re-evaluative practices should incorporate "innovative approaches and solutions," signaling a transition from traditional to a current "more holistic, health-centered approach" in city planning and design, geared towards future sustainability considerations [33].

An integrated digital and HeWe (Health and Welbeing) infrastructure could be a successful path towards a resilient city. Even in order to reshape closed and open space in urban areas, the pandemic has revived the health- and wellbeing-related benefits of the open space [34]. Inadequate, densified, uneven accessibility in cities of many global-south countries like Dhaka, Lagos, Mumbai, Ankara etc. shows multiple spatial challenges in re-designing open spaces whereas adequate open space areas in global north countries shows innovative ideas and practices like 15 m city planning, corridor planning, sustainable urbanism. To establish enduring design solutions, it is imperative to augment the quantity of open spaces, green infrastructures, and climate-resilient structures, thereby fortifying the health and well-being infrastructure, including hygiene and sanitation, within built environments [3, 33]. Consequently, post-COVID-19, there has been a heightened demand for open and green spaces in densely populated cities worldwide, compelling a variety of transformations in open space planning to foster resilience. The call for pandemic-resilient spatial planning underscores a critical necessity to address future crises, encompassing both densely populated global-south and global-north nations.

Concerning enclosed spaces, the advent of COVID-19 has reshaped how individuals inhabit both residential and commercial real estate, prompting a shift in the design and redesign of buildings to incorporate health, safety, and resilience principles [34]. This evolving demand for enclosed spaces in global north countries introduces novel challenges, necessitating adherence to new building principles and design considerations. Apartments with limited square footage face heightened challenges, particularly in terms of insufficient or absent external visual contact. Moreover, non-residential buildings are required to adhere to new guidelines that prioritize public health considerations. The transformation of office spaces in major urban centers reflects substantial changes in their utilization, favoring the optimization of existing small office spaces over larger ones.

In the context of global south countries such as Bangladesh, India, Cambodia, the Philippines, and others, the evolving demand for enclosed spaces post-COVID-19 poses unique challenges primarily due to financial constraints and the prevailing high-density built environment.

3.1 Financial Incapability

3.1.1 Limited Resources

Many global south countries face economic challenges that limit their resources for implementing comprehensive changes to building designs, hindering the incorporation of advanced technologies or resilient building materials.

3.1.2 Affordability of Modifications

Retrofitting existing structures to meet new health and safety standards can be financially burdensome for residents and businesses, especially in densely populated urban areas.

3.2 *High Density Built Environment*

3.2.1 Space Constraints

Densely populated urban areas lack spacious land for new constructions, complicating efforts to introduce social distancing measures, ventilation improvements, or create additional open spaces.

3.2.2 Limited Open Spaces

Scarcity of open spaces in densely populated regions restricts possibilities for creating open-air alternatives, exacerbating challenges associated with designing spaces adhering to COVID-19 protocols.

3.3 *Informal Settlements*

3.3.1 Unregulated Structures

Many global south countries have significant informal settlements characterized by unregulated and makeshift structures, making compliance with new building principles complex due to informality and lack of oversight.

3.3.2 Community Challenges

Informal settlements are tightly knit communities where social distancing is inherently difficult, requiring tailored solutions considering unique social and economic.

3.4 *Public Health Awareness and Education*

3.4.1 Limited Awareness

Lack of awareness necessitates not only infrastructural changes but extensive public health education to ensure effective compliance.

3.5 *Government Intervention*

3.5.1 Policy Implementation

Implementing new building principles relies heavily on the commitment and capability of governments to enforce and regulate changes, posing challenges in global south countries with less robust regulatory frameworks.

Addressing these challenges requires a multi-faceted approach involving architectural and infrastructural considerations, considering the socio-economic realities of these regions. Innovative, cost-effective solutions balancing health and safety requirements with financial and technical capacities are essential which most of the global south nations are missing.

4 A Modelling Approach to Synthesize the Diversity in Spatial Transformation

Amidst the challenges posed by the COVID-19 pandemic, innovative modelling approaches have emerged as a useful tool for understanding the impact of current and proposed adaptations on urban spaces. In particular, the stochastic models can be leveraged to analyze changes in citizens' mobility patterns and predict how they may evolve over time [5]. Additionally, the increasing adoption of open data policies provide an opportunity to test and explore new modelling approaches that can help inform effective urban responses to the pandemic context. By utilizing these innovative tools and taking advantage of open data, it is possible to gain valuable insights for perceiving the changing needs of our urban spaces and develop responsive strategies that prioritize the health and well-being of our communities.

As COVID-19 has impacted cities in different ways, it is crucial to study these variances and extract valuable lessons that can inform global urban responses to the pandemic. However, data availability differs across socio-political contexts of the global south and global nortj countries, which is an important consideration for using modelling to understand and inform urban responses [35]. To gain a better understanding of the impacts on urban spaces, the studies should consider case studies across the spectrum of the global south and global north economies. These cities can provide insights into how to use urban spaces effectively during and beyond the pandemic. It is critical to understand the relationship between COVID-19 response measures, such as physical distancing, and the impact on urban spaces. As such, one must ask a number of research questions, including: (1) How can stochastic modelling be used to estimate the effects of COVID-19 responses on urban spaces? (2) Which open data sources can be utilized to develop models in different national contexts? (3) What physical characteristics of existing public spaces can be identified as safe and accessible through modelling? (4) How can modelling results be made accessible to urban communities and policymakers? (5) To what extent do modelling results inform contemporary planning and design principles for cohesive and adaptive urban spaces? Answering these questions can offer valuable insights and develop effective strategies to address the challenges posed by the pandemic on the built environment.

The transformation of urban spaces in response to the COVID-19 pandemic has highlighted the need for innovative approaches to understanding the potential effects of physical distancing regulations on public space usage [36]. Stochastic modelling is one such approach that can be utilized to simulate crowd movement and analyze the impact of regulations on urban space context. Some specific examples are: (1) Conducting stochastic analysis by generating scenarios and simulating crowd movement in urban spaces using mobility data. (2) Using big data analytics from open sources to triangulate secondary data on urban spaces such as the average volume of people and daily mobility patterns. (3) Integrating geospatial indicators from open-source geodata services to capture the spatial dimensions of urban spaces. (4) Performing a multi-criteria analysis that incorporates both qualitative and quantitative indicators to assess the quality of public spaces in terms of usability and safety. This analysis should involve diverse stakeholder's opinions from local professionals, municipal officials, and planning experts. (5) Utilizing a geographically weighted regression model to analyze the spatial variability in regression modelling that is sensitive to spatial causality.

The open-sourced big data is adopting to develop such multi-criteria analysis, stochastic dynamic model, and scenarios. The global north multi-criteria analysis considers open-source information, key variables, consequences, and perspectives to aid in defining the primal stochastic model. The stochastic model uses a scenario-based approach to account for the uncertainty of variables, such as physical distancing and crowd volume restrictions, as modelling parameters. Based on the primal stochastic dynamic model, the proposed adaptive stochastic dynamic model is synthesized, which fine-tunes itself by automatically updating its parameters with respect to open-sourced data [37]. The adaptive stochastic model updates the statistical characteristics of generated scenarios based on secondary data, making it adaptable to dynamic changes in urban spaces. This model addresses the challenges of uncertainty and adaptability in modelling the future waves of COVID-19.

A conceptual framework has the potential to contribute to discussions on urban space development dynamics in relation to COVID-19 and other pandemics. It is not only limited to the scientific domain but also applicable to practical applications and exploring challenges for broader social benefit through innovation in planning and design principles. Further tasks include identifying relevant strategic guidelines for new urban planning processes, creating a replicable model adaptive to other communities and governance, making an assessment toolkit available to build a long-term vision of smart urbanism, providing reliable information for evaluating decision-making processes, and comparing the adaptation potential of innovation in urban open data for broader social benefit.

5 Conclusions

This chapter delves into the intricate transformation dynamics of urban spaces in response to the COVID-19 restrictions, emphasizing adaptability and cohesiveness across the diverse landscapes of global south and developed nations. A conceptual landscape is attempted to portray integrating a stochastic dynamic decision model, providing insights into how COVID responses reshape the utilization of urban spaces.

In the context of global south nations, the ability to implement comprehensive changes is often hindered by financial constraints. The critical challenge involves striking a balance between affordability and effective transformation. Densely populated areas, common in global south nations, face challenges in creating open spaces due to space limitations, exacerbating difficulties in adhering to social distancing measures. Additionally, informal settlements present unique challenges, demanding tailored solutions that consider the specific socio-economic dynamics.

Conversely, global north nations possess greater financial capacities to implement advanced technologies and resilient materials. However, challenges persist in reorganizing existing structures to align with new health standards. The shift towards smaller office spaces and home-based work poses challenges in maintaining the collaborative and interactive nature of urban spaces.

A way forward involves a context-sensitive modeling approach that generates scenarios to inform planners and civic societies. The proposed stochastic dynamic decision model provides a foundation for anticipating potential effects, guiding the (re-)organization and (re-)design of urban spaces. This framework, adaptable to the varied landscapes of global south and global north nations, aims to foster resilient, inclusive, and responsive urban environments for future pandemics. How can global south nations overcome financial constraints to implement the proposed stochastic dynamic decision model for (re-)organizing and (re-)designing urban spaces in response to future pandemics? A deeper cooperation is needed between the global south and global north countries on matters of data sharing and technologies to maintain cross borders safety and maintain coherent transformation across the borders. While virtual jobs and services are at an extreme peak, inter country cooperation is extremely critical to create the right opportunities for transforming urban spaces and evolving employment opportunities.

References

1. Nahiduzzaman KM, Holland M, Sikder S, Shaw P, Hewage K, Sadiq R (2021) Urban transformation toward a smart city: an e-commerce–induced path-dependent analysis. J Urban Plan Dev 147(1). https://doi.org/10.1061/(ASCE)UP.1943-5444.0000648
2. Engle S, Stromme J, Zhou A (2020) Staying at home: mobility effects of covid-19. https://papers.ssrn.com/sol3/papers.cfm?abstract_id=3565703. Accessed 15 Jun 2022
3. Ahsan MM (2020) Strategic decisions on urban built environment to pandemics in Turkey: lessons from COVID-19. J Urban Manag 9(3):281–285. https://doi.org/10.1016/j.jum.2020.07.001

4. Campisi T, Nahiduzzaman KM (2023) Beyond COVID-19: planning the mobility and cities following "15-minute city" paradigm. In: Campisi T, Nahiduzzaman KM (eds) The city in an era of cascading risks: new insights from the ground. Springer Nature
5. Campisi T, Nahiduzzaman KM, Al-Rashid M, Tesoriere G (2022) COVID-19 and impacts of the evolving city models on mobility. AIP Conf Proc 2611, International conference of computational methods in sciences and engineering, 4–7 September 2021, Heraklion, Greece
6. Ahsan MM, Jerin SS, Islam MR (2023) Digital divide in education sector of Bangladesh during COVID-19. Eur J Sci, Innov Technol 3(3):96–112
7. Sikder SK, Nagarajan M, Mustafee N (2023) Augmenting EV charging infrastructure towards transformative sustainable cities: an equity-based approach. Technol Forecast Soc Change 196:122829. https://doi.org/10.1016/j.techfore.2023.122829
8. Shawon A (2020) Offices open again after the two months lock down. The Dhaka Tribune. https://www.dhakatribune.com/bangladesh/211376/offices-reopen-after-two-months-of-shutdown. Accessed 21 Feb 2023
9. Woodward A, Su R (2020) Anthony Fauci says there's 'virtually no chance' the coronavirus will disappear—3 charts predict what future waves of cases will look like. https://www.businessinsider.com/second-wave-of-coronavirus-infections-may-peak-in-fall-2020-5. Accessed 5 Jan 2022
10. Spotswood EN, Benjamin M, Stoneburner L, Wheeler MM, Beller EE, Balk D, McPhearson T, Kuo M, McDonald RI (2021 December) Nature inequity and higher COVID-19 case rates in less-green neighbourhoods in the United States. Nat Sustain 4(12):1092–1098. https://doi.org/10.1038/s41893-021-00781-9
11. Berdejo-Espinola V, Zahnow R, Suárez-Castro AF, Rhodes JR, Fuller RA (2022, March 17) Changes in green space use during a COVID-19 lockdown are associated with both individual and green space characteristics. Front Ecol Evol 10:804443. https://doi.org/10.3389/fevo.2022.804443
12. Pan J, Bardhan R, Jin Y (2021 July) Spatial distributive effects of public green space and COVID-19 infection in London. Urban For Urban Green 62:127182. https://doi.org/10.1016/j.ufug.2021.127182
13. Zhang H, Liu Y, Chen F, Mi B, Zeng L, Pei L (2021 December) The effect of sociodemographic factors on COVID-19 incidence of 342 cities in China: a geographically weighted regression model analysis. BMC Infect Dis 21(1):428. https://doi.org/10.1186/s12879-021-06128-1
14. Ahsan MM (2022) Urban governance in Turkish global competitive cities: a study on the right to the city perspective. J CivIzation Stud 7(2):14–27
15. Nahiduzzaman KM, Lai SK (2020) What does the global pandemic COVID-19 teach us? Some reflections. J Urban Manag 9(3). https://doi.org/10.1016/j.jum.2020.08.004
16. Ahsan MM, Sadak C (2021) Exploring housing market and urban densification during COVID-19 in Turkey. J Urban Manag 10(3) https://doi.org/10.1016/j.jum.2021.06.002
17. Taylor G, Anderson K, Cramer D (2021) Municipal digital infrastructure and the COVID-19 pandemic: acase study of Calgary, Canada. J Digit Media & Policy 12:137–157. https://doi.org/10.1386/jdmp_00052_1
18. Tanrıvermiş H (2020) Possible impacts of COVID-19 outbreak on real estate sector and possible changes to adopt: a situation analysis and general assessment on Turkish perspective. J Urban Manag 9(3):263–269. https://doi.org/10.1016/j.jum.2020.08.005
19. Stephen Johns (2020) COVID-19 in Brazil: research reveals how epidemic spread around country. https://www.imperial.ac.uk/news/200453/covid-19-brazil-research-reveals-epidemic-spread. Accessed 15 May 2022
20. Reuters (2021) Australia enjoys Easter with no new local coronavirus cases. https://www.reuters.com/business/healthcare-pharmaceuticals/australia-enjoys-easter-with-no-new-local-coronavirus-cases-2021-04-04. Accessed 14 May 2023
21. Dhaka Tribune (2021) Urban space during COVID-19 lockdown in birds eye view of the Sapla Chattar at Motijheel-the very busy commercial hub of capital city Dhaka, Bangladesh. https://www.dhakatribune.com/bangladesh/health/250916/timeline-government%E2%80%99s-efforts-to-curb-covid. Accessed 22 Dec 2022

22. Greif_Contzen (2021) Cologne's office space market: COVID-19, politics and market development in the months ahead. https://www.greif-contzen.de/en/press/detailseite-en/News/detail/der-koelner-bueromarkt-pandemie-politik-und-die-marktentwicklung-der-kommenden-monate.html. Accessed 19 Jul 2023
23. The Rockefeller Foundation (2021) Integrating digital technology and public health to fight Covid-19: linking China's experiences to the world. https://www.rockefellerfoundation.org/case-study/covid-19-linking-chinas-experiences-to-the-world. Accessed 16 Apr 2022
24. Seattle Times (2021) A man crosses a nearly deserted street in the city center, in Ankara, Turkey, Saturday, Turkish security forces are patrolling main streets and setting up checkpoints at entry and exit points of cities, to enforce Turkey's strictest COVID-19 lockdown to date. Burhan Ozbilici/The Associated Press. https://www.seattletimes.com/seattle-news/health/coronavirus-daily-news-updates-may-2-what-to-know-today-about-covid-19-in-the-seattle-area-washington-state-and-the-world. Accessed 18 Mar 2023
25. Intelligencer (2021, May 2) India sees 400,000 new cases in a day: COVID-crisis updates. https://nymag.com/intelligencer/article/the-ongoing-covid-19-crisis-in-india-updates.html. Accessed 12 Jan 2023
26. Adam D (2020) Special report: the simulations driving the world's response to COVID-19. Nature. https://doi.org/10.1038/d41586-020-01003-6. Accessed 3 Jan 2020
27. An der Heiden M, Buchholz U (2020) Modellierung von Beispielszenarien der SARS-CoV-2-Epidemie 2020 in Deutschland. Unter Mitarbeit des Robert Koch-Instituts, Berlin
28. Nahiduzzaman KM, Aldosary A, Mohammed I (2019). Framework analysis of e-commerce induced shift in the spatial structure of a city. J Urban Plan Dev 145(1). https://doi.org/10.1061/(ASCE)UP.1943-5444.000051
29. Salama AM (2023) Coronavirus questions that will not go away: interrogating urban and socio-spatial implications of COVID-19 measures. Emerald Open Res 1(5). https://doi.org/10.1108/EOR-05-2023-0006
30. Forsyth A (2020) What role do planning and design play in a pandemic. News–Harvard's Graduate School of Design. https://www.gsd.harvard.edu/2020/03/what-role-do-planning-and-design-play-in-a-pandemic-ann-forsyth-reflects-on-covid-19s-impact-on-the-future-of-urban-life. Accessed 29 Mar 2022
31. Honey-Roses J, Anguelovski I, Chireh VK, Daher C, Konijnendijk van den Bosch C, Litt JS, Mawani V, McCall MK, Orellana A, Oscilowicz E, Sánchez U (2020) The impact of COVID-19 on public space: an early review of the emerging questions—design, perceptions and inequities. Cities & Health 5(1):263–279. https://doi.org/10.1080/23748834.2020.1780074
32. Klaus I (2020) Pandemics are also an urban planning problem. CityLab Design. https://www.bloomberg.com/news/articles/2020-03-06/how-the-coronavirus-could-change-city-planning. Accessed 16 Mar 2023
33. Nahiduzzaman KM, Ahsan MM (2023) Asian cities in and beyond COVID-19. In: Hu R (ed) Routledge handbook of Asian cities, 1st edn. Routledge, pp 38–48
34. Galven OL, Nikolić N, Folić B, Vitošević B, Mitrović A, Kosanović S (2022) COVID-19 and city space: impact and perspectives. Sustainability 14(3):1885. https://doi.org/10.3390/su14031885
35. Sikder SK (2020) Transition to the new normal: lens on the urban spatial data science agenda. J Urban Manag 9(4):396–397. https://doi.org/10.1016/j.jum.2020.08.003
36. Berglund O, Britton J, Hatzisavvidou S, Robbins C, Shackleton D (2023) Just transition in the post-pandemic city. Local Environ 28(6):753–767. https://doi.org/10.1080/13549839.2023.2173732
37. Paköz MZ, Işık M (2022) Rethinking urban density, vitality and healthy environment in the post-pandemic city: the case of Istanbul. Cities 124:103598. https://doi.org/10.1016/j.cities.2022.103598
38. Penntoday (2020) Reflections public spaces age COVID protest, USA. https://penntoday.upenn.edu/news/reflections-public-spaces-age-covid-protest. Accessed 16 Apr 2022

Dr. Sujit Kumar Sikder is a Research Associate at the Leibniz Institute of Ecological Urban and Regional Development in Dresden, Germany. His research focuses on data-driven approaches to promote spatial sustainability transformations, particularly in the integration of land use, mobility, and energy systems. Dr. Sikder extensive professional journey and education across both global northern and southern regions have endowed him with valuable experience and a wide-reaching network. He actively participates in national and international open science and research data infrastructure communities, including the German National Research Data Infrastructure (NFDI) and GO-FAIR. Additionally, he serves on the editorial boards of the international journals Data & Policy (Cambridge Core) and SN Social Sciences (Springer Nature). He earned a PhD from the University of Bonn, a Master of Science (MSc) from the Technical University of Munich (TUM), and a Bachelor of Urban and Rural Planning (BURP) from Khulna University, Bangladesh.

Dr. Md Moynul Ahsan holds a Bachelor of Urban and Rural Planning degree from Khulna University, Bangladesh, and a master's degree in public policy from Universiti Brunei Darussalam. He received PhD in Urban and Environmental Sciences from Ankara University, Ankara, Türkiye. He is now serving as an Assistant Professor at the Department of Real Estate Development and Management at the Faculty of Applied Sciences in Ankara University, Ankara, Türkiye. He worked as a "Senior Research Officer" at the Bangladesh Centre for Advanced Studies (BCAS). He served as a "Town Planner" in different land and housing projects in Bangladesh. He worked as a "Research Assistant" in the Universiti Brunei Darussalam and as a "Teaching Assistant" at the Department of Urban and Regional Planning in the Bangladesh University of Engineering Technology. His current research interests include urban planning, strategic planning, sustainable cities/real estate, public policy, housing, disaster, and climate change.

Prof. Hassan Radoine is an architect-planner and the Director at the School of Architecture, Planning & Design at UM6P. He received his PhD and M.Sc. in architecture and city planning from the University of Pennsylvania, USA. He is an expert in UN-habitat, ICOMOS, ICROM, UNESCO and Agha Khan Award. He holds key roles in various research projects. For instance, he was co-Principal Investigator of the Urban Metabolism research project with MIT, and he is currently the Principal Investigator of the Territorial Innovation and Performance Lab, and Principal Investigator of urban/territorial design Project. His expertise ranges from territorial innovation and smart performance, resilient and sustainable urban planning, urban regeneration, sustainable urban conservation, integrated urban/territorial design to smart cities.

Dr. Omar Al-Mahadi is the Chairman of the Architecture and City Design Department at King Fahd University of Petroleum and Minerals (KFUPM). He holds a PhD in Architecture from IIT Chicago, an MArch from the University of Waterloo, Canada, and a BSc in Architecture from KFUPM. With a passion for urban design and placemaking, Dr. Al-Mahdy specializes in research areas including walkability, livability, and place identity. He has been instrumental in launching the BSc program in Smart and Sustainable Cities at KFUPM, aligning with Saudi Vision 2030. As a dedicated educator and industry expert, Dr. Al-Mahdy actively engages in research and collaborations, fostering a symbiotic relationship between academia and practice. His work emphasizes data-driven approaches and innovative solutions in architecture and urban design, reflecting his commitment to advancing the field and enhancing urban living through quality of life and vibrant public spaces.

Prof. Kh Md Nahiduzzaman is Nordic trained Canadian urban planner who received PhD from the Royal Institute of Technology (KTH), Sweden, and MPhil from Norwegian University of Science and Technology (NTNU), Norway. Prior to joining Mohammed VI Polytechnic University (UM6P), Morocco, Prof. Nahiduzzaman held faculty positions at the University of British Columbia Okanagan (Canada), King Fahd University of Petroleum & Minerals (Saudi Arabia), and the Department of Urban Planning and the Environment in the Royal Institute of Technology (Sweden). His core research interest is urban resilience through the lenses of digital twins, urban transformational planning, smart cities, and climate risks and vulnerability assessment. He is a recognized researcher in the field. He is heavily engaged on collaborative research projects with the key governments and industrial partners, and (indigenous) communities across British Columbia, Canada and Morocco. So far, he has secured reputed research grants worth more than $3 million from the prestigious national and international funding agencies. He has published more than 100 scientific manuscripts in the top ranked journals, books and other reputed proceedings. Many of his works such as non-structural flood mitigation, floodplain maps and land use planning strategies in BC have direct impact on the provincial policies. He is an "Executive" and "Book Review" Editor of the Journal of Urban Management, Editor-in-Chief of "City Development: Issues and Best Practices" journal, and Editorial Board Member of the Journal of Urban Planning and Development. He is also seating on many national and international scientific committees, fora and think-tanks. Prof. Nahiduzzaman is the editor of the book "Making Sense of Planning and Development for the post-pandemic Cities" by Springer. He is also a co-chair of "Smart and Connected Cities" track in the prestigious US-Africa Frontiers program to be held on February 18-20, 2025, in Kigali, Rwanda.

Sustainable Urban Transformation as an Approach to Solving the Paradoxical Challenges of Contemporary Cities

Wail Ismail Bakhit

Abstract Cities throughout history have faced and dealt with a myriad of urban challenges, yet the ones we are witnessing now seem multifaceted and paradoxical, with contradictory drivers and divergent targets. At the forefront of these contemporary challenges is the battle for sustainability, but it exists within a complex web of endogenous and exogenous factors. This introductory article critiques our understanding of the nature of these challenges by introducing the Paradoxical Challenges Model (PCM) that depicts what contemporary cities really face; a model by which sustainability is framed, and within which it is positioned. The article stresses that the only viable method to resolve the contradictory factors of this model and to successfully address sustainability is to embark on Sustainable Urban Transformation (SUT). The article concludes with the need to develop a Sustainable Urban Transformation Framework (SUTF) to better respond to the variables and changing situations surrounding PCM construction for different cities.

Keywords Urban transformation · Paradoxical challenges · Urban sustainability · Contemporary cities · Sustainable transformation

1 Introduction

We live, year over year, in a world of dynamic changes with increasing uncertainty, rising complexity, and escalating risks; a situation in which cities are prone to unprecedented challenges. Although cities throughout history have faced and dealt with a myriad of urban challenges, the ones we are witnessing now seem multifaceted and paradoxical, with contradictory drivers and divergent targets. At the forefront of these contemporary challenges is the battle for sustainability. While the path towards

W. I. Bakhit (✉)
Urban Planning and Policy Consultant, Khobar, Saudi Arabia
e-mail: wailbakhit@hotmail.com

K. M. Nahiduzzaman (ed.), *Making Sense of Planning and Development for the Post-Pandemic Cities*, City Development: Issues and Best Practices,
https://doi.org/10.1007/978-981-97-5481-6_5

sustainable cities may appear clear and straightforward from a reductionist perspective, it is characterized nevertheless by complex ramification with many entangled endogenous and exogenous factors.

The challenges of contemporary cities and the complexities of sustainability, along with the inherent risks, have been increasingly raised in recent years. For instance, Husar et al. [1] refer to the spatial problems facing European cities and societies, from climate crises and economic developments to socio-political instability, describing them as wicked problems to highlight the complexity of issues faced in the field of urban planning. Whereas Burch et al. [2] emphasize the breadth and complexity of sustainability that transcends the climate dimension to a more entangled group of environmental, social, and economic goals.

Moreover, Childers et al. [3] acknowledge the challenges of cities worldwide that range from exploding population and deteriorating infrastructure to environmental disruptions and advise that the pressing needs of the twenty-first century are to understand urban sustainability as well as to enhance policymakers' ability to achieve sustainable management. The Organization for Economic Co-operation and Development (OECD) [4] also highlight the increasing damage caused by several major systemic risks around the world in recent decades, due to significant changes in the nature of those risks, which are also expected to undergo further changes. OECD defines systemic risks as those affecting the systems such as health, environment, transport and the like on which societies, and cities necessarily, depend on.

This introductory article critiques our understanding of the nature of these challenges surrounding contemporary cities and the complexities underlying the ongoing endeavours to achieve urban sustainability. It attempts to provide an explanation for why sustainability becomes such a vexing issue and how it cascades with so many other challenges in a way that complicates decision-making processes, both locally and globally.

Arguments and conceptual ideas presented in this article constitute an ongoing observation resulting from a wide-ranging practice across many sectors, bringing together urban planning and public policy domains from critical, holistic, and realistic perspectives. Thus, the following discourse will benefit from other evidence-based research and best practices for further development.

2 Contemporary Cities and the Paradoxical Challenges

In their pursuit to develop, prosper and stabilize in a globalized world and turbulent times, cities struggle with many negative hindrances. From the one hand, they are confronting escalating environmental and health issues that are cross-border, such as the ever-worsening climate change that resulted in severe weather disruptions and led to voluminous losses and costly damages. The Covid-19 that has swept the world since 2020 is yet another global catastrophe that upended our world and threatened our current lifestyle practices! At the same time, cities are facing local urban problems that are deeply rooted in their growth and urbanization processes such as urban sprawl, traffic congestion, air pollution, social inequality, and urban poverty.

From the other hand, cities are required to adopt and show commitment to global agendas and conventions such as the 2030 Agenda for Sustainable Development, Paris Agreement, and the New Urban Agenda (NUA) that seek a less rugged path towards peace and prosperity for people and the planet. Moreover, many cities also find themselves exposed to various threats that they have no control over, such as being in a landlocked country which makes them cut off from world markets especially for seaborne trade, or lack of developable lands or sustainable water sources that make development plans more challenging.

Beside these negative hindrances, the cities' pursuit of development has a counter side of positive motives. From the one hand, a growing number of countries are pursuing their promising national visions and development plans towards a prosperous economy and a decent quality of life. In addition, cities can benefit from global interconnectedness by reaping the benefits of globalization and free trade to strengthen their countries and uplift their economies. From the other hand, many cities have some effortless characteristics such as their unique locations that help connect faraway regions or control a vital trade route.

These endogenous and exogenous factors that include both positive motives and negative hindrances, and vary from being created to imposed, represent the current challenge of contemporary cities. This challenge can be expressed through the Paradoxical Challenges Model (PCM), which reflects the many facets of this challenge in a way that complicates achieving the desired goal by a particular city or reaching an agreed-upon roadmap among the global community. The interconnection among these factors has a reciprocal influence in a way that affects the goal that each city aspires. The paradoxical nature of these challenges appears in their internal conflict and mismatch in a way that seems irrational but can prove true when examined and clarified.

Paradoxes are an inherent feature of urban life. Many authors have raised the concept of 'urban paradoxes', which points out that there are contradictions surrounding many aspects of the urban life. Carlsen and Leknes [5] highlight the paradox found in large cities, especially in developed countries, where people keep moving to those cities despite their low score on indices of happiness or life satisfaction compared to other cities in their countries. Tijanić [6] also links urban characteristics to apparent paradoxes where the determinants of better life in cities, such as wealth and accessible transport, are contrasted with negative factors, such as dissatisfaction and traffic congestion. In his argument that urbanization is an environmentally beneficial trend, Kelbaugh [7] supports the idea of the environmental paradox of cities, where urban areas consume a large number of resources and produce huge waste yet are greener than sprawling suburbs that consume and produce less resources and waste. It is therefore quite reasonable that the contemporary challenges facing cities today display similar self-contradiction (Fig. 1).

The Paradoxical Challenges Model is composed of three dichotomous dimensions; endogenous—exogenous, positive—negative, and created—imposed factors.

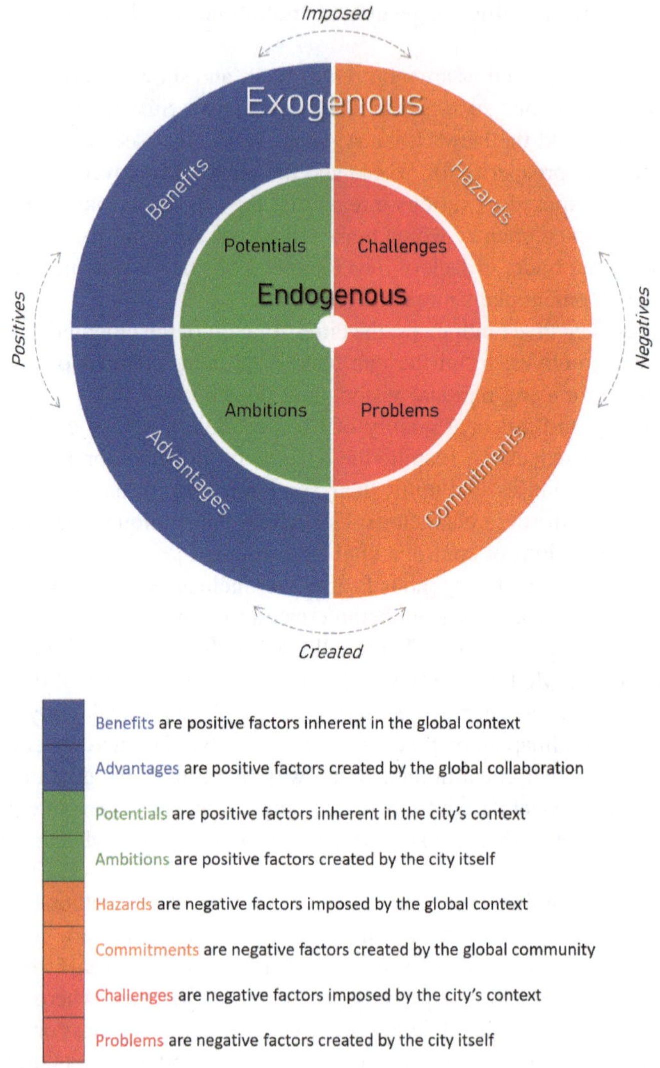

Fig. 1 The Paradoxical Challenges Model affecting contemporary cities

It assumes that each contemporary city is facing, in one way or another, a complicated challenge that consists of:

- Endogenous factors that originate from the city itself, which can be positive such as potentials and ambitions, or negative such as challenges and problems. These endogenous factors also differ from factors that have been created by the city, such as ambitions and problems, to those that have been imposed on the city, such as potentials and challenges.

- Exogenous factors that emerge from the outside, which can also be positive such as benefits and advantages or negative such as hazards and commitments. They also vary between those created by the global community such as commitments and advantages to those imposed by the global context, such as benefits and hazards.

Although not all cities are exposed to all these factors with the same degree of magnitude and diversity, however, there is no city that is completely devoid of these endogenous and exogenous factors. A city's model of possible paradoxical challenges that it may face depends on its geographic, political, environmental, urban, economic, and socio-cultural contexts. A key characteristic of the model, in general, is its internal inconsistency that brings obvious contradiction, and which requires either sacrificing some factors at the expense of the others or reaching an equilibrium through a compromise approach between all factors.

Obvious conflict can be seen between an exogenous factor (commitment), such as the SDGs, and an endogenous factor (ambition), such as national visions of countries. Although Chimhowu et al. [8] highlighted that national development plans are witnessing a global resurgence since 2006, and this trend is accelerating from 2015 due to the need to adopt SDGs, nevertheless national visions often tend to be not fully aligned or in full compliance with the global agendas! Developing countries, for instance, with emerging economies that have large reserves of natural resources (e.g. oil, gas, or coal) are likely to use this wealth in their pursuit of industrialization and modernization regardless of what SDG 13 calls for. The 2021 Production Gap Report by UNEP has alarmed that even leading developed countries are planning to increase their oil and gas production over the next two decades [9].

Endogenous characteristics of cities that pose positive motives and negative hindrances can also be in real conflict. A city ambition or goal to increase its current population, in order to capitalize on large labour market for greater economic benefits, can negatively impact its urban mobility or housing affordability. And the opposite is similarly true; an urban problem of a city such as traffic congestion can severely hinder its goal of becoming a livable city with a higher quality of life. Thus, it looks like an unbalanced equation with paradoxical variables. This situation complicates policy decisions, exacerbates global disputes, and makes the ultimate result vaguer or even with a zero-sum output. In fact, this is a multifaceted puzzle that portraits a wicked phenomenon.

The perfect equilibrium of this (PCM) that brings benefits to both individual cities and the global community, and putting sustainability at the fore, would be through maximizing the impact of positive endogenous and exogenous factors while at the same time controlling and minimizing the negative impact of endogenous and exogenous factors respectively. This is an ambitious target that is easy to theorize and difficult to achieve in practice, which again highlights the wickedness of the challenges we are facing.

3 Why Is Achieving Sustainability So Complicated?

The dilemma of achieving sustainability arises from the place where it has been positioned in the paradoxical sustainability model. This can vary from considering sustainability as an endogenous ambition (positive factor) to dealing with it as an exogenous commitment (negative factor). The global commitment to implement the 2030 Agenda for Sustainable Development and its 17 SDGs is an indispensable step and a lofty mission indeed, nevertheless when these goals are imposed from outside party without much relevance to the internal situation, they may be less welcome. And if they represent a challenge against the priorities of a country or city they will definitely be pushed aside. In contrary, when sustainability emerges as an internal ambition or goal embedded in the vision of a county or city, the attempts to achieve this goal will be more serious and the path towards it will gain more momentum. This may explain why the United Nations and other developmental agencies usually promote the idea of embracing and integrating SDGs within national visions and development plans.

There are many issues underlying the complex nature of the political ecosystem of sustainability. First, we encounter the dichotomous nature of planning between being regarded as 'normative', which focuses on what should be done from an idealistic theoretical approach and being considered as 'descriptive' or 'explanatory', which emphasizes what is being done on ground from a realistic practical view [10, 11]. Derived from and influenced by this dichotomy, sustainability is approached in most current endeavors from an idealistic non-practical perspective which looks to what ought to be; ignoring a variety of on-ground realities that should be considered to make the transition viable. An ultimate proof of this is the realization that 105 of the 169 SDGs targets cannot be achieved without adequate engagement and coordination with regional and local governments [12], those who better know particulars on ground.

Second, global cooperation and coordination, though highly needed and reflect great solidarity, require the acknowledgment that they exist in a highly contested arena. Actors in this arena take their decisions intuitively based on the likely reaction of other players, a situation that economists and political scientists prefer to interpret through game theory strategies. When achieving sustainability, at global level, requires a set of commitments that countries and cities feel uncomfortable with since they will get a higher payoff by not committing; a choice that prioritizes individual (country/city) rationality over group (the global community) rationality [13], the result will be not to cooperate despite their public agreement and regardless of the multiple rounds of the game. Thus, a look at the commitments to reduce greenhouse gas emissions reveals that developing countries, among the others, cannot easily abandon dependence on fossil fuels to meet their energy needs, especially those with large oil reserves. Another side concerns the global monitoring efforts of the greenhouse gas emissions, which relies so far on national reporting by countries. Under this situation, no country can be charged if its emissions exceed the agreed-upon commitment while officially reporting full compliance.

Third, each government is confronted, when establishing its policy agendas, with a wide range of issues that need to be tackled and solved. Within the constraints of uncertainty and bounded rationality that surround the policy environment, and according to John Kingdon's multiple-streams framework for agenda setting [14], a few of these issues will have the attention of policymakers, and very few may be properly addressed. With this in mind, sustainability can be seen as only one issue competing with many others, and due to the multiple goals for sustainability in each agency within the same government, the decision to prioritize sustainability policy may wane at the agenda setting stage.

4 Sustainable Urban Transformation

The world has finally started to seriously consider sustainability as a real and pressing target. Nevertheless, sustainability problems are considered wicked [15], and so does urban sustainability [16]. Wicked problems as introduced by Horst Rittel in 1967, are those with many characteristics among them are the lack of a definitive formulation, the good-or-bad nature of solution rather than true-or-false, and the uniqueness of every wicked problem [17]. In such problems, traditional, linear, and simple ways of thinking and acting are inappropriate.

As more than 56% of the world's population now lives in cities, with a continuous concentration in urban centers and the rapid expansion of metropolitan areas, while at the same time urban lands outpace population growth by as much as 50% [18], we can claim that most cities have far been developed, in terms of urban lands consumed, and went beyond their early stage of growth. This situation clearly highlights the issue of 'transformation' rather than 'development'. The main struggle of contemporary cities is the issue of urban transformation not urban development; it's how to transform their existing urban forms and practices to desirable ones. It is, therefore, critical to realize that "sustainable urban development is primarily about development in urban areas while sustainable urban transformation is about development or change of urban areas" [19].

This brings the issue of 'urban transformation' to the fore of this contemporary urban battle. This article argues that, in the pursuit of the most ambitious goal we are heading; the sustainability, confronting the paradoxical challenges and resolving the wicked unbalanced equation that cities are facing nowadays can only be achieved through embarking on urban transformation. UN Habitat [20] warned that maintaining the 'business as usual' approach that characterized the pre-pandemic era will eventually lead to a pessimistic scenario and that only changing course to a sustainable one can lead to an optimistic scenario.

Mere transformation however is not the ultimate goal. Due to the inherent characteristics of the 'wicked phenomenon' that surrounds the cities of today, as well as the global campaign to fight climate change and to promote sustainable future, it is highly and indisputably required to adopt the concept of Sustainable Urban Transformation (SUT).

There is an increasing importance and growing attention towards the issue of 'sustainability' as well as the topic of 'transformation'. Nevertheless, the link between the two has just recently made [21]. Despite the existence of a general common definition for both 'sustainability' and 'sustainable development', there exists no clear definition of what urban transformation is [21], which has resulted in acknowledged confusion and looseness about sustainable urban transformation. This ambiguity can be traced back to different approaches to the use of urban transformation:

- First, the term 'urban transformation' has been used as a synonym to other intersecting terms such as urban development, urban change, and urban conservation. This has been changing in planning theory and planning practice through history [22].
- Second, this term is also being used to explain the urbanization phenomenon, reflecting the migration of people from rural to urban areas or the general change in the built environment as well as in urban cultures [21].
- Third, the implementation gab between theory and practice which highlights the discrepancy between conceptual papers and empirical cases [21].

Urban Transformation has emerged as a response to the recognition that a radical and systemic change towards sustainable and resilient cities is needed [23]. Complex sustainability pressures at both local and global levels [15] as well as persistent urban problems that resulted from demographic, environmental and economic changes have derived the necessity for a comprehensive and fundamental transformation. The holistic approach to this transformation is called upon due to the failure of sectoral approaches to these sustainable challenges, and thus urban transformation is introduced as a cross-sectoral and far-reaching process [21].

According to McCormick et al. [19], Sustainable Urban Transformation "refers to structural transformation processes—multi-dimensional and radical change—that can effectively direct urban development towards ambitious sustainability goals" (p. 1). They further defined SUT according to two dimensions: drivers of radical change, such as governance and innovation, and multi-dimensional sustainable urban structure, such as spatial environment and public space. Ernst et al. [16] argue that achieving sustainable urban places is part of sustainable urban transformation, "which encompasses both sustainable urban structures and environments and (radical) economic, social, cultural, organizational, governmental and physical change processes" (p. 1).

Although the importance of urban transformation started to grow since 2010, current development processes in cities are still behind advances in related studies [21]. This is supported by McCormick et al. [19], who highlighted that growing awareness of climate change as well as sustainable development have not resulted in powerful initiatives that transform urban development into a sustainable, resilient, and low-carbon track. Moreover, the current urban transformation discourse still does not include the complex challenges highlighted in the relevant policy debates and global agendas of urban sustainability [15].

5 'Transformation' or 'Transition'?

There exists a considerable debate and apparent confusion in the use and differences between 'transformation' and 'transition'. Although these two terms have been interchangeably used more often than not, some scholars contend that 'transformation' is more encompassing concept, well connected with sustainable development and indicates both the process and the outcome, whereas 'transition' remains to be the preferred term in socio-technical system studies and only indicates the process [15]. Other authors consider 'transformation' as a subset of 'transition' [24]. A cursory look at language dictionaries can highlight the subtle differences between these two terms. The Cambridge Dictionary defines 'transition' as "a change from one form or type to another, or the process by which this happens",[1] while defines 'transformation' as "a complete change in the appearance or character of something or someone, especially so that that thing or person is improved".[2] This can give a glimpse into the difference between these two terms. It appears that the linguistic meaning of 'transformation' is more concerned with the complete and radical change which is compatible with the use of the terminological meaning stated above.

6 A Step Forward: The Need for Sustainable Urban Transformation Framework

Although we have come to suggest so far, that the only viable way for resolving the contradictory variables of the unbalanced equation explained through the Paradoxical Challenges Model and to appropriately address sustainability is to embark on Sustainable Urban Transformation, nevertheless, there is a need to develop a Sustainable Urban Transformation Framework (SUTF) to proceed on that makeover.

The importance of such a framework arises from several concerns. First, the wickedness and fuzzy nature of those interrelated problems associated with sustainability and urban transformation necessitates this framework. The multi-dimensional interconnectedness of those problems assumed by the Paradoxical Challenges Model makes the situation even more complicated to be tackled without a proper framework. Second, the apparently unsuccessful global and national endeavours in making the needed transformation, which resulted from their approach to the existing context of cities, requires this framework. Those agendas and plans sound as if we are dealing with simple problems of new cities being established on vacant tracts! The reality is however totally different and more complex; we are dealing with existing cities located within limited areas with excessively consumed resources and facing wicked problems. Third, the uniqueness of the model for each city due to the changing nature of the problems and how each city defines a certain factor (e.g. as an endogenous

[1] https://dictionary.cambridge.org/dictionary/english/transition.

[2] https://dictionary.cambridge.org/dictionary/english/transformation.

created problem or as an exogenous imposed hazard) also entails the establishment of (SUTF). As mentioned earlier, sustainability can be viewed differently by different cities, and depending on where it is placed in the (PCM), the features of the model and its desired equilibrium may vary accordingly.

Sustainable Urban Transformation Framework (SUTF) should draw upon the dynamics and realities of contemporary cities and their evolving challenges assumed by the Paradoxical Challenges Model. It also needs to consider the inter-disciplinary approach through which sustainable urban transformation operates, recognizing and acknowledging the role of public policy in embodying ambitious visions and goals, and overcoming gaps and shortage in current attempts. The desired (SUTF) is expected, therefore, to help cities balance their paradoxical challenges within an ever-contested world in a way that makes all parties, including our planet, better off.

7 Conclusion

This introductory article tried to understand the nature of the challenges that face contemporary cities and the complexities underlying the ongoing endeavours to achieve urban sustainability. The modern era brings a complex web of challenges characterized by entangled endogenous and exogenous factors that appear multi-faceted and even paradoxical. This situation complicates decision-making processes at the local and global levels in terms of achieving the desired goal by a particular city or reaching an agreed-upon sustainability roadmap among the global community. The article, therefore, introduced the Paradoxical Challenges Model (PCM) that depicts what contemporary cities really face; a model by which sustainability is framed, and within which it is positioned. Based on the understanding provided by the PCM, the article advised the need to embark on Sustainable Urban Transformation (SUT) as the only viable way to achieve a dual outcome: resolving the contradictory variables of the PCM and appropriately addressing the global sustainability agenda. Given the wickedness of the challenges, unsuccessful approach to transformation, and the uniqueness of the model for each city, the article concludes with the need to develop a Sustainable Urban Transformation Framework (SUTF) that should take into account the many aspects surrounding cities and their changing environment.

References

1. Husar SCV et al (2023) What's next? Some priorities for young planning scholars to tackle tomorrow's complex challenges. Eur Plan Stud 31(11):2368–2384
2. Burch S, Hughes S, Romero-Lankao P, Schroeder H (2018) Governing urban sustainability transformations: the new politics of collaboration and contestation. In: Elmqvist T (ed) Urban planet: knowledge towards sustainable cities. Cambridge University Press, Cambridge, pp 303–326

3. Childers DL et al (2014) Advancing urban sustainability theory and action: challenges and opportunities. Landsc Urban Plan 125:320–328
4. OECD (2003) Emerging systemic risks in the 21st century: an agenda for action. OECD, Paris
5. Carlsen F, Leknes S (2022) The paradox of the unhappy, growing city: reconciling evidence. Cities, pp 1–9
6. Tijanić L (2019) Urban paradoxes—"The other side" of urban growth and quality of life in European cities. Ekonomska Misao i Praksa 28(1):309–328
7. Kelbaugh D (2014) The environmental paradox of the city, landscape urbanism, and new urbanism. Consilience J Sustain Dev 13(1):1–15
8. Chimhowu AO, Hulme D, Munro LT (2019) The 'new' national development planning and global development goals: processes and partnerships. World Dev 120:76–89
9. SEI, I. O. E. a. U. (2021) The Production Gap Report 2021. Stockholm Environment Institute, Stockholm
10. Fainstein SS, DeFilippis J (2016) Introduction: the structure and debates of planning theory. In: Readings in planning theory, 4th edn. Wiley Blackwell, West Sussex, pp 1–18
11. Ladra' T (2013) Modern trends in urban planning theory: from the generalities of normative theories to the specifics of wisdom-based practice in reality. Courrier Du Savoir 16:107–124
12. OECD (2020) A territorial approach to the sustainable development goals: synthesis report. OECD Publishing, Paris, OECD Urban Policy Reviews
13. Axelrod R (1980) Effective choice in the prisoner's dilemma. J Conflict Resolut 24(1):3–25
14. Beland D, Howlett M (2016) The role and impact of the multiple-streams approach in comparative policy analysis. J Comp Policy Anal 18(3):221–227
15. Wolfram M, Frantzeskaki N, Maschmeyer S (2016) Cities, systems and sustainability: status and perspectives of research on urban transformations. Curr Opin Environ Sustain 22:18–25
16. Ernst L, de Graaf-Van Dinther RE, Peek GJ, Loorbach DA (2016) Sustainable urban transformation and sustainability transitions: conceptual framework and case study. J Clean Prod 112:2988–2999
17. Lönngren J, Van Poeck K (2021) Wicked problems: a mapping review of the literature. Int J Sust Dev World 28(6):481–502
18. The World Bank (2022) Urban development overview [Online]. Available at: https://www.worldbank.org/en/topic/urbandevelopment/overview
19. McCormick K, Anderberg S, Coenen L, Neij L (2013) Advancing sustainable urban transformation. J Clean Prod 50:1–11
20. UN Habitat (2022) World cities report 2022: envisaging the future of cities. UN Habitat, Nairobi
21. Koch F, Kabisch S, Krellenberg K (2018) A transformative turn towards sustainability in the context of urban-related studies? A systematic review from 1957 to 2016. Sustainability 10(1):58
22. Gülersoy NZ, Gürler E (2011) Conceptual challenges on urban transformation. A|Z ITU J Fac Archit 8(1):10–24
23. Hölscher K, Frantzeskaki N (2021) Perspectives on urban transformation research: transformations in, of and by cities. Urban Transformation 3(2):1–14
24. Nielsen J, Farrelly MA (2019) Conceptualising the built environment to inform sustainable urban transitions. Environ Innov Soc Trans 33:231–248

Wail Ismail Bakhit is an urban planning and policy consultant with more than 20 years of experience in urban planning, design management and policy domains. Throughout his extensive career, he has worked in various entities in the public, private and academia s sectors. His main research interests range from the dilemma of development between national policies and local plans, to the role of urban design and urban policy in achieving a decent quality of life, to the sustainable urban transformation amidst the changing dynamics of urban societies. Wail holds a BSc in architecture and a Master of City and Regional Planning (MCRP).

Reimagining Travellers' Airport Experience: Innovative Approaches for a Post-COVID-19 Pandemic Future

Johannes Bhanye, Mareli Hugo, Abraham Matamanda, and Kgosi Mocwagae

Abstract As the aviation industry recovers from the COVID-19 pandemic, the ability of African airports to adapt and innovate has become even more critical. This chapter presents the travellers' airport experiences in the post-COVID-19 pandemic period. It also presents practical solutions to improve passengers' airport experiences in the post-COVID-19 pandemic period. Data for the study was collected through desktop research and rapid airport ethnography by observing and interviewing travellers at Bram Fischer International Airport (South Africa), Robert Gabriel Mugabe International Airport (Zimbabwe), and Oliver Tambo International Airport (South Africa) from check-in, departure concourse, security checks and boarding. Our findings reveal that travellers' airport experiences in the post-COVID-19 pandemic period were mixed but largely negative, with reports of high uncertainty among travellers, concerns about COVID-19 infections, flight delays and cancellations, stressful experiences, luggage loss, and bad treatment by airport personnel. We conclude that accelerated recovery of the air transport sector across the African continent is vital in the post-COVID-19 pandemic period. Although African airports have been striving for improved passenger volumes and expanding their capacities, the concept of analysing and improving passenger experience is yet to be widely improved among African airports. Given that the passenger is central to the success of any airport, enhancing passenger experience should be at the heart of airport improvement initiatives in Africa. Governments should step up efforts to provide financial and regulatory support for the industry. Airports should also strive to enable seamless travel experiences where digitalisation, sustainability, and newer technologies can help ease the passenger journey.

J. Bhanye (✉)
African Climate and Development Initiative (ACDI), University of Cape Town, Cape Town, South Africa
e-mail: Joebhanye@gmail.com

M. Hugo · A. Matamanda · K. Mocwagae
Department of Urban and Regional Planning, University of the Free State, Bloemfontein, South Africa
e-mail: mocwagaeks@ufs.ac.za

K. M. Nahiduzzaman (ed.), *Making Sense of Planning and Development for the Post-Pandemic Cities*, City Development: Issues and Best Practices,
https://doi.org/10.1007/978-981-97-5481-6_6

Keywords Airports · Travellers' experience · COVID-19 pandemic · Technology and innovation · Customer service · Africa

1 Introduction

The recent health emergency caused by the COVID-19 pandemic forced people to change their mobility habits, with the reduction of non-essential travel and the promotion of online activities. During the first phase of the emergency in 2020, governments considered several mobility restrictions to avoid pandemic diffusion [1–3]. The goal behind mobility restrictions was to slow down the spread of the virus [4]. However, it also led to the stand still of transportation industries [5]. Air transport was most affected by these restrictions, with scholars claiming that COVID-19 was the greatest crisis ever experienced by the global aviation industry. With international travel bans in place, lockdowns and travel restrictions, the aviation travel demand decreased significantly [6]. This led to many airlines cancelling flights and ultimately seizing operations. With no flow of income, many airlines laid off staff to cut expenses [7]. However, following more than two years since the World Health Organization (WHO) declared the COVID-19 outbreak a global pandemic, a certain "normality" has started to emerge. The momentum created by the reopening had a positive and immediate impact on global air travel demand recovery. While many indicators point toward recovery, the aviation industry also faces significant headwinds [8].

There is a growing number of publications on the aviation industry and the COVID-19 pandemic, primarily focusing on the impacts of the COVID-19 pandemic on the aviation industry. Studies pointed out the heavy toll on global aviation, which resulted in rating downgrades, liquidation and bankruptcy of several airlines and airports due to severe cash burn instigated by travel restrictions [2, 9]. Other studies focused on job losses in aviation and related industries [7], while some studies focused on the fall in GDP supported by aviation in the region, with International Air Transport Association (IATA) previously estimating a US$28 billion decline in GDP [10]. However, there is a paucity of research on the aviation industry during the post-COVID-19 pandemic era. As the world is recovering, airports are under increased pressure to cut costs and claw back revenue. Still, they also have the challenging task of restoring passengers' confidence in air travel, getting them back to the Airport and encouraging them to spend. Critical questions in the post-COVID-19 pandemic are: How can airlines and airports turn air travel back into an experience for passengers instead of a dull necessity? How can the efficiency of airports be improved?

Airports have become complex systems in modern society. This complexity arises from the various components of the Airport, all of which have different requirements. Components include various systems, procedures, stakeholders and artefacts necessary for the operation of an airport [11]. Stakeholders include the parties interested in the running of the Airport, such as private interests (the airport owner, shareholders, the airlines), government bodies (customs and security); travellers (passengers and visitors); and agencies (such as the International Air Transport Association [IATA] and various government agencies at different levels) [11]. As travellers' demands continue to change, airports are no longer travel infrastructure only but an experience hub for passengers. They act as a gateway that connects passengers for social interactions, business opportunities, and leisure. Hence, it is critical for airports to make their passengers feel safe, engage them, and improve the 'passenger experience.' However, as passenger numbers increase, so too do the challenges faced by airports. IATA [12], for example, projected the global passenger traffic to double by 2037. Accommodating this fascinating growth is a significant challenge for the air transport industry and governments. It will require new standards, harmonised regulations and adequate infrastructure [12].

This chapter looks at the airport travellers' experience in the post-COVID-19 pandemic era, with a specific focus on Bram Fischer International Airport (South Africa), Robert Gabriel Mugabe International Airport (Harare), and Oliver Tambo International Airport (South Africa). As the world is emerging from the COVID-19 pandemic, the aviation industry faces new challenges. How we travel has changed, and airports must adapt to ensure the safety of passengers and staff. However, the pandemic also presented an opportunity to reimagine airport travel and create a more efficient and enjoyable experience for travellers. While there is some research on the impact of the pandemic on the airline industry and airport operations, there is limited research on how it has affected travellers' experiences and what can be done to address the challenges they face. This chapter aims to fill this gap by providing insights into the post-pandemic travellers' experiences and identifying practical solutions for improving their experiences at airports.

2 Airport Travellers' Experiences

The sequential nature of the passenger process has remained unchanged since the 1990s [13]. This sequential nature can be seen in Fig. 1, showing the 14-step journey established by the International Air Travel Association (IATA). Passengers usually arrive between 1 and 2 h before the departure of the actual flight, in order to prepare for the flight. Preparing for a flight includes ticket issuance, check-in, document and ticket checks, a separate station for authentication checks, baggage check-in and drop-off, and exit control. After all these steps, the passenger is still not on the plane. Security checks are done before boarding the plane, and passengers may have to wait a while to board. Upon arriving at their destination, entry control takes place, after which passengers can collect their baggage before heading to customs. All these steps add to the overall travel time and can confuse uninformed travellers.

Airports have many types of travellers: airlines, retailers, service providers, and, of course, passengers. To have the most significant impact on all the Airport's travellers, the passenger should be the starting point for innovation [14]. But, airports pose unique traveller experience challenges. First, their sophisticated, multilayered operations resemble miniature cities [15]. Second, a traveller's total experience is the product of many different airport stakeholders—security, baggage handling, and

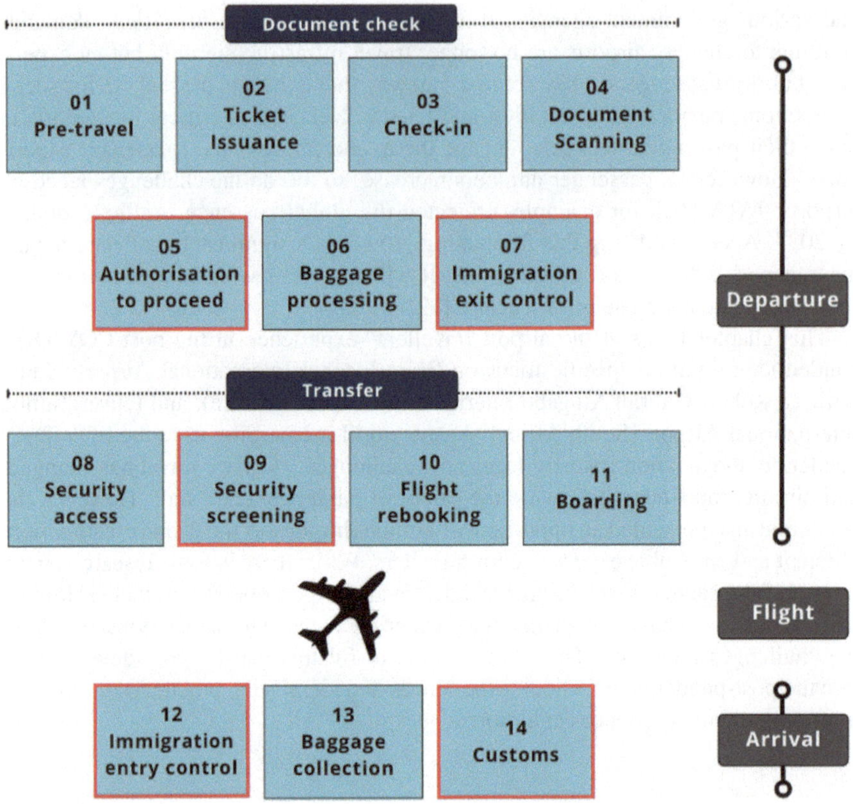

Fig. 1 The passenger journey defined. *Source* Adapted from the International Air Transport Association [13]

airlines included—necessitating a community approach dedicated to delivering a delightful, seamless passenger experience [16–18]. Airports also cater to diverse passengers with different profiles, trip purposes and expectations of what constitutes a stellar experience [16]. Flexibility and adaptability must become a strategic part of any successful passenger experience [19]. Finally, the airport traveller experience embodies how travellers perceive their interactions with an airport. These interactions can be delivered in person, online or via other channels many days before and/or after an actual travel event. These diverse and sporadic experiences add to critical moments of truth that create the overall traveller experience [18, 19].

It is becoming increasingly essential to ensure passengers enjoy a smooth, trouble and stress-free journey, not only within the Airport and during the flight but from the moment they make their booking [17, 20]. Passengers want to feel relaxed and confident and enjoy the travel experience, whether travelling for business or pleasure [16, 20]. A satisfied passenger is vital if an airport wishes its revenues to increase, traffic to grow, and reputation for remaining esteemed. This is achieved by providing travellers with a high-quality passenger experience. 'Seamless travel' is frequently used to describe the optimum passenger experience, meaning a traveller can complete the entire airport journey without experiencing delays, disruption, or confusion [21, 22]. Self-service technology, predictive analysis, artificial intelligence, real-time information, and data-sharing are some of the digital concepts airports worldwide are now going for in an attempt to achieve a seamless passenger experience [21, 23].

With demand rising and expectations high, airports and airlines are striving to find new ways to keep passengers happy and operations as efficient as possible [20]. Competition is high within the transport sector, and the importance of traveller satisfaction cannot be overlooked. Improving the airport experience for passengers can significantly affect the value of the airport services and its reputation [16, 22]. Increased passenger satisfaction can also translate into opportunities for airports to increase revenue, both aeronautical and non-aeronautical. Studies show that measuring passenger experience is a provable way to increase revenue and grow their airport business. In fact, according to AICE research, every 1% increase in the global passenger satisfaction mean generates an additional 1.5% in non-aeronautical revenue [24]. Leading airports are now creating Traveller Experience departments dedicated to ensuring that every action the Airport takes has the traveller's best interest in mind [17, 24]. Modern airports interested in improving the traveller experience need to understand the traveller's perspective on the needs and wants of the people passing through the terminals and gates.

With the shocks of the COVID-19 pandemic, the expectations of travellers and, by extension, aviation passengers have changed dramatically [25]. Airports and airlines no longer get credit for delivering on the basics,they must exceed expectations through innovative, useful, and usable traveller-centric solutions [25, 21]. Airports increasingly compete with each other and alternative transport modes for passengers,therefore, understanding airport traveller experience in the post-COVID-19 period is essential right now.

3 Methodology: Rapid Airport Ethnography

After entering the post-COVID-19 pandemic period and passing through many airports, a question came to our minds, having witnessed the increased volumes of people, to understand the experiences of travellers at airports in the post-COVID-19 pandemic period through ethnographic lenses. Airports are somewhat complex and new places of study. However, researchers are increasingly adopting diverse approaches to studying airport travellers' experiences, including surveys, case studies, and experimental, qualitative, and ethnographic approaches. Bogicevic et al. [26] did a quantitative study on the impact of traveller-focused airport technology on traveller satisfaction. The study revealed a positive relationship between airport self-service technologies and travellers' confidence benefits and enjoyment, positively affecting overall traveller satisfaction. Munoz et al. [27] also adopted a quantitative approach for a study modelling air travellers' experience based on service quality stages related to airlines and airports. Guerreiro et al. [28] adopted focus group discussions to study airport accessibility and navigation assistance for people with visual impairments. Mithilesh Kumar [29] adopted a case study on Indira Gandhi International Airport in India, while Awad et al. [30] adopted a survey approach to study travellers' perception of service quality at Dubai International Airport. The studies equally demonstrate how airports have become one of the most critical infrastructures in the background of globalisation. Airport spaces are now gateways associated with fast mobility and crossing borders by diverse travellers, and like Laghidze [15] highlighted, modern airports are now like micro-cities.

Scholars are also increasingly adopting ethnographic studies [31] often involving spending significant time in airports, observing and interacting with travellers, and documenting their experiences through field notes, photographs, and video recordings. Brenda Chalfin [32] ethnographically explored a peripheric airport in Ghana. Zuskáčová [33] also ethnographically took on the role of a Business Class passenger to empirically reflect on the issues of placeness and non-placeness while routinely passing through one of the world's busiest airports. Airport ethnography is a research method that involves observing and analysing people's behaviour, interactions, and experiences in and around airports. It is a form of ethnographic research that aims to gain a deep understanding [34] of the cultural practices and social dynamics unique to airports and the people who use them.

In this study, we adopted rapid airport ethnography considering airport travellers' busy and mobile nature. Rapid airport ethnography is a research methodology involving quick and focused observation of airport travellers to gain insights into their experiences and behaviours. We adopted rapid airport ethnography at three airports: O.R. Tambo International Airport, South Africa (1 July 2022, 4 July 2022, 8 July 2022, 28 August 2022, 1 September 2022); Robert Gabriel Mugabe International Airport (1 July 2022, 4 July 2022, 8 July 2022, 30 July 2022); and Bram Fischer International in Bloemfontein (28 August 2022, 1 September 2022). In the first step of the rapid airport ethnography, we defined the research question and objectives. The key research question for the study was: How has the COVID-19 pandemic impacted airport travellers' experiences, and what are some practical solutions for improving their experiences in the post-COVID-19 era? The research objectives were to observe and document airport travellers' behaviours and experiences, to identify patterns and trends in their behaviors and experiences, and to identify practical solutions for improving their experiences.

The second step was to select participants for the study. After securing their consent, we conveniently selected 15 participants at each Airport for brief interviews. Participants were selected from different parts of the Airport, such as check-in, security, and boarding gates, to understand their experiences comprehensively. Data collection was also done through observations and note-taking. We observed participants' behaviours, interactions, and experiences at different stages of their airport journey. We also took note of changes in procedures and protocols that have been implemented due to the COVID-19 pandemic. Researchers can also conduct brief interviews with participants to gain additional insights into their experiences. Data analysis was done by coding and categorising the notes taken during interviews and observations. We identified patterns and trends in the data and used them to generate insights into participants' experiences. These insights were used to identify practical solutions for improving travellers' experiences. We also assured participants of their privacy and confidentiality in writing the final manuscript by anonymising their names using pseudonyms.

Reflecting on our rapid airport ethnography methodology, we encountered several challenges in implementing the study. One significant challenge was the dynamic and fast-paced nature of airport environments. Airports are bustling with activity, and travelers are often on tight schedules, making it challenging to capture their experiences comprehensively within limited timeframes. Additionally, the ever-changing COVID-19 protocols and procedures added another layer of complexity, requiring constant adaptation and flexibility in our approach.

To address these challenges, we implemented several solutions:

- **Focused observation**: Given the limited time available for each observation session, we focused on specific areas within the airport where significant traveler interactions occur, such as check-in counters, security checkpoints, and boarding gates. This focused approach allowed us to maximize our observations within the constraints of time.
- **Convenient participant selection**: We adopted a convenient sampling method to select participants for brief interviews. While this approach may have limitations in terms of representativeness, it enabled us to quickly gather insights from a diverse range of travelers across different parts of the airport.
- **Flexibility in data collection**: Recognizing the fluid nature of airport environments, we remained flexible in our data collection methods. In addition to interviews, we relied on observations and note-taking to capture real-time behaviors and experiences. This multi-method approach provided a more holistic understanding of travelers' experiences.
- **Adaptation to COVID-19 protocols**: With COVID-19 protocols constantly evolving, we ensured that our methodology remained adaptable to these changes. We closely monitored and documented any adjustments in airport procedures related to the pandemic and incorporated them into our observations and analysis.
- **Privacy and confidentiality assurance**: To maintain the trust and cooperation of participants, we prioritized their privacy and confidentiality. We anonymized participants' names using pseudonyms in the final manuscript, ensuring that their identities were protected while still contributing valuable insights to the study.

Through these strategies, we were able to navigate the challenges of conducting rapid airport ethnography and gather valuable insights into travelers' experiences during the post-COVID-19 era.

To answer the second research question of the study: "What are the practical solutions to improve passengers' airport experiences in the post-COVID-19 pandemic period?" we reviewed secondary sources of information. The desktop research for this study was performed between 1 August 2022 and 14 September 2022. Literature sources ranged from credible journals, online books, newspaper articles, and airline blogs. Search terms for the literature search included: practical solutions, passengers' airport experiences, post-COVID-19 pandemic period, technology and innovation, wayfinding, traveller service, airport passenger flow, airport amenities, ambience and atmosphere, airport parking, and non-aeronautical activities. Data search was improved through the use of truncation and wildcard symbols ('*', '?' or '$'), Boolean operators (AND/OR/NOT), and field codes (i.e. [TI] vs [TIAB] vs [TW]). Using literature sources with some degree of credibility was very essential as this influences the final accuracy and validity of the information presented in the chapter.

The data were analysed according to the themes guided by the research questions. Collected data was then scanned regarding its usefulness to the research questions. The analysis focused on the existing text and was linked to the themes identified by the research questions. Empirical data's overall plausibility and credibility are essential criteria for validating research findings [35]. Various descriptive and comparative units of the text were systematically identified and allowed for interpretations based on the research questions. Data were visualised using Microsoft word, SmartArt and NVivo.

4 Case Studies: Bram Fischer International Airport, Robert Gabriel Mugabe International Airport, and Oliver Tambo International Airport

4.1 Bram Fischer International Airport

Bram Fischer International Airport (Afrikaans: *Bram Fischer Internasionale Lughawe*) is a primary airport located in Bloemfontein, the capital city of the Free State province of South Africa. It was established in 1961 and was formerly known as Bloemfontein Airport. Figure 2 is the map of Bram Fischer International Airport.

Figure 3 is the entrance of Bram Fischer International Airport.

Cargo flights are also available to and from the Airport. Bram Fischer International Airport facilities include a tourist information desk, VIP passenger, business lounge, and conference room; basic retail outlets include a restaurant/coffee shop and ATMs [36]. Bram Fischer Airport provides domestic passenger flights to Durban, Johannesburg, Cape Town and George through airlines like Airlink, South African Airways, CemAir, Mango and South African Express, handling over 300,000 passengers a year. Figure 4 shows the annual passenger traffic at Bram Fischer International Airport, demonstrating a fall in passenger flow during the COVID-19 pandemic and a sudden rise in passenger flow in the post-pandemic period.

Table 1 shows Bram Fischer International Airport's main airlines and destinations.

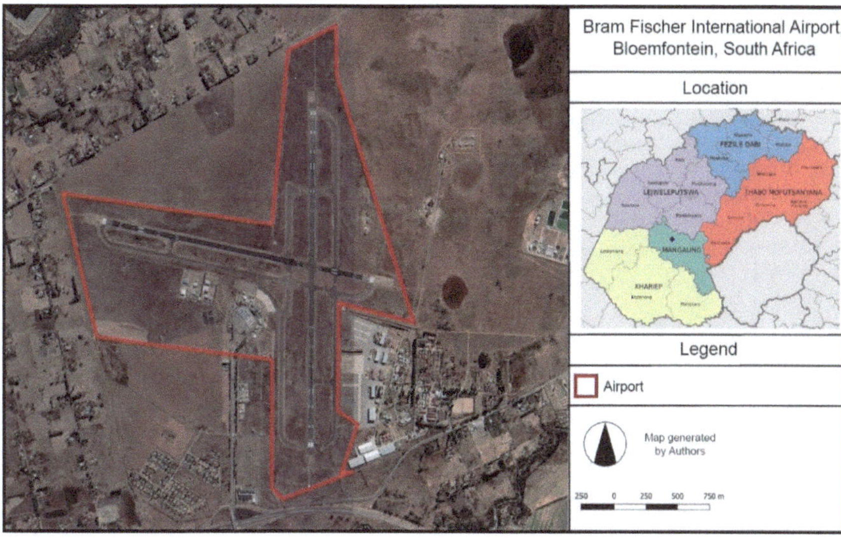

Fig. 2 Map of Bram Fischer International Airport. *Source* Authors

Fig. 3 Bram Fischer International Airport. *Source* Authors

4.2 Robert Gabriel Mugabe International Airport

Robert Gabriel Mugabe International Airport (RGIA) (Fig. 5), formerly Harare International Airport, is Zimbabwe's largest international Airport. The Airport is operated by the Civil Aviation Authority of Zimbabwe (CAAZ).

Over the years, annual airport traffic flow at Robert Gabriel Mugabe International Airport has been ever increasing. Figure 6 shows the gradual annual passenger traffic flow increase at Robert Gabriel Mugabe International Airport from 2004 to 2016.

Thus, the Civil Aviation Authority of Zimbabwe embarked on a project to upgrade the Airport. The project involves the upgrade of an international airport by increasing

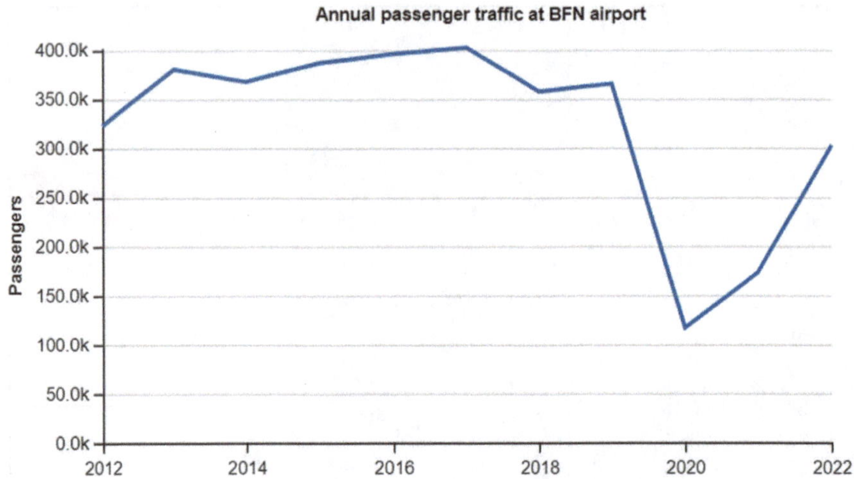

Fig. 4 Annual passenger traffic at Bram Fischer International Airport. *Source* Bram Fischer International Airport [36]

Table 1 Bram Fischer International Airport main airlines and destinations

Airlines	Destinations
Airlink	Cape Town, Johannesburg–O.R. Tambo
CemAir	Durban, George, Johannesburg–O.R. Tambo
FlySafair	Cape Town, Johannesburg–O.R. Tambo

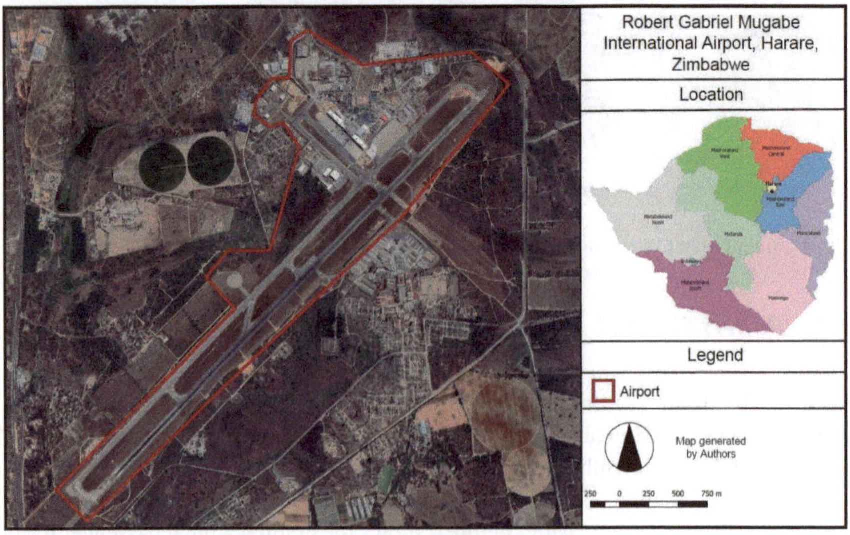

Fig. 5 Map of Robert Gabriel Mugabe International Airport. *Source* Authors

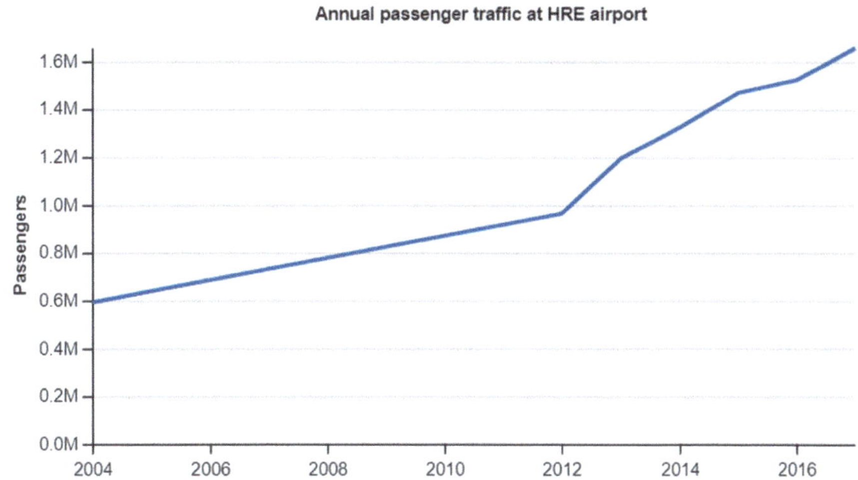

Fig. 6 Annual passenger traffic flow at Robert Gabriel Mugabe International Airport[1]

Fig. 7 Aerial view of Robert Gabriel Mugabe International Airport before the expansion project. *Source* Taderera [38]

the handling capacity from 2.5 million passengers a year to 6 million a year. It includes the construction of four new bridges and a secondary radar system, the transformation of a car park into a multi-story facility, the expansion of the runway to allow for multiple landings at the same time, and the installation of communication, navigation and surveillance system and airfield lighting system and safety systems [37]. Figure 7 shows the aerial view of Robert Gabriel Mugabe International Airport before the expansion project.

RGIA serves a variety of airlines and destinations (Table 2), including Airlink to Johannesburg's O.R Tambo airport, Air Namibia to Windhoek- Hosea Kutako airport, Air Tanzania to Dar es Salam, British Airways, Emirates, Ethiopian Airlines, Kenya

[1] Figures on annual passenger traffic flow at Robert Gabriel Mugabe International Airport beyond 2016 were not available.

Table 2 Robert Gabriel Mugabe International Airport main airlines and destinations

Airlines	Destinations
Air Tanzania	Dar es Salaam
Air Zimbabwe	Bulawayo, Dar es Salaam, Johannesburg–O.R. Tambo, Victoria Falls
Airlink	Cape Town, Durban, Johannesburg–O.R. Tambo
Emirates	Dubai–International
Eswatini Air	Manzini (begins 14 April 2023)
Ethiopian Airlines	Addis Ababa
Fastjet Zimbabwe	Bulawayo, Johannesburg–O.R. Tambo, Kariba, Victoria Falls
Kenya Airways	Lusaka, Nairobi–Jomo Kenyatta
Malawian Airlines	Lilongwe, Lusaka
Proflight Zambia	Lusaka
Qatar Airways	Doha
RwandAir	Cape Town, Kigali
South African Airways	Johannesburg–O.R. Tambo
TAAG Angola Airlines	Luanda
Zambia Airways	Lusaka

Airlines, Malawian Airlines, Proflight Zambia, South African Airways and TAAG Angola airlines for international flights. The Airport also caters to local/domestic flights under Air Zimbabwe to Bulawayo and Victoria Falls.

4.3 Oliver Tambo International Airport

Oliver Tambo International Airport (Fig. 8) in Johannesburg serves as the primary Airport for both domestic and international travel to and from South Africa and is reputed as the busiest Airport in Africa. The Airport was founded in 1952 as Jan Smuts International Airport, then Johannesburg International Airport, and O.R. Tambo International Airport in 1994.

O.R. Tambo is one of the few airports in the world that has direct flights to all six inhabited continents on earth, with a capacity to handle up to 28 million passengers annually. The Airport has over 140 retail stores open from 6 in the morning until ten at night. These times also apply to banks, bureau de change, dry cleaners, pharmacy and the spa at the Airport. The duty-free stores also stock products exclusively available only at O.R Tambo International Airport. There are six terminals at the Airport, but these can be broken down into three major areas: international, domestic, and transit. Terminals A1 and A2 handle international passengers, while the other two terminals handle domestic passengers. Due to the Airport's design, departure and arrivals terminals are considered separate terminals. The Central Terminal under construction will be named Terminal A3 and will be used for both international

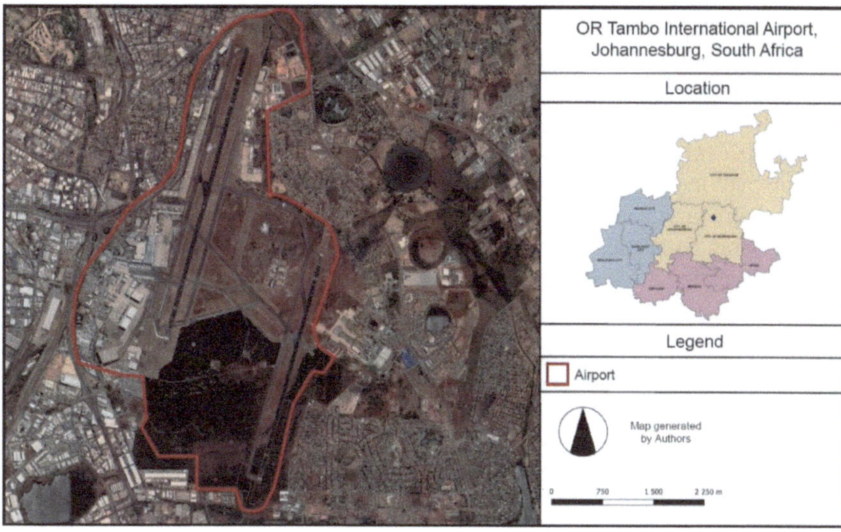

Fig. 8 Map of O.R. Tambo International Airport. *Source* Authors

and domestic passengers. The two terminals, Terminal A and Terminal B have been restructured. The Airport also houses a travel clinic, open 24 h a day with full medical staff. Figure 9 is an aerial view of O.R. Tambo International Airport.

Several major airlines operate from Johannesburg International Airport, including SAA (or South African Airways), British Airways, Singapore Airlines, Emirates, Airlink and Virgin Atlantic. Several Domestic airlines like Flysafair, Kulula and

Fig. 9 Aerial view of O.R. Tambo International Airport. *Source* Tshuma [39]

Mango Airlines operate multiple flights at the Airport daily. The Airport is also capable of handling 650,000 tons of cargo per year. Table 3 shows some of the airlines and destinations at O.R. Tambo International Airport

O.R. Tambo International Airport recorded 21.2 million passengers in 2017–2018, up from 20.7 million passengers the year before. Of those passengers, 9.2 million were international and 11 million domestic, with the remainder being classified as "regional" or "unscheduled". 220,644 aircraft traffic movements were recorded; the majority being domestic services. Figure 10 shows annual passenger traffic for O.R. Tambo International Airport.

Table 3 Some of the airlines and destinations at O.R. Tambo International Airport

Airline	Destination
South African Airways	Accra, Blantyre, Cape Town, Nairobi–Jomo Kenyatta, Durban, Harare, Kinshasa–N'djili, Lagos, Lilongwe, Lusaka, Maputo, Mauritius, Port Elizabeth, Victoria Falls, Windhoek–Hosea Kutako
RwandAir	Kigali, Lusaka
TAAG Angola Airlines	Luanda
Turkish Airlines	Istanbul
Uganda Airlines	Entebbe
United Airlines	Newark
Virgin Atlantic	London–Heathrow
Ethiopian Airlines	Addis Ababa
Emirates	Dubai–International
Zambia Airways	Livingstone, Lusaka
Kenya Airways	Nairobi–Jomo Kenyatta
British Airways	London–Heathrow
Qatar Airways	Doha
Swiss International Air Lines	Zürich

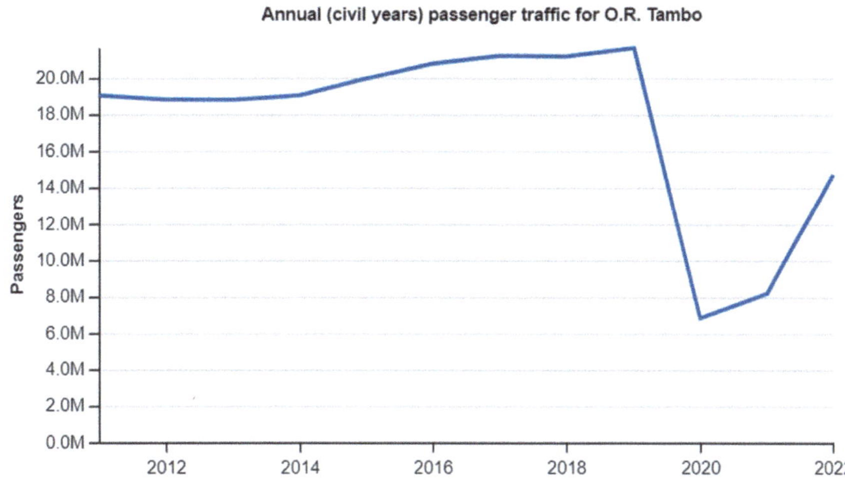

Fig. 10 Annual passenger traffic for O.R. Tambo International Airport

5 Travellers' Airport Experiences in the Post-COVID-19 Pandemic Period

Our study findings indicate diverse travellers experiences at the three airports: Bram Fischer International Airport, Robert Gabriel Mugabe International Airport, and Oliver Tambo International Airport in the post-COVID-19 pandemic period. The word cloud in Fig. 11 gives a snapshot of the travelers' airport experiences in the post-COVID-19 pandemic period.

We categorised these experiences into positive experiences and negative experiences.

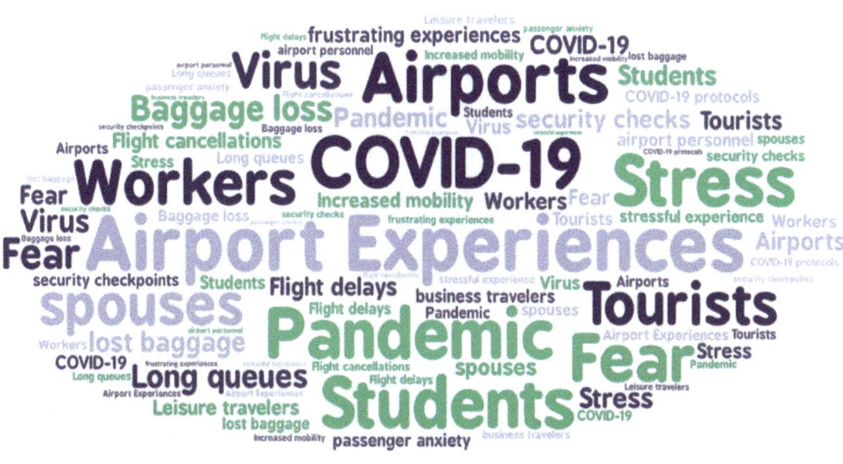

Fig. 11 Word cloud of passengers' airport experiences in the post-COVID-19 pandemic period

5.1 Positive Travellers' Airport Experiences in the Post-COVID-19 Pandemic Period

Positive travellers' airport experiences in the post-COVID-19 pandemic period ranged from airports leveraging on COVID-19 innovations, relief from tedious COVID-19 airport protocols, and elimination of fear of contracting the COVID-19 virus (Fig. 12).

5.1.1 Leveraging on COVID-19 Innovations

Some travellers admitted that the airport experience has improved because of drastic innovations adopted during the COVID-19 pandemic. During the COVID-19 pandemic, the aviation industry used technology to minimise the interaction between travellers and staff. Two years later, the post-COVID era is still benefiting from these technologies. During the pandemic, for example, online check in's become more popular, and this innovation is still in place, minimising traveller stress during check-in. The technological innovations assist in reducing operational costs and addressing the issue of staff shortage (allowing the industry to bounce back financially) and improving customer experience and satisfaction.

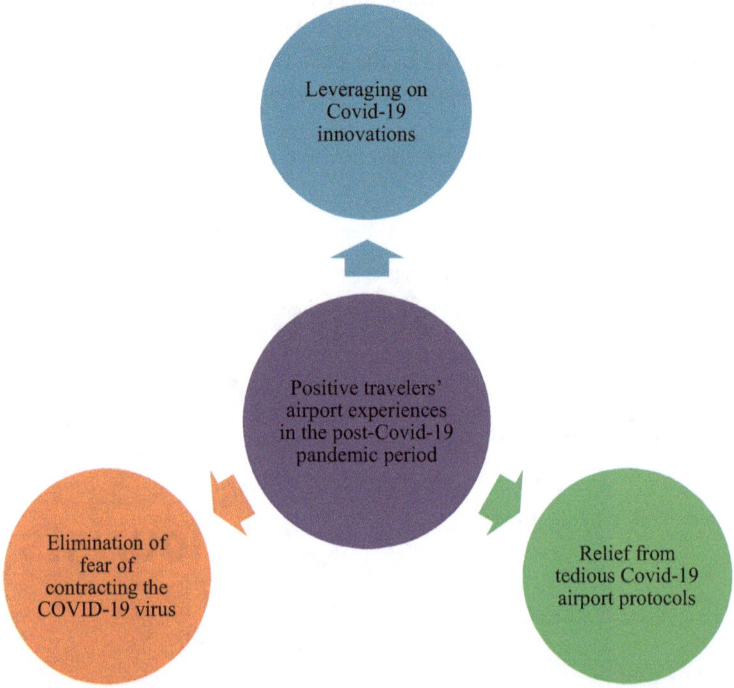

Fig. 12 Positive travellers' airport experiences in the post-COVID-19 pandemic period

5.1.2 Relief from Tedious COVID-19 Airport Protocols

With the sort of return to normalcy after the COVID-19 pandemic, many travellers indicated frictionless travel at the airports compared to the COVID-19 pandemic period. During the COVID-19 pandemic, air travel was tightly controlled, with travellers being thoroughly scrutinised. The South African government, for example, developed a risk categorisation model for different international travellers. The model classified international travellers according to a high, medium and low risk scale. High-risk travellers came from countries with higher COVID-19 infections and reported deaths compared to South Africa. Medium-risk travellers were from countries with a relatively equal number of infections and death toll to South Africa, and low-risk travellers originated from countries with fewer infections of COVID-19 and death toll than South Africa. Leisure travellers from high-risk countries were not permitted. The exception was business travellers with scarce and critical skills, including diplomats, repatriated persons, investors and people participating in professional sporting and cultural events, who would undergo the same health protocol screenings. Travellers from medium and low-risk countries were only allowed into the country subject to the prevailing visa requirements. While long-term visa holders who visit the country for business purposes were allowed to travel to South Africa, they were also subject to health screenings for COVID-19 symptoms at the port of entry [40]. Figure 13, taken at O.R Tambo International Airport, demonstrates how airports started to operate normally post-COVID-19 pandemic.

We observed increased volumes of passengers at all three airports (O.R. Tambo International Airport, Robert Gabriel Mugabe International Airport and Bram Fischer International Airport) that we passed through after the opening of international borders. Passengers ranged from students returning to school, business people, expatriates, spouses relocating, and tourists. Most of the different categories of passengers on the move that we interacted with at the airports indicated that, because of the relief from tedious COVID-19 airport protocols, they had a better experience in the post-COVID-19 pandemic phase compared to the COVID-19 pandemic period. Figure 14 shows the relaxed mood of travellers at Robert Gabriel Mugabe International Airport and Bram Fischer International Airport.

The first case involves a Business Mogul who said, "I am a business mogul. With the opening up of borders, it is time to fly again. Due to the pandemic, I had completely abandoned flying. Now that the economies are open and there are no more travel restrictions, I will be a regular traveller passing through O.R. Tambo airport. I like airports and flying, but I did not enjoy the long queues and strict COVID-19 checks at the airports."[2] The second case was of a Zimbabwean migrant worker who narrated: "I am glad the borders are now open; I have been locked down in Zimbabwe for nearly two years. I did not want to be vaccinated so that I could travel. I go to the apostolic church, and we do not believe in vaccinations. I rather waited, and now that international travel is open, I am returning to South Africa."[3]

---

[2] Interview with a Business Mogul at O.R. Tambo International Airport, 1 July 2022.

[3] Interview with Zimbabwean migrant worker at O.R. Tambo International Airport, 4 July 2022.

Fig. 13 Airports operating normally post-COVID-19 pandemic—O.R. Tambo International Airport. *Source* Authors

Fig. 14 Relaxed mood of travellers at Robert Gabriel Mugabe International Airport and Bram Fischer International Airport. *Source* Authors

The third case was of a local South African student who said: "I am an academic with the University of Free State. I am going to an academic conference in Mexico. This is my first conference in 3 years. I was being invited to several academic conferences during the COVID-19 pandemic, however, I could not attend any of them physically because of the COVID-19 travel restrictions. All my international academic engagements were online. I am so happy to meet other academics after a long physical isolation physically."[4] Another international student also said: "I a Nigerian PhD student with the University of Cape Town in South Africa. Because of COVID-19 travel restrictions, I decided to stay in Nigeria for some time. Now that the borders have opened and there are no more travel restrictions, I must fly back to South Africa to proceed with my studies."[5] The last case involves a White tourist travelling with his family, who said: "The COVID-19 screening measures at the airports were so frustrating. Now that they have removed all the screening measures, we can move and explore the world. I am on my way to Victoria Falls with my family. We have a bucket list of other places we want to visit now that we are no longer mandated to produce too many documents."[6]

5.1.3 Elimination of Fear of Contracting the COVID-19 Virus

For some travellers, the good airport experience was beyond removing the frustrating COVID-19 protocols. It was also associated with eliminating of fear of contracting the COVI9-19 virus. Figure 15 shows travellers travelling without fear of contracting the COVID-19 virus.

At O.R. Tambo international airport, for example, we observed that passengers were more relaxed, moved freely and spent more time exploring various outlets. With the massive vaccination programmes globally and the resultant complete haul of the COVID-19 pandemic, travellers felt air travel was becoming less stressful. On our flight back to South Africa with the Fast Jet Airline, we interviewed an elderly South African lady who indicated that, because she had underlying health conditions (diabetes, to be specific), she could not enjoy air travel during the COVID-19 period. She narrated: "I am diabetic, and because of this health condition, I was so vulnerable to COVID-19. I completely abandoned air travel. Airports were epi centres of the spread of the virus. But now I can travel anywhere in the world. I think air travel in the post-COVID-19 pandemic is a good experience for me."[7]

[4] Interview with a Local South African Student at Bram Fischer International Airport, 28 August 2022.

[5] Interview with a Nigerian Student at O.R. Tambo International Airport, 8 July 2022.

[6] Interview with a Tourist at O.R. Tambo International Airport, 8 July 2022.

[7] Interview on air with an elderly South African lady, 28 July 2022.

Fig. 15 Travellers travelling without fear of contracting the COVID-19 virus *Source* Authors

5.2 Negative Travellers' Airport Experiences in the Post-COVID-19 Pandemic Period

Negative travellers' airport experiences in the post-COVID-19 pandemic period ranged from high levels of uncertainty among travellers, concerns about COVID-19 infections, flight delays and cancellations, stressful experiences, luggage loss, and bad treatment by airport personnel (Fig. 16).

5.2.1 High Levels of Uncertainty Among Travellers

Our observations and interviews pointed out that there were high levels of uncertainty among travellers in the post-COVID-19 pandemic phase. Travellers faced more travel stress than pre-COVID, influencing how they approach air transportation and travel trips. This is due to the uncertainty they face around check-in, flight delays, security issues, luggage issues, long waiting times, and parking, among other airport services. During the study, other travellers also highlighted that regulations around COVID-19 were still different at airports in other countries, leaving travellers uncertain about travelling to those countries. Other travellers were also concerned about another lockdown, leaving them in a different country or town. Besides these uncertainties, many travellers are still concerned about possible infection during travel or abroad in other countries. We present this below.

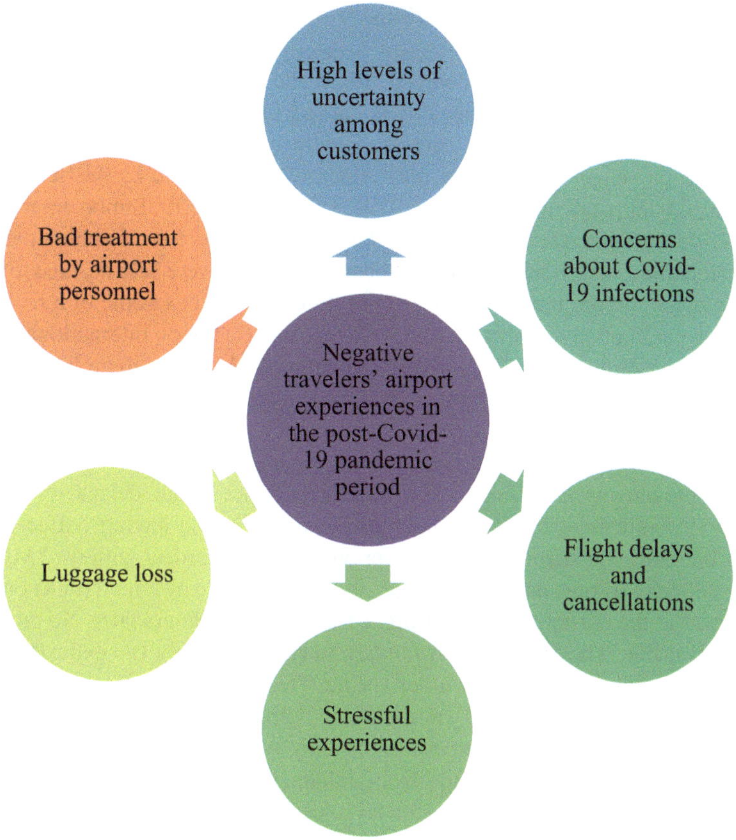

Fig. 16 Negative travellers' airport experiences in the post-COVID-19 pandemic period

5.2.2 Concerns About COVID-19 Infections

For some travellers, there are still fears of the spreading of the virus at the airports in the post-COVID-19 pandemic period. For these passengers, the removal of COVID-19 protocols gives them goosebumps that the spread of the virus is now much easier. One of our interviewees at O.R. Tambo international airport said: "We are never sure if the COVID-19 pandemic is completely over. Yes, I am travelling, but the removal of COVID-19 travel protocols makes me feel uncomfortable. Why? People are now roaming around without masks, no sanitising, no social distancing. Many airlines have also removed the mask mandate. This is concerning!"

5.2.3 Flight Delays and Cancellations

Passengers do not want to be delayed. From the start of the passenger journey, delays become stressful. Whether this is caused by an unexpected traffic issue, a problem with car parking at the Airport, a security hold-up or a flight delay, it all adds up for passengers [20]. Other passengers were frustrated by the delays in the post-COVID-19 pandemic phase. We observed that, at O.R. Tambo International Airport, immigration queues were long, delaying travellers from catching their next connecting transport. Other travellers indicated that, they were held hostage for hours by their airline before they could board the flight. A good example of a frustrating delay was of a Zimbabwean friend whom we met at O.R. Tambo International Airport on our way to Zimbabwe. We had coffee together at Mugg & Bean while waiting for our flight at 15.30 and his flight at 18.00. After we arrived in Zimbabwe, he later told us that he had languished in the cold for hours because his flight was delayed to 21:00.

Other passengers also shared their bad experiences with delays in the post-COVID-19 pandemic phase. A young man we met while having coffee at O.R. Tambo International Airport shared his previous week's ordeal with us: "My flight was cancelled two hours before departure, and I was notified by e-mail (that I only saw when I was at the Airport). The next flight was eight-plus hours later. No rebooking to another airline. They suggested refunding only the base fare (no extras/baggage), and then I can pay my fare on the other airline. This is not an acceptable response as they could have helped us avoid waiting at the Airport for 8 h.[8]

Another traveller reported a terrible flight delay experience with one of the airlines: "I recently flew on Fast Jet airline from Johannesburg to Harare. The flight was delayed by almost an hour due to technical issues on the aircraft. They made us wait for almost one hour on the bus while they fixed the aircraft. I did not feel safe at all while flying, and I was highly disappointed. While up in the air, the aircraft was making a really strange squeaky noise which was very unsettling considering that the aircraft had some technical issues. It was a terrible experience![9]

In another case, a couple also reported terrible delays at O.R. Tambo International Airport: "My husband and I were travelling from Johannesburg to Harare, we got to the Airport 3 h before the time only to wait almost an hour to check-in. The staff finally arrived to open up the counters but made the passengers wait another half an hour before proceeding to assist. Fastjet has a promotion where you allowed two checked-in bags; however, we were told to leave two bags behind on standby and collect them a day later from Harare airport; this type of service is unacceptable. They should not advertise something which they cannot live up to.[10]

---

[8] Interview with Julius at O.R. Tambo International Airport, 1 July 2022.

[9] Interview with Lucia at O.R. Tambo International Airport, 30 July 2022.

[10] Interview with the Kruger Couple at O.R. Tambo International Airport, 30 July 2022.

5.2.4 Stressful Experiences

While the majority of the travellers were very delighted about the removal of COVID-19 travel protocols, returning to the skies meant dealing with the same old air travel stress: long flight delays, cancellations, extensive security checks, and lost baggage, among other frustrating experiences. With the increased volumes post-COVID-19, some passengers expressed having stressful airport experiences. The long lines at check-in, ticketing, and security checkpoints triggered stress. Especially queueing is one of the biggest causes of passenger anxiety within the Airport. One of the travellers narrated, "I do not like queues, and I hate being late! Because of the overflow of people post-COVID-19 pandemic, moving through airports now feels like a continuous process of moving from one queue to the next: baggage drop, security, passport, and gate. The experience is very stressful."[11] Another traveller narrated: After the COVID-19 pandemic, airports have become places of stress, urgency, and inconvenience. I work a stressful job, but airports are now more stressful than work."[12] Ineffective communication from airport staff, such as how long the wait time for screening at a checkpoint will take, also added to travellers' frustrations. Another stressful experience travellers had was the confusion of navigating maze-like terminals like O.R. Tambo International Airport and the seemingly endless waiting that comes with checking baggage and screening through security checkpoints. These moments where passengers do not have control cause even more stress that they might not make that flight or the next flight.

5.2.5 Luggage Loss

Another bad airport experience we observed in the post-COVID-19 period was luggage loss. On our way to Zimbabwe, we checked in luggage at O.R. Tambo International airport to collect it at Robert Gabriel International Airport. Unfortunately, upon arrival, the luggage was nowhere to be found. The airline had deliberately placed the luggage on the delayed flight that our friend was to come with at 21:00. Upon our arrival; we observed hundreds of other bags that were misplaced and mishandled. We had to come and collect the bags the following day after several calls and follow-ups. Other passengers also had the same bad experiences with their luggage. One lady at Robert Gabriel Mugabe International Airport said: "I lost my check-in baggage from Harare to Johannesburg. I nearly missed my connecting flight trying to recover my bag. To make matters worse, the plane was delayed for more than 2 h without any communication from the team on the ground.[13] Another young man we met the day we recovered our bags at Robert Gabriel Mugabe International could not recover his bag for over a week: "I lost my luggage, and up to now, I have not yet received any

[11] Interview with Maria at O.R. Tambo International Airport, 30 July 2022.

[12] Interview with Portia at Robert Gabriel Mugabe International Airport, 8 July 2022.

[13] Interview with Judith at Robert Gabriel Mugabe International Airport, 8 July 2022.

meaningful assistance. I had the most horrible experience with Fast Jet. They lost my luggage over a week ago, and no one is willing to assist."[14]

5.2.6 Bad Treatment by Airport Personnel

Other passengers expressed dissatisfaction with the treatment they got from some airport personnel. Travellers expressed that they expect smooth and easy travel experiences where they feel valued and important. However, as one traveller lamented, there are still huge disparities in the treatment of travellers: "Aviation companies are not designing equally for all travellers. Most of the aviation experiences are designed to cater for business or first-class passengers first and foremost. The economy class passengers are treated as garbage. Whilst it is true that not all travellers have an equal level of short-term value—many that are being deprioritised have great lifetime value. All the travellers should be treated equally."

These negative airport experiences and overall dissatisfaction could potentially result in travellers choosing another mode of transport or other better airports and airlines. These bad travellers' experiences could be easy fixes that might not even call for grand renovations of airport terminal layout designs and expensive upgrades. Below, we present some practical solutions that can be adopted to improve passengers' airport experiences in the post-COVID-19 pandemic period.

6 Practical Solutions to Improve Passengers' Airport Experiences in the Post-COVID-19 Pandemic Period

Providing a good airport experience can make the difference between travellers regularly choosing to fly out of one hub over another and increasing revenue for the Airport. There are several practical solutions that can be adopted to improve passengers' airport experiences in the post-COVID-19 pandemic period. These include technology and innovation, airport cleanliness, clear and efficient signage, good traveller service, improving airport passenger flow, improving airport amenities, improving airport ambience and atmosphere, investing in restaurants and retail, investing in activities for passengers, improving airport parking, and capturing data on travellers experiences. The word cloud (Fig. 17) gives a visual snapshot of the practical solutions to improve passengers' airport experiences in the post-COVID-19 pandemic phase.

[14] Interview with Lameck at Robert Gabriel Mugabe International Airport, 8 July 2022.

Fig. 17 Practical solutions to improve passengers' airport experiences in the post-COVID-19 pandemic period. *Source* Authors

6.1 Technology and Innovation

As the world gradually recovers from the COVID-19 pandemic, the aviation industry is seeing an increase in travellers. To ensure a smooth and efficient journey for passengers, airlines can use technology to improve the overall traveller journey [16, 21, 26]. How can airlines use technology to improve the overall traveller journey as travellers take off in growing numbers in the post-COVID-19 pandemic period? Investing in disruptive technology is very critical for a positive passenger experience. Self-service opportunities today permeate all areas of a passenger's air-travel journey. These opportunities begin before arriving at the Airport and continue through the airport departure process, inflight services, airport transfer points, and ultimately through the airport departure process [26, 41]. Figure 18 presents the many journey steps at which self-service opportunities may be offered. Self-services can be offered to passengers to decrease travel time and enhance their overall travel experience. Self-check-in can be completed in the comfort of the traveller's home through an online app. Passengers are assigned an ID token that they need to gain access to the airport. In cases where passengers cannot self-check-in, assistance check-in is still made available at the airport, and an ID token is received. Passengers can make their way from the check-in services to baggage drops without any assistance or human interaction. By means of face authentication, passengers can go through immigration and security. Transfer passengers will also receive ID tokens and board alongside those from the check-in services. ID tokens make the process much faster and remove the need for constant human interaction, which in the case of COVID-19, protects passengers and workers from getting the virus.

In a Technology Opinion writing, the chief commercial officer for Cinq North America, Juan Hoyos [42], highlighted three areas that airlines could focus on to

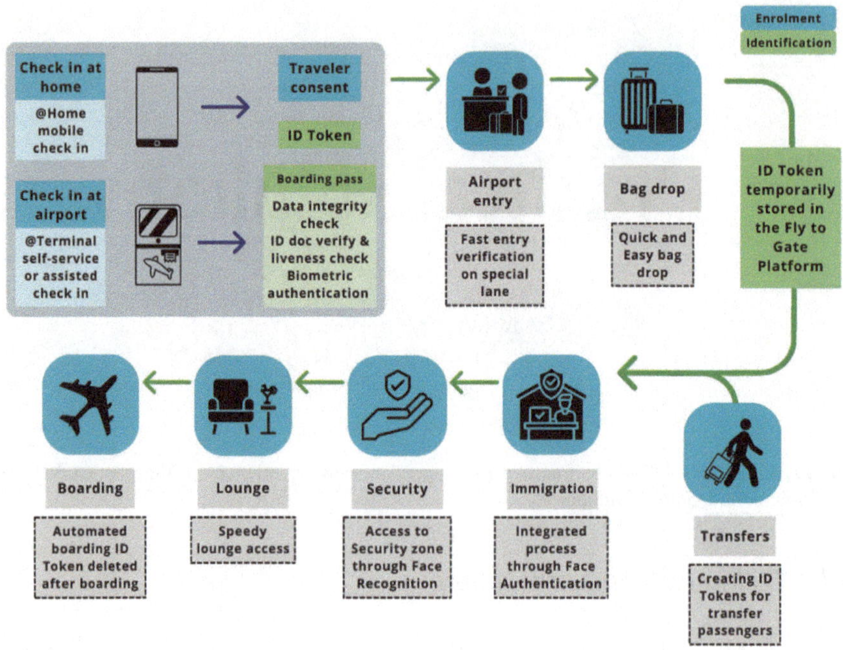

Fig. 18 Fly to Gate: a touchless biometric journey improving passenger experience and airport efficiency. *Source* Authors

improve the passenger experience—increasing speed, better communications, and securing luggage. In increasing speed, streamlining the various checkpoints in the traveller's journey can go a long way to improving traveller satisfaction. Too many checkpoints are a major pain point for travellers. Instead of having agents and security personnel scan boarding passes and checking IDs to drop luggage, check-in, go through border control, and board, airlines should use kiosks and smart devices to ease the process [21, 42]. These advanced systems handle everything from luggage tags to identity management, facilitating faster speeds throughout the traveller's journey [16]. The advanced biometric-enabled kiosks also allow a passenger's face to serve as their ID while seamlessly connecting to mobile devices [43]. This significantly reduces the need to handle documents, touch shared devices, or continually interact with staff. Studies show these kiosks can speed up boarding times by up to 30%, getting passengers to and from their destinations faster [42, 43].

When researching fares and schedules, travellers want the latest information accessible at their fingertips. Thus, good communication is very important. From ticket purchases to gate changes, keeping passengers informed on the latest fares and flight changes is critical. In the post-COVID-19 pandemic phase, airfare systems should provide airlines with all the data they need to make pricing decisions in real-time [21]. This helps airlines keep fares competitive and gives passengers the latest pricing information to make a more informed buying decision. Once they have a ticket, updating passengers on delays and flight changes is the priority [42]. Uninformed passengers can quickly become impatient and frustrated. Airports should also use modern ground-to-air communication platforms to connect to apps to provide passengers with real-time flight and travel information [26, 41]. Airlines can integrate these platforms into their systems to send passengers emails, texts, and push notifications as soon as possible. With automation and artificial intelligence, these fast and powerful apps can keep passengers better informed throughout the journey [42].

There is nothing that ruins a fantastic trip like lost luggage. With more travellers in the air in the post-COVID-19 pandemic phase and a significant shortage of airport workers, concerns over missing bags are rising. As discussed in the passengers' airport experiences section, not having one's luggage can create anxiety and frustration. Airports should introduce efficient baggage tracking systems to reduce the odds of losing passengers' bags at arrival and during connections [42]. Airlines can use technology such as Radio Frequency Identification (RFID) tags and baggage tracking systems to provide passengers with real-time updates on their luggage. This can help to reduce the stress and inconvenience of lost or delayed baggage.

The COVID-19 pandemic has undoubtedly prompted a faster uptake of technologies such as biometrics and queue-less journeys [44]. It has forced airports to rapidly adopt better design and technology features that may have taken years to implement. Examples of some of the critical technologies disrupting the airport industry include Contactless Fast Track Bookings and Queue-Less Journeys, Biometrics, Airport eCommerce, Robotics and automation. At the height of the pandemic, many airports implemented, fast-tracked and upscale these disruptive technologies in response to the immediate situation [21]. However, as we move into the recovery period, it is essential to consider how these technologies fit into airports' long-term, five and 10-year plans. Post-pandemic airport technology must be sustainable, futureproof and serve long-term benefits as part of a complete digital transformation strategy [44].

6.2 Airport Cleanliness

In the post-COVID-19 pandemic period, ensuring a clean and hygienic airport environment is critical in building passenger confidence and ensuring a safe and healthy travel experience. Cleanliness—not just in bathrooms but also in terminals, lounges, and other public areas of an airport—can improve airport revenue because it establishes a level of comfort and positivity between the Airport and the consumers [16, 45]. It also lends to the Airport's prestige which benefits them in the long run with loyal passengers. It is beneficial for airports to establish an atmosphere and ambience that encourages passengers to become repeat travellers. Passengers will likely remember an unclean and filthy airport bathroom experience and associate that negative experience with the airport hub. After all, if one bathroom is not maintained properly, there is a good chance that the rest of the bathrooms at the Airport are just as unclean. Airports should increase the frequency of cleaning and disinfection, especially in high-traffic areas such as check-in counters, security screening areas, and boarding gates [16, 45]. Cleaning should include all surfaces that passengers and staff may encounter, such as seats, armrests, and touchscreens. Ample hand hygiene facilities, such as hand sanitiser and hand-washing stations, should also be provided throughout the airports [16]. These facilities should be placed in high-traffic areas and easily accessible to passengers and staff. Airports can also implement touchless technology, such as automatic doors, touchless faucets, and soap dispensers, to reduce the spread of germs [21]. Airports can also monitor the air quality in the Airport to ensure that it meets or exceeds the recommended standards. This can be achieved through the use of air filters and regular air quality testing.

6.3 Clear and Efficient Signage

Clear and efficient signage is an essential aspect of airport design and operations. It improves the passenger experience, enhances safety and security, reduces delays and improves efficiency, facilitates accessibility, and promotes brand reputation [46, 47]. Figure 19 is an example of clear signage at Frankfurt International Airport, Germany.

Digital signage technology helps passengers quickly locate their gate and time for arrival or departure in even the busiest airports [46, 48]. Airport terminals and waiting areas should display digital messages with updates as needed for gate changes, shuttle

Fig. 19 Signage at Frankfurt International Airport, Germany. *Source* Authors

delays and wait times. Keeping information updated helps passengers get to where they are going efficiently. The flow of passenger travel is essential in an airport, especially in large hubs with multiple terminals like O.R. Tambo International Airport. By investing in high-quality signage, airports can create a more positive and seamless travel experience for passengers, improving their overall satisfaction and loyalty. Airports can be confusing and overwhelming for passengers, especially those unfamiliar with the airport layout or travelling to a new destination. Unlike airport staff, passengers do not always have a thorough understanding of the Airport's layout, so clear and frequent signage is necessary for wayfinding. This also helps improve airport passenger flow by preventing congestion throughout terminals and helps keep foot traffic moving smoothly [46, 48]. Clear signage help passengers navigate through the Airport easily, reducing stress and anxiety [47]. In an emergency or security incident, clear signage help direct passengers to the nearest exit or shelter, ensuring their safety. Efficient signage can also help airport staff respond quickly to an emergency, reducing the risk of injury or harm to passengers. Efficient signage also allows passengers to find their way to check-in, security, and boarding gates quickly and easily, reducing the risk of delays and missed flights [46, 47]. This can also help airlines and airports operate more efficiently, improving customer satisfaction. Passengers with disabilities or limited mobility can also navigate the Airport independently, making air travel more accessible and inclusive. Finally, clear and efficient signage enhances the overall brand reputation of an airport. A well-designed and easy-to-navigate Airport can leave a positive impression on passengers, improving their overall travel experience and increasing the likelihood of repeat business.

6.4 Good Customer Service

In the same vein as having airport staff available to assist with wayfinding, it is crucial to provide top-tier customer service to travellers [17, 18, 22]. Rethinking airports as consumer hubs can be beneficial as customer service is a pillar of successful businesses [17, 18]. Customer service is essential to improving the airport experience. Passengers need to feel confident that airport staff know the airport layout and can accurately guide them to where they need to go. In addition to being knowledgeable, staff must treat passengers as travellers by being courteous and respectful [48]. The key to smooth-running airport foot traffic is passenger flow. The idea is to have passengers seamlessly pass from check-in to security to the gate without needing to wait hours in between. While the staff is not usually involved in the planning process when it comes to designing airport layouts, they can help to improve passenger flow without the need for construction by understanding where terminals, gates, and other destinations are located [17, 48].

Cape Town International Airport (CTIA) is an excellent African example offering good customer service. In 2023 CTIA won their eighth consecutive award for 'Best Airport in Africa' and third consecutive award for 'Best Airport Staff in Africa' at the prestigious Skytrax World Airport Awards. The accolade is proof of the outstanding

Fig. 20 Capetown International Airport Check-in Area. *Source* Authors

standards and level of service the Airport offers. Figure 20 shows the organised check-in area at Capetown International Airport.

In the post-COVID-19 pandemic, providing good customer service is more important than ever in ensuring a positive travel experience for passengers. Airports should provide clear and concise communication to passengers regarding their travel experience, including updates on flight schedules, changes in procedures due to the pandemic, and health and safety guidelines [46, 47]. Airports staff should also be trained to provide friendly, helpful, and compassionate service to passengers, especially in situations where passengers may be experiencing stress or anxiety due to the pandemic. Airports facilities should also be accessible to all passengers, including those with disabilities, by providing clear signage, accessible bathrooms, and wheelchair accessibility [22]. Airport customer service is also improved by offering amenities like comfortable seating, charging stations, and food and beverage options catering to different dietary needs and preferences. Digital technologies, such as mobile apps and self-service kiosks, also improve the check-in and boarding process and reduce wait times [18, 21]. Airports can also provide passengers with a mechanism to provide feedback on their travel experience, such as surveys or feedback forms. This can help airports identify areas for improvement and address concerns on time.

6.5 *Airport Passenger Flow*

Airports should also focus on creating a smooth passenger journey from the start [16, 21]. This allows operators and airlines to make a difference by providing as smooth an experience as possible once passengers approach the Airport. Good airport design goes a long way in enhancing airport passenger flow. Figure 21 shows the open airport design at O.R. Tambo International Airport, facilitating faster passenger flow.

Implementing Smart Booking integrations, for instance, can help arrange advance parking and give real-time travel advice [20, 21]. Advanced check-in or baggage handling systems can also reduce initial processing times [20, 21]. Cutting queue time is the best way to move waiting crowds and avoid running into friction points throughout the terminal. There are several ways to cut queue times: Automation (Online check-ins or self-serve kiosks), good communication and proper signage [21, 43].

In the post-COVID-19 pandemic period, managing airport passenger flow is critical to ensure a safe and efficient travel experience. Increased use of technology, such as mobile apps and self-service kiosks, reduces the need for physical interaction with staff and waiting times [43]. Airports can also use real-time data to monitor passenger flow and adjust operations accordingly [21]. This can include adjusting security checkpoint staffing and opening additional boarding gates as needed. Airports can also improve wayfinding by providing clear signage and digital displays to guide

Fig. 21 The open airport design at O.R. Tambo International Airport facilitates faster passenger flow. *Source* Authors

passengers to their destinations and reduce confusion and congestion [46, 47]. Security screening procedures must also be improved to reduce wait times and improve the passenger experience. This can include implementing new screening technologies and providing clear instructions to passengers on what to expect during the screening process [21]. Airports can also manage gate areas to reduce congestion and improve passenger flow. This can include using boarding groups or assigning seats to specific gates to reduce crowding [16]. Increasing staff presence is also essential. Airports can increase staff presence in high-traffic areas, such as security checkpoints and boarding gates, to assist and guide passengers and reduce confusion and delays.

6.6 Improving Airport Amenities

Airports need to adapt and upgrade to suit travellers' unique needs, which are changing with the times [16, 21, 43]. The seating area is very important. One of the most frustrating things about air travel can be the inability to find available or adequate seating, especially if one has a long wait ahead. For people travelling in groups or with family, finding seating together at some airports can be even harder [48]. Airports can offer comfortable seating, including lounge chairs and charging stations, to provide passengers with a comfortable place to rest or work while they wait for their flights [49]. Comfortable seating throughout the terminal and gates allow passengers to take a load off and take a breath after the hustle and bustle of security checks and finding the right gate. Ample comfortable seating takes the stress out of airports, allowing passengers to begin their vacation or business trip in earnest. [16, 48]. Airports can provide access to wellness amenities, such as yoga rooms, meditation areas, and massage services, to help passengers relax and reduce stress. Offering play areas for children to help families with young children pass the time while waiting for their flights is also essential.

Airports can also enhance their shopping options by providing various retail options, including duty-free shopping and local souvenirs [16, 49]. Airports can also provide various dining options to cater to different dietary needs and preferences. This can include healthy, vegetarian, vegan, and food options for people with allergies or food sensitivities. Passengers also need electrical sockets [16]. Nowadays, travellers do not leave home without smartphones and are always looking for an electrical socket to charge electronic devices. Ensuring enough outlets for charging phones, tablets, and laptops is an indirect way to help passengers relax [48, 49]. A charged device means they can entertain themselves with games or ebooks while waiting for their flights or get some work done before business travel. Those who travel for work will find it beneficial not to wait to board the aeroplane before resuming work [48]. When it comes to improving the layout of airports, instead of access to only a handful of electrical outlets throughout the terminal, airports should ensure that they have to charge capability at every seat or more evenly distributed throughout the gates [48].

6.7 *Improving Airport Ambience and Atmosphere*

In the post-COVID-19 pandemic period, improving the Airport's ambience and atmosphere is essential to create a welcoming and comfortable environment for passengers. Every place has its unique ambience and can make travellers feel in various ways, including calm and serene or stressed and anxious [16, 49]. Part of improving airports is by making small changes, such as switching from harsh, unflattering lightbulbs to softer, more calming ones [50]. Adding plants or even an entire indoor garden is another way to improve airports by creating a peaceful environment. Similarly, installing artwork can be a great way to encourage people to spend extra downtime browsing the terminal [16, 49].

Another way to improve airports' ambience and atmosphere is through enhancing lighting and décor. Airports can enhance their lighting and decor to create a more welcoming and relaxing environment [50]. This can include natural lighting, greenery, and artwork. Figure 22 shows the welcoming ambience and atmosphere at Frankfurt International Airport, Germany.

Providing comfortable seating, including lounge chairs and charging stations, to provide passengers with a comfortable place to rest or work while they wait for their flights also improves airport ambience and atmosphere [16, 49]. Airports can also play music and audio content to create a relaxing and pleasant atmosphere. This can

Fig. 22 Welcoming ambience and atmosphere at Frankfurt International Airport, Germany. *Source* Authors

include curated playlists or audio content that promotes local culture and attractions. Airports can also use scents and aromas to create a more pleasant atmosphere [50]. This can include diffusing natural fragrances or using air purifiers to improve air quality. Lastly, airports can also provide access to green spaces, such as outdoor gardens or indoor green walls, to help passengers relax and reduce stress.

6.8 *Investing in Restaurants and Retail*

Investing in restaurants and retail is crucial to improving the airport experience for passengers in the post-COVID-19 pandemic period. Airports should incorporate a selection of eateries and retail shops for travellers [49, 51, 52]. With the extra time on their hands, travellers can do shopping and dining. Figure 23 shows authors having coffee and working at O.R. Tambo International Airport, awaiting their connecting flight to Zimbabwe. At O.R. Tambo International Airport, restaurants and fast-food outlets include Nando's, Cafe Dulce, Debonairs Pizza, Kauai, KFC, News Cafe, Steers, Wimpy, Spur, and more. This service can also be extended to other growing African airports like Robert Gabriel International Airport in Zimbabwe.

Shopping and dining are at the top of the list regarding how airports can increase revenue [48, 49]. Various eateries and stores spread throughout the terminal

Fig. 23 Authors having coffee and working at O.R. Tambo International Airport. *Source* Authors

Fig. 24 Popular shopping brands (Rolex and Wempe) at Frankfurt International Airport, Germany. *Source* Authors

encourage travellers to spend money by enjoying a nice meal or purchasing souvenirs or other items on the go [16, 49]. Restaurants and retail also help prevent heavy congestion in other airport areas, such as the gate seating areas and provide something for people to do while waiting for their flight [48, 52]. Figure 24 shows some popular shopping brands (Rolex and Wempe) that attract travellers at Frankfurt International Airport in Germany.

There are various ways airports can invest in their restaurants and retail. Recently expanding Airports like Robert Gabriel Mugabe international Airport should attract local vendors to set up restaurants and retail shops at the Airport. This can give passengers a taste of local culture and cuisine and create a unique airport experience [17]. Airports should also offer healthy food options, including vegetarian and vegan options and food options for people with allergies or food sensitivities, to cater to different dietary needs and preferences [49, 52]. Providing various dining options like fast-casual dining, sit-down restaurants, and cafes to cater to different tastes and preferences is also important [51]. The use of technology to improve the shopping experience is now timely. Airports can use technology, such as augmented reality and mobile apps, to improve the shopping experience for passengers [21, 26]. This can include providing virtual try-on experiences for clothing and accessories or using beacons to send targeted promotions to passengers. Airports should also offer loyalty programs to encourage passengers to shop and dine at the Airport. This can include offering program members discounts, special offers, and exclusive perks [16].

6.9 *Investing in Activities for Passengers*

Investing in activities for passengers is another excellent way for airports to enhance the airport experience and provide a more enjoyable and entertaining stay for passengers in the post-COVID-19 pandemic period. One of the worst things about airports is inevitably waiting. While a lot of it can be alleviated with solutions to steady passenger flow, there are other times when it is unavoidable, such as while waiting at the gate before a flight or if a flight is delayed [16, 21]. Shopping and dining presented above are two ways to occupy travellers who have a long wait ahead of them, but other activities can also be entertaining [49, 51, 52]. Airports can provide entertainment options, such as live music, art exhibitions, and movie theatres, to keep passengers entertained while waiting for their flights [16]. Airports can create family-friendly areas, including play areas and interactive exhibits, to provide children with a fun and safe space to play while their parents wait for their flights. Indoor gardens or art exhibits are also other forms of entertainment activities airports should consider for travellers [49, 50]. Airports can also offer exercise and wellness options, such as yoga rooms and fitness centres, to allow passengers to stay active and healthy while waiting for their flights. Airports can also organise airport tours, allowing passengers to learn more about the Airport's history, architecture, and operations [16]. Virtual reality experiences, such as virtual tours of the city or immersive games, provide passengers with an engaging and interactive experience and also provide travellers with a good experience at airports. Airports can also offer cultural experiences, such as workshops and exhibitions showcasing local arts and crafts, to provide passengers with a taste of local culture and heritage [49, 50].

6.9.1 Improving Airport Parking

Often, parking lots are forgotten once built, with airports focusing more on improving the indoor airport experience for travellers inside. However, the parking lot is one of the first and last things passengers will notice and is critical for improving the airport parking experience [50]. Some ways to improve the airport parking lot experience include automation, shuttle service, wayfinding and covered parking [48, 53]. If there are long lines or congestion at the entrance or exit of a parking garage, upgrading to an automated payment system might be an ideal solution. This enables passengers to avoid waiting in long lines and quickly and securely pay for parking as they make their way to the airport exit (VPS). Airports with off-site parking can offer shuttle buses to the terminals, with the cost of this service integrated into the total parking fee. Covered parking, whether surface or rooftop, offers added comfort and security through protection from catastrophic weather elements and extreme heat [48]. With covered parking lots, passengers who need to leave their cars in long-term airport parking can do so with peace of mind. During the hottest months of the year, they will also return to a more comfortable vehicle, as adequately covered parking solutions can help significantly decrease the internal temperature of a vehicle [48, 50].

6.9.2 Capturing Data on Traveller's Experiences

Capturing data on traveller's experiences is essential for airports to understand their customers and improve their overall experience in the post-COVID-19 pandemic period. A key part of increasing travellers' experience is understanding what they want from their journey [16, 22]. This means asking and listening by giving them a platform to air their views and making them understand their critical role in your operational success [17, 18]. Airports should invest in data gathering, assessments and analysis to improve passenger experience and increase passenger flow efficiency at airports. Using integrated and in-cloud solutions for operational management can effectively automate reporting and analysis, enabling a comprehensive view of the Airport in real-time [21, 43]. By gaining this view, airport management teams can pinpoint potential delays in advance, identify issues before they become problems and maintain passenger flow while improving resource management on the ground. Capturing data on travellers' experiences can help airports identify areas for improvement. This data can provide insights into what passengers like and dislike about their airport experience and highlight areas that need attention [54]. Capturing data on travellers' experiences also help airports personalise their services. Analysing passenger behaviour and preferences data, airports can tailor their services to individual passenger needs and provide a more personalised experience. Through understanding the factors contributing to passenger satisfaction, airports can change their services and facilities to enhance the passenger experience [16, 22]. Capturing data on travellers' experiences also help airports increase revenue. By understanding passenger behaviour and preferences, airports can identify opportunities to offer new services and products to generate additional revenue. Finally, by tracking passenger feedback over time, airports can see the impact of their efforts and adjust their strategies accordingly [54].

7 Conclusion and Way Forward

The COVID-19 pandemic presented significant challenges for the aviation industry but also presented an opportunity to reimagine airport travel. By adopting new technologies, improving communication, enhancing hygiene measures, offering more flexibility, and prioritising the customer experience, airports and airlines can create a more efficient and enjoyable experience for travellers in the post-COVID-19 era. Airports are key platforms for transporting people and goods regionally, nationally and internationally. Understanding traveller experiences post-COVID-19 is critical to solving common Airport operating problems, adapting appropriate new technologies, and introducing innovations into the airport industry. While there have been significant improvements in passenger experience in the aviation industry over the years, the post-COVID-19 pandemic period ushered in a new phase where airlines should consider treating travellers better. With travel operations returning to normal from the lows of the pandemic, airports need to restore consumers' confidence in travelling again by establishing a travel experience that evokes safety and trust. Why is this important? The COVID-19 pandemic sparked a lot of change; passengers are no longer the same as they were before. Behaviours have evolved, and priorities have changed; thus, airports must revisit their plans and strategies to ensure they align with the post-pandemic passenger needs.

Though African airports have been striving for improved passenger volumes and expanding their capacities, the concept of analysing and improving passenger experience is yet to be widely improved among African airports. Given this scenario, it is instructive to state here that the passenger is central to any airport's success; enhancing passenger experience should be at the heart of airport improvement initiatives in Africa. By starting with the passenger experience and working backwards, the airport industry in Africa can create operational efficiencies, better retail experiences, and a smoother journey. Thus, accelerated recovery of air transport across the continent is vital in the post-COVID-19 pandemic period. Governments should step up efforts to provide financial and regulatory support for the industry. Airports should strive to enable seamless travel experiences where digitalisation, sustainability, and newer technologies can help ease passenger journeys. Airports should also use data capture of airport passenger behaviours to aid implementation. There is also a need to diversify away from retail to offer an airport waiting experience that improves health and mental well-being. Airport planners and designers should also consider designing future airport areas to increase natural light and improve the acoustic environment. Finally, airports should also adjust to the personalisation of the journey process.

The chapter provided insights into how airports and airlines can adapt to the new reality of travel in a post-COVID-19 world. However, several areas of future research could further be explored to enhance the airport experience. Further research can explore the effectiveness of new technologies, such as biometric scanning and facial recognition in improving the airport experience. While airports have introduced several health and safety measures in response to the pandemic, there is still room for improvement. Further research could explore the effectiveness of different measures in reducing the spread of infectious diseases and how these measures could be implemented more efficiently and streamlined. The airport experience can be challenging for people with disabilities, and there is a need for greater accessibility in airport design and services. Further research could explore how airports can better cater to the needs of people with disabilities, such as by providing accessible facilities or personalised assistance.

Finally, reflecting on the adoption of airport ethnography, this methodology may pose various challenges and require ethical considerations inherent in studying such complex spaces. The dynamic and fast-paced nature of airports often present difficulties in capturing comprehensive data within limited timeframes, while ensuring the privacy and confidentiality of participants may require careful navigation. Additionally, the diverse and transient nature of airport travelers can also present challenges in participant selection and data collection. Despite these challenges, the adoption of airport ethnography methodology offers numerous advantages and opportunities for innovation. Through immersing researchers in the airport environment, deep insights into traveler experiences, behaviors, and interactions can be gained, which traditional research methods may overlook. This approach allowed us to uncover nuanced patterns and trends in traveler experiences post-COVID-19, informing practical solutions for enhancing the airport journey. Moreover, airport ethnography provides a unique opportunity to study the cultural practices and social dynamics unique to airports, contributing to a richer understanding of these complex spaces. Embracing insights from innovative research methodologies like airport ethnography, airports and airlines may better adapt to the evolving needs and expectations of travelers, ultimately enhancing the overall airport experience in the African continent.

References

1. Bhanye J (2023) "Nimble sociality and belonging": an ethnography of migrants' responses to bans on associational life during the COVID-19 pandemic. In: Urban Forum. Springer Netherlands, Dordrecht, pp 1–27. https://doi.org/10.1007/s12132-023-09503-0
2. Bonaccorsi G, Pierri F, Cinelli M, Flori A, Galeazzi A, Porcelli F, Schmidt AL, Valensise CM, Scala A, Quattrociocchi W, Pammolli F (2020) Economic and social consequences of human mobility restrictions under COVID-19. Proc Natl Acad Sci 117(27):15530–15535
3. Fazio M, Pluchino A, Inturri G, Le Pira M, Giuffrida N, Ignaccolo M (2022) Exploring the impact of mobility restrictions on the COVID-19 spreading through an agent-based approach. J Transp Health 25:101373. https://doi.org/10.1016/j.jth.2022.101373
4. Bhanye J, Bhanye A (2023) A blessing or a curse?: the role of social media during the COVID-19 pandemic in Africa. In: The Palgrave handbook of global social problems. Palgrave Macmillan, Cham. https://doi.org/10.1007/978-3-030-68127-2_367-1
5. Zanin M, Papo D (2020) Travel restrictions during pandemics: a useful strategy? Interdisc J Nonlinear Sci 30:1–8. https://doi.org/10.1063/5.0028091
6. Deveci M, Çiftçi ME, Akyurt İZ, Gonzalez ED (2022) Impact of COVID-19 pandemic on the Turkish civil aviation industry. Sustain Oper Comput 3:93–102
7. Law CC, Katekaew R (2022) COVID-19: ASEAN aviation policy and the significance of intra-regional connectivity. J Asian Econ Integr 4(1):1–23. https://doi.org/10.1177/26316846221075476
8. Airport Council International (ACI) (2022) The impact of COVID-19 on airports—and the path to recovery. https://aci.aero/2022/06/28/the-impact-of-COVID-19-on-airportsand-the-path-to-recovery/. Accessed 10 Sept 2022
9. Dube K, Nhamo G, Chikodzi D (2021) COVID-19 pandemic and prospects for recovery of the global aviation industry. J Air Transp Manag 92:102022. https://doi.org/10.1016/j.jairtraman.2021.102022
10. International Air Transport Association (IATA) (2020) Impact of COVID-19 on African aviation and economies is worsening. https://www.iata.org/en/pressroom/pr/2020-08-13-03/. Accessed 9 Sept 2022

11. Popovic V, Kraal B, Kirk P (2010) Towards airport passenger experience models. In: Sato K, Desmet P, Ludden G, Mathew A, Hekkert, P (eds) Proceedings of the 7th international conference on design and emotion 2010. IIT Institute of Design, United States, pp 1–11
12. International Air Transport Association (IATA) (2022) Passenger experience & facilitation. https://www.iata.org/en/programs/passenger/. Accessed 9 Sept 2022
13. International Air Transport Association (IATA) (2015) Fast travel strategy. http://iata.org/whatwedo/passenger/fast-travel/Pages/index.aspx. Accessed 9 Sept 2022
14. Kwik B (2021) How airports can innovate by obsessing over passenger experience. https://aws.amazon.com/blogs/industries/how-airports-innovate-obsessing-over-passenger-experience/. Accessed 12 Sept 2022
15. Laghidze D (2021) Ethnography of modern airports. Heinrich Böll Stiftung. https://ge.boell.org/en/2021/06/29/aeroportebis-etnograpia. Accessed 8 Sept 2022
16. Batouei A, Iranmanesh M, Mustafa H, Nikbin D, Ping TA (2020) Components of airport experience and their roles in eliciting passengers' satisfaction and behavioural intentions. Res Transp Bus Manag 37:100585
17. Hwang J, Lee J (2019) A strategy for enhancing senior tourists' well-being perception: FOCUSING on the experience economy. J Travel Tour Mark 36(3):314–329
18. Majra H, Saxena R, Jha S, Jagannathan S (2016) Structuring technology applications for enhanced customer experience: Evidence from Indian air travellers. Glob Bus Rev 17(2):351–374
19. Forrester (2021) A quest to improve real-time traveller experience. https://www.forbes.com/sites/forrester/2021/07/21/a-quest-to-improve-real-time-traveller-experience/?sh=2a960a0e35a4. Accessed 11 Sept 2022
20. AICE (2020) Three ways to transform passenger experience. https://www.a-ice.aero/three-ways-to-transform-passenger-experience/. Accessed 10 Sept 2022
21. Macnaughton P (2022) Building smart airports: technology solutions to improve health, enhance experience and increase revenue. J Airport Manage 16(2):125–133
22. Orakani SN, Smith K, Weaver A (2021) Reframing the experiences of travellers with mobility impairments: enhancing the leisure constraints model. J Hosp Tour Manag 47:84–92
23. International Airport Review (2022) Passenger experience. https://www.internationalairportreview.com/core_topic/passenger-experience/. Accessed 9 Sept 2022
24. Coll D (2022) Small steps to success: improving passenger experience with complimentary data collection. https://blog.aci.aero/small-steps-to-success-improving-passenger-experience-with-complimentary-data-collection/. Accessed 5 Sept 2022
25. Lees E, Freibrun S (2022) Reimagining the passenger experience: the importance of building a traveller-centric strategy. https://www.icf.com/insights/transportation/reimagining-the-passenger-experience. Accessed 10 Sept 2022
26. Bogicevic V, Bujisic M, Bilgihan A, Yang W, Cobanoglu C (2017) The impact of traveler-focused airport technology on traveler satisfaction. Technol Forecast Soc Chang 123:351–361
27. Munoz C, Laniado H, Córdoba J (2019) Modeling air travelers' experience based on service quality stages related to airline and airports. Mod Appl Sci 13(11):37–53
28. Guerreiro J, Ahmetovic D, Sato D, Kitani K, Asakawa C (2019) Airport accessibility and navigation assistance for people with visual impairments. In: Proceedings of the 2019 CHI conference on human factors in computing systems, pp 1–14
29. Kumar M (2017) Infrastructure, labor, and government a study of Delhi Airport. Doctoral dissertation, Western Sydney University, Australia
30. Awad M, Alzaatreh A, AlMutawa A, Al Ghumlasi H, Almarzooqi M (2020) Travelers' perception of service quality at Dubai International Airport. Int J Qual Reliab Manage 37(9/10):1259–1273
31. Bhanye J (2023) Temporality, translocality, and sedentariness: 'complex and varied' perceptions for the future among African irregular migrants on the margins. In: The Palgrave handbook of global social change. Springer International Publishing, Cham, pp 1–21. https://doi.org/10.1007/978-3-030-87624-1_396-1

32. Chalfin B (2008) Sovereigns and citizens in close encounter: airport anthropology and customs regimes in neoliberal Ghana. Am Ethnol 35(4):519–538. https://doi.org/10.1111/j.1548-1425.2008.00096.x
33. Zuskáčová V (2023) 'Where are you?': (Auto) ethnography of elite passage and (non)-placeness at London Heathrow Airport. Mobilities:1–16. https://doi.org/10.1080/17450101.2023.2171805
34. Bhanye J (2023) "Strategic, dual sense of place" among middle-aged migrants in the diaspora. Int J Migr Health Soc Care 19(3/4):376–390. https://doi.org/10.1108/IJMHSC-07-2022-0075
35. Prinsloo J (2015) Service learning in custodial settings in South Africa. J Psychol Afr 25(2):151–155. https://doi.org/10.1080/14330237.2015.1021532
36. Bram Fischer International Airport (2023) Bram Fischer International Airport. https://bramfischerairport.co.za/. Accessed 8 Apr 2023
37. Global Data (2018) CAAZ—Robert Gabriel Mugabe international airport upgrade—Zimbabwe. https://www.globaldata.com/store/report/caaz-robert-gabriel-mugabe-international-airport-upgrade-zimbabwe/. Accessed 2 Apr 2023
38. Taderera E (2020) Robert Mugabe International Airport. https://www.zimlegends.com/2020/11/29/robert-mugabe-international-airport/. Accessed 1 Apr 2023
39. Tshuma, N (2022) SARS intercepts crystal meth shipment worth R236 million at OR Tambo. https://www.iol.co.za/capeargus/news/sars-intercepts-crystal-meth-shipment-worth-r236-million-at-or-tambo-2201a21b-9cfe-434d-a4eb-883889d51dde. Accessed 1 Apr 2023
40. SA Gov (2020) Minister Naledi Pandor: Re-opening of borders and ports of entry for international travellers during Coronavirus COVID-19 epidemic lockdown level 1. https://www.gov.za/speeches/minister-naledi-pandor-re-opening-borders-and-ports-entry-international-travellers-during. Accessed 12 Sept 2022
41. Barich F, Ruiz L, Miller J (2015) Enhancing the passenger experience through an integrated approach to self-service opportunities. J Airport Manage 10(1):49–63
42. Hoyos J (2021) Airlines must embrace tech to improve passenger experience. https://www.phocuswire.com/How-airlines-can-use-to-tech-improve-the-traveler-journey. Accessed 13 Sept 2022
43. Taufik N, Hanafiah MH (2019) Airport passengers' adoption behaviour towards self-check-in Kiosk Services: the roles of perceived ease of use, perceived usefulness and need for human interaction. Heliyon 5(12):e02960
44. Wallace V (2022) How to create a positive passenger experience at the airport? https://www.rezcomm.com/resources/blog/ecommerce/create-a-positive-passenger-experience. Accessed 13 Sept 2022
45. Del Chiappa G, Martin JC, Roman C (2016) Service quality of airports' food and beverage retailers. A fuzzy approach. J Air Transp Manag 53:105–113
46. Carutasu NL, Adir V, Pascu NE, Arion GC (2018) Signage and ergonomics in packaging systems. Ann DAAAM Proc 1:0173–0180
47. Saronge B (2017) A comparative analysis of vehicle traffic flow at Jomo Kenyatta and Julius Nyerere international airport. Doctoral dissertation, Maseno University
48. VPS (2020) How to improve the airport experience? https://www.vpslp.com/about/insights/how-to-improve-the-airport-experience. Accessed 12 Sept 2022
49. Prentice C, Kadan M (2019) The role of airport service quality in airport and destination choice. J Retail Consum Serv 47:40–48
50. Straker K, Wrigley C (2016) Translating emotional insights into digital channel designs: opportunities to enhance the airport experience. J Hosp Tour Technol 7(2):135–157
51. Brida JG, Moreno-Izquierdo L, Zapata-Aguirre S (2016) Customer perception of service quality: The role of Information and Communication Technologies (ICTs) at airport functional areas. Tourism Manage Perspect 20:209–216
52. Fodness D, Murray B (2007) Passengers' expectations of airport service quality. J Serv Mark 21(7):492–506

53. Akai K, Aoki K, Ueda Y, Kanetsuku K (2021) Redesigning parking facilities using a parking service satisfaction survey at Izumo local airport in Japan. Transp Res Interdisc Perspect 11:100433
54. Graham A, Budd L, Ison S, Timmis A (2019) Airports and ageing passengers: a study of the UK. Res Transp Bus Manag 30:100380

Johannes Bhanye is a Postdoctoral Research Fellow at the African Climate and Development Initiative (ACDI), University of Cape Town. He holds a PhD in Social Sciences - Migration and Land Settlement, B.Sc. (honors) degree in Rural and Urban Planning and an M.Sc. in Social Ecology, both from the University of Zimbabwe. He is a former Research Fellow at the University of the Free State, the Andrew W. Mellon Foundation and the Centre for Applied Social Sciences (CASS). His research interests cut across land and peri-urbanity, migration, urban informality, and climate change adaptation and resilience. His recent BOOK publication by Springer is titled: "COVID-19 Lockdowns and the Urban Poor in Harare, Zimbabwe: Emerging Perspectives and the Morphing of a Sustainable Urban Future. Dr Bhanye has been invited as a speaker at several International Conferences and Seminars in countries like Senegal, Zambia, China, Germany, South Africa, Lesotho, Morocco, Belgium, Ghana, and Switzerland.

Mareli Hugo is a PhD student at the University of the Free State, in South Africa specializing in Urban and Regional Planning. Her research interests encompass a diverse range of topics, including mobility and transport planning, eco-industrial parks, and Geographic Information Systems (GIS). With her dedication in her field, Mareli has been recognized as an exceptional emerging researcher and scholar, earning invitations to speak at numerous local and international academic conferences.

Abraham Matamanda (PhD Urban & Regional Planning) is an NRF Y2-rated Urban and Regional Planner. Abraham lectures at the University of Free State (UFS) and is also affiliated with the SARChI Chair on City Region Economies in the faculty of Economic and Management Sciences at UFS. Currently, he serves as the editor of the Town Planning Journal published by UFS and serves on the academic editorial board of Plos Water Journal. His research focuses on urban governance and planning, climate change adaptation and resilience, informal Global South urbanism, urban land markets and housing studies. Abraham has authored and edited several books published with Springer Nature including:

- Urban Geography in Postcolonial Zimbabwe: Paradigms and Perspectives for Sustainable Urban Planning and Governance;
- Housing and Technology: Special Focus on Zimbabwe;
- Urban Infrastructure in Zimbabwe: Departures, Divergences and Convergences; and
- Secondary Cities and Local Governance in Southern Africa.

Kgosi Mocwagae is the Academic Head of Department of the Department of Urban and Regional Planning at the University of the Free State. Dr Kgosi (Pr. Pln. A/2896/2019) is a qualified professional town planner with the following qualifications: BSc (IT) Management, B.Spatial Planning (Hons), Master and PhD in Urban and Regional Planning. His career in urban and regional planning started at the University of the Free State in 2014 as a research assistant, but has grown to lecturing, research, supervision, private consultations and membership of Municipal Planning Tribunal.

Picturing Pandemics as Urban Disaster: Enumerations of Post-Covid Life in Informal-Windhoek, Namibia

Uchendu E. Chigbu, Penehafo Ricardo, Cathrine Marenga, and Malcon L. Mazambani

Abstract Historically, pandemics have had critical influences on the development of human societies. It can impact social cohesion, politics, economic productivity, and population growth (among many other factors). Following the outbreak of the COVID-19 pandemic between March 2020 and December 2021, Namibia faced a tumultuous and wide-ranging shock. Due to its high morbidity and mortality rates, curfews and states of emergency were declared in the country. Although the pandemic is now in its post-outbreak period, its effect remains poorly researched. This chapter evaluates the post-COVID situation in Namibia's city of Windhoek. It situates the COVID-19 pandemic as a disaster combining biological threats with various vulnerabilities, such as the organizational and response capacity of socio-spatial systems, including urban informality. The chapter used visual illustrations to show Windhoek's pandemic situation as a reflection of a one-city two-system city consisting of formal and informal experiences. It also used vignettes from Namibia's Newspapers to depict the post-COVID life of the informal settlements of Windhoek. It prescribed measures for making COVID-stressed cities more resilient to pandemic disasters.

Keywords Covid-19 · Community · Enumerations · Informal settlements · Namibia · Okahandja · Pandemic · Town planning · Urban disaster · Windhoek

1 Background and Approach to the Study

The World Health Organization (WHO), on the 5th of May 2023, ended the global emergency status for COVID-19 more than three years after its original declaration. There is no better period to make a renewed sense of planning and development for post-pandemic cities than now. Cities have been identified as the epicentres of new

U. E. Chigbu (✉) · P. Ricardo · C. Marenga · M. L. Mazambani
Department of Land and Spatial Sciences, Namibia University of Science and Technology, Windhoek, Namibia
e-mail: echigbu@nust.na

K. M. Nahiduzzaman (ed.), *Making Sense of Planning and Development for the Post-Pandemic Cities*, City Development: Issues and Best Practices,
https://doi.org/10.1007/978-981-97-5481-6_7

pandemics, with the "worst fears of unprecedented health havoc" caused in informal settlements worldwide [1, p. 115].

The number of people living in urban areas is not only increasing around the world. The population of Namibians that reside in urban areas, incredibly informal settlements, is ever-growing. Tackling this challenge through good urban governance has always posed a challenge in Namibia.

When the outbreak of the COVID-19 pandemic emerged in 2020, it was a big concern to all-urban planners, policymakers, city administrators and informal settlers worldwide. The pandemic affected Namibia—one of Sub-Saharan Africa's least populated countries—with almost 50% of the population now living in urban areas. Due to the seriousness of the pandemic, scholars in different fields have studied its nature, occurrence, and impacts [2–5]. Within this brief period, studies on the outbreak of COVID-19 or Coronavirus have "evolved in the aspects of information update on deaths and casualties, and the search for medically certified preventive and curative measures" [6]. In urban studies, it has extended into informal settlement upgrading. However, studies on the impact of the COVID-19 pandemic on urban residents, especially informal settlement dwellers, "remains poorly explored" in Namibia. The study presented in this chapter aims to contribute to knowledge on how to better proof cities to be resilient to future pandemic disasters. The focus is on informal settlements in Windhoek, Namibia.

Not much has been written about Windhoek's or Namibia's COVID-19 pandemic experience in the context of planning and development for post-pandemic cities. This chapter fills this gap by assessing the COVID-19 situation in the informal section of the city of Windhoek. By way of approach, the chapter is a descriptive qualitative urban community evaluation of life under COVID-19 in Windhoek, Namibia, post-COVID. It used conventional literature to argue and establish that the COVID-19 pandemic is a disaster, just like earthquakes, volcanoes and tsunamis. It used visual illustrations to show Windhoek's pandemic situation as a reflection of a one-city, two-system city consisting of formal and informal experiences. It used grey literature to evaluate the COVID-19 pandemic post-COVID situation in Namibia's city of Windhoek. The grey literature used is Namibia's national newspapers, where extracts or vignettes were used to depict residents' experiences in Windhoek's section. It also used vignettes from Namibia's Newspapers to describe the post-COVID life of the informal section of Windhoek. A vignette is a short piece of writing that does not have a beginning, middle, or end but focuses on a specific moment.

The chapter is divided into five sections. The following section (that is, Sect. 2) uses existing literature to frame the COVID-19 pandemic as a viral disaster. This conceptual framing is crucial to understanding why the pandemic is only one of many diseases that have devastated the world in the history of humanity. It also shows that pandemic situations are nothing new to humans. Still, the human response to it has been the problem. Section 3 deals with the issue of urban informality and COVID-19 in Namibia. This is followed by a presentation of Namibia's post-COVID experience

(Sect. 4). The experience reflects on the informal section of Windhoek as a pandemic-stressed city. Based on Windhoek's experience, the last section (Sect. 5) prescribes measures for making COVID-stressed cities (such as Windhoek) more resilient to disease disasters.

2 Grasping COVID-19 as a Viral Disaster: Natural or Human-Caused?

Positively or negatively, disease outbreaks play a crucial role in shaping human societal development [7]. Just like other natural and human-made disasters, diseases can cause displacements, poverty, inadequate housing, environmental degradation, poor sanitation and vulnerability to future disasters, among other havoc capable of impeding the achievement of the development objectives. Natural disasters that cause setbacks in spatial development include hurricanes, earthquakes, landslides, tsunamis, droughts, tornadoes, flooding, violent volcanoes, avalanches, and many others [8]. Human-made disasters capable of causing similar effects, including gas leaks or use of oil spills, biological weapons, forest fires or wild-fires, wars, terrorism, gaseous explosions, and nuclear meltdowns, among many others [9]. This study considers a disease to be a disaster.

A summary of significant disease disasters recorded in medical and public health literature [10–12], as compiled in Fig. 1, indicates that disease disasters have been part of humankind, from earliest times until today.

The critical inferences from the disease disaster compilation are sixfold. First, disease disasters, like other natural disasters, have been named in the past based on their geographical origins. Second, they have often occurred in ways like earthquakes,

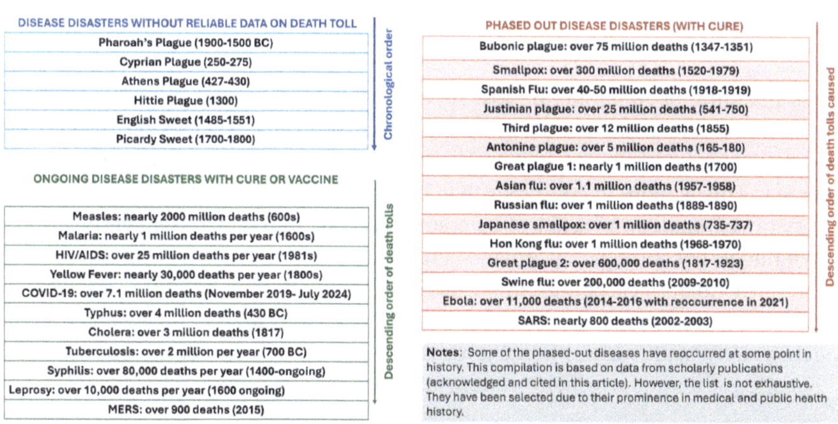

Fig. 1 Outlook of prominent disease disasters in the history of humanity. *Source* Authors' illustration

volcanoes, floods, and other natural disasters. That is, they ravage their geography and leave behind human casualties. Third, the magnitude of deaths over time depends on the availability of preventive, mitigative and curative means or resources. Fourth, they can occur and reoccur. For instance, Cholera has hit the world more than six times, and the great plagues hit the world more than two times [13, 14]. Fifth (with a focus on death tolls), some earlier disease disasters (such as the Bubonic plague, Smallpox, Spanish flu, and the Justinian plague) make the Ebola and the HIV—human immunodeficiency virus, which can lead to AIDS or acquired immunodeficiency syndrome—outbreaks seem medically tolerable. According to the United States President's Emergency Plan for AIDS Relief [15], 8.45% of Namibia's population live with HIV/AIDS. The disease was responsible for about 3165 deaths in 2022. This means that HIV/AIDS remains a viral disaster in Namibia post-COVID. Sixth, unlike the later disease disasters (in the past 100 years), the earlier disasters were more regional than global in their impacts. Another crucial experience discernible from all the earlier disease disasters is that people or societies do not have to be at the origin of a disease to be affected by ill health or discomforts caused by the disaster it brings. This is exemplified by the 2019 novel coronavirus disease (COVID-19) outbreak in Wuhan (China), which spread quickly worldwide. On this basis, COVID-19 can be considered a viral disaster.

There have been speculations, and even debates, on whether the COVID-19 pandemic is human-manufactured or naturally occurring [16, 17]. Findings from the Scripps Research Institute[1] have confirmed that it has natural origins and was not made in a laboratory or artificial setting, as some people have speculated [18]. Whether the COVID-19 pandemic was a natural or unnatural disaster is not the interest of this chapter because it does not matter. In the context of this study, what matters is the havoc it caused and how it has affected the lives of people in Windhoek.

3 The Informal-Windhoek and COVID-19 in Namibia

Urban planning in Namibia, to most scholars, can be summarised to mean putting the country's future into consideration for development [19]. This broad planning perspective, consciously or unconsciously, embraces segregation as part of urban planning. This is more unconscious than conscious in Namibia due to the country's history of segregation, a legacy of its colonial apartheid past. Consequently, the city of Windhoek has an inherent urban segregation pattern, reflected in a feature that excludes or hinders certain groups from accessing services, activities and spaces. Therefore, several scholars have described this Windhoek-style (un)planned urban system pattern with various nomenclatures or phrased descriptions. Friedman [20, p. 1] called it a system which "entrenches economic and political inequality" and

[1] Scripps Research is ranked the most influential institution in the world for its impact on innovation. The have advanced profound innovations that improve wellbeing, including breakthrough studies that address the world's most pressing health concerns.

helps guarantee "people's economic and social marginalisation." Chigbu [21, p. 276] describes it as a place that needs to foster a "sense of place in the planning and designing."

The city is a damningly unequal space for Namibians. Jauch [22] considered it an urban system whose social fabric remains racially disjointed, partly due to the separateness in planning and its associated effect on human interactions. Mphambukeli [23, p. 1] identified it as a planning system rooted in separation, racial segregation and the exclusivity of one section of the society (and the space they occupy) over the others. Kohima et al. [19] called it "one-city two-system planning." This study broadly refers to it as a formal-and-informal system—meaning that the city is demarcated in formal and informal patterns—promoting urban segregation and space exclusivity.

3.1 Overview of Informal-Windhoek: Characteristics and Conditions

The City of Windhoek was founded in 1890 and now inhabits a quarter of Namibia's 2.6 million population. It has been the hallmark of Namibia's evolution to urbanisation since 1928 [24]. It has become the commercial, industrial, and administrative hub of Namibia in its 130-year history. It is the political, administrative, social, and economic capital of Namibia. Like many sub-Saharan African cities, it is a city "which is a product of the colonial era, and its development was initiated by Whites" [25, p. 1]. The urban spatial anthropology of Windhoek allows for an understanding of how Windhoek's colonial city governance has evolved. A look at the current socio-spatial structure of Windhoek reflects its formal and informal landscape.

From an urban planning perspective, the city of Windhoek has continuously developed along "formal" and "informal" lines. The formal section of the city areas is duly planned and serviced according to the city's planning specifications. The informal part of the city is where residents live in unplanned and unserviced quarters (i.e., informal settlements). Like in any other sub-Saharan African country, the informal settlements in Namibia have always been the hubs of poor health and sanitation. Efforts to tackle informal settlement growth in Namibia have had historical difficulties.

Frames a–d (in Fig. 2) comprise four (4) photos and depict the risky environment where most informal settlements are located in Okahandja within the Windhoek area—which is characterised by a gross lack of planning, infrastructure and other essential urban services. Okahandja is one of the settlements where the land and housing insecurity level is high around Windhoek. According to one of Namibia's national newspapers, *The Namibian* [26], during the COVID-19 pandemic, the Okahandja Medical Centre had a capacity of fifteen (15) medical beds and a mortuary capacity of 35 bodies, with less than ten (10) doctors and nurses.

Frame a: Picture depicting waste disposal in the Ekunde 1 Informal Settlement. Note that the waste disposal site is directly within the living and open spaces ©Photograph by authors.

Frame b: Picture showing a distant view of the environmental landscape of the Okahandja Park Informal Settlement. The settlement is sandwiched between erosion and landslide-prone foothills. Note that the houses are made of corrugated iron sheets — ©Photograph by authors.

Frame c: Picture showing a closer view of the state of housing in the Okahandja Park Informal Settlement. Note its poorly serviced substandard housing in a highly unhealthy and unsafe landscape — ©Photograph by authors.

Frame d: Picture showing a view of an open bush for defecation due to a lack of toilets in the informal settlement of Ekunde 1. Note the aerial view of the state of housing in the settlement — ©photograph by authors.

Fig. 2 Pictorial representations (Frames a–d) of Okahandja Park and Ekunde 1 Informal Settlements in Windhoek

The informal settlements in Windhoek and its suburbs are the areas with the highest growth rates in town in Namibia [27]. They also constitute the low and ultra-low incomes areas of the Windhoek area.

Figure 3 is the map of Windhoek showing the current city layout and a socio-spatial characterisation of the city plan. An aerial view of the city still clearly depicts the high-density western and low-density eastern parts. On the upper left of the map are Katutura and Khomasdal, and the old location of the former black-only neighbourhood later relocated to Katutura. During Namibia's apartheid and colonial periods, segregationist policies were enforced to ensure that whites' and 'non-whites' did not mix. Hence, urban planning was used to make spaces and infrastructure serve as objects of segregation in the city.

The informal-Windhoek represents the section of the city of Windhoek (that is, the Western part of Windhoek in Fig. 2) where informal settlements (depicted with green circles in Fig. 2) dominate the spatial and aspatial landscape of the city. The east (formal side of the city) still serves as the place of employment for most city residents. In its current spatial form, the main road (known in Namibia as *B1*) cuts the city into two parts, indirectly creating two cities in one [19]. The east side functionally serves the upper/middle class, and the western side serves the middle/lower class and constitutes the section of informal settlement proliferation.

3.2 Informal Settlements in Informal-Windhoek

Informal settlements have been defined in numerous ways. The common description that is rare in the most scholarly cited definitions relates to the housing standard issue. Most definitions of informal settlements recognise them as "areas within cities" which consist of "substandard housing" in "poorly serviced and/or overcrowded, and therefore unhealthy, unsafe, and socially undesirable" [28, p. 383]. From a planning perspective—informal settlements constitute a planning problem)—informal settlements emerged because of poor urban management, leading to socio-spatial exclusions. Many governments have moved from eviction-related approaches to more focused, inclusive methods [29]. The more inclusive practices relate to self-help community development strategies, gendered techniques, in-situ upgrading and relocation, and many others [30–32].

The reasons behind the expansion of informal settlements in Windhoek are historical and structural. From the structural dimensions, informal settlements in Windhoek are due to uncontrolled urban expansion or sprawl, the increasing population within an unorganised physical urban structure exacerbated by poor infrastructure [19, 33]. From a historical development perspective, informal settlement proliferation in Namibia is attributable to the country's transition from a "rural-based society to one based largely in urban areas" [34, p. 1].

From a structural dimension, the informal settlement is characterised by a lack of land ownership by settlers, the temporary nature of building materials, a lack of infrastructure and poor hygienic conditions. Corrugated iron is the primary building

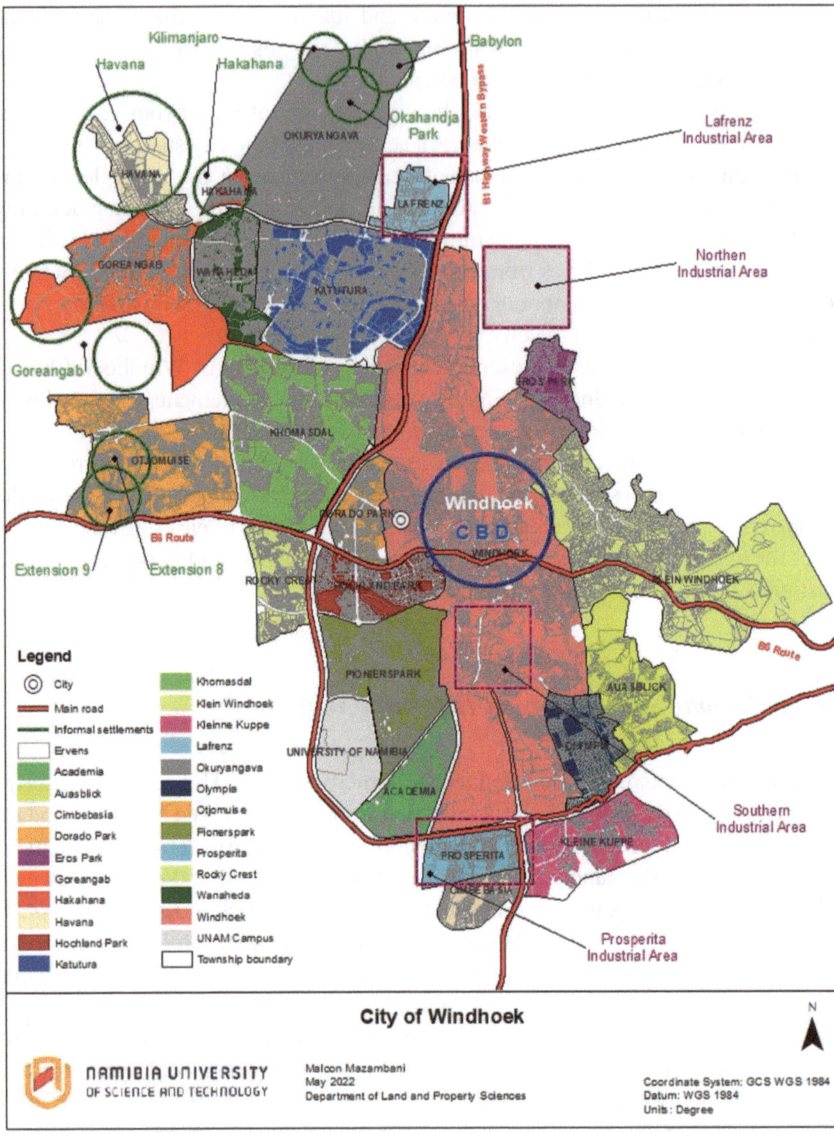

Fig. 3 Map of Windhoek depicting the formal and informal suburbs. *Source* Authors' illustration

material used in these settlements for houses or shacks because it is cheap to source [35]. They are easy to erect within a day or two. They can also be easily moved. The materials can also pitch a new structure in a different location.

Most importantly, "local authorities also tolerate corrugated iron in informal settlements, where the use of permanent construction materials for houses is often prohibited" [34, p. 25]. Tenure insecurity (lack of tenure security) is critical in informal settlements. People in informal settlements in Windhoek have no tenure security on the land they are using [36]. This means that residents of these settlements "are not eligible for freehold title and do not have formal tenure security over the land on which they are built [34, p. 26]. They have no access to energy for cooking. Many of them depend on firewood for their cooking needs.

In most cases, they must often fetch firewood from naturally wooded areas around their settlement. Private and shared flush toilets are also prevalent in the settlements. For most of them, using a bush for a toilet or open defecation is a reality. These have negative hygiene or sanitary implications on the health of informal settlement residents. As one woman described the hygiene situation in the Namibian Newspaper [37], going to the bush is so dirty and dangerous that "I used to make my children do number two (poop) in a plastic bag, and throw it away (the flying toilets)."

Despite their deplorable socio-spatial conditions, the informal section of Windhoek (spatially located on the west side of Windhoek) is an active socio-economic space because it comprises populations of people from different works of life. Like many informal settlements elsewhere, it is constantly involved in livelihood-related activities [38]. However, these activities are impeded by inadequate socio-economic and appropriate spatial planning and tenure security on land or property rights [39]. Efforts to improve people's living conditions in informal settlements (informal settlement upgrading) have shifted over time.

Informal settlements emerged in Namibia due to unplanned occupation by people eager to migrate to cities. This led to people living in places with no recognised street addresses or any form of officially recognised addresses. Based on this situation, the informal settlements are excluded from spatial developmental plans by local authorities. Although inclusive approaches to informal settlement upgrading are now being applied in Namibia, there have been incidences of evictions. In Katima Mulilo, a town council demolished structures in an area identified for upgrading [40]. This incident raised a public outcry that led to Namibia's national government issuing a directive to all local authorities to desist from using evictions as a solution to informal settlement upgrading. By heeding this directive, Windhoek's development and upgrading policy incorporates informal settlements within its development structures, presenting different planning areas to be assisted through various stages of upgrading [41]. This has also led to the more collaborative implementation of informal settlement upgrading in Windhoek. For instance, Kohima et al. [40, p. 4] noted that "in 2021, the government implemented the informal settlement upgrading project in partnership with the National Housing Enterprise and City of Windhoek." Despite this effort, this project had its limitations. It only considered those with access to title deeds and ignored others—even though they might have had other forms of land documents to prove their occupation and use of land. However, this is changing. Municipal authorities now apply land recordation approaches as a precondition for informally issuing certificates of recognition to residents occupying the land. This practice is essential because land documents and records prove land occupation in informal settlements.

The concept of the informality of settlements in Namibia is linked to the country's colonial apartheid system, which encouraged spatial inequality. However, Namibia's proliferation of informal settlements is linked to its post-colonial practice of reception areas in Windhoek. Between 1991 and 1994, the City of Windhoek established so-called 'reception areas' to accommodate poor in-migrants temporarily to relocate them to permanent areas once they were available [42, p. 23]. Section 91(A) of the Local Authority Act No. 23 of 1992 gives power to municipalities and town councils to set aside reception areas [43, p. 92] by stating that:

> (1) A municipal council and a town council may, in its local authority area, set aside reception areas for the construction and erection of informal housing structures or buildings. (2) for the purposes of subsection (1), "informal housing structures or buildings" means structures or buildings of a temporary nature which are not dwelling houses, incremental houses or initial self-help dwellings as contemplated in the National Housing Development Act, 2000.

The implementation (or application) of the prescription of this legislation has not been efficient within the context of informal settlement proliferation in Windhoek. Weber and Mendelsohn [34] found that higher demand overgrew land supply. This meant that the local authority, which has the mandate to allocate and service land, has struggled to meet the challenges posed by the increasing demand for land. There has been a higher demand for housing "due to the internal growth of the households and the increased urbanisation" [40, p. 5]. To address the challenges of informal settlements, controlled entry to the informal settlements was introduced. Controlled access was considered a strategic intervention to minimise land invasion by allowing households with a genuine need for accommodation in designated buffer zones—called reception areas—permission to enter informal settlements based on the availability of land for residential purposes [24]. Windhoek developed several formal low-income housing schemes between 1991 and 1999.

During this period, the serviced plots provided by the city became unaffordable to most people experiencing poverty while the influx of poor urban migrants continued. In response to this, the municipality developed three reception areas that were intended to be temporary. According to the World Bank [44], in 1992, the first reception area (then known as *Big Bend* but now called *Havana*) was established. Another reception (*Okuryangava Extension 6*, locally known as *Babylon* and *Kilimandjaro*) was later established. Later (in 1998), a third reception area (*four blocks*) was developed in a suburb of Windhoek called *Goreangab*. The set-up of these reception areas was a top-down emergency initiative for minimising the proliferation of informal settlements. "These areas consisted of tracts of land where earth roads were cut to a rudimentary layout, lifeline water supply was provided and, in some cases, communal toilet facilities were also provided" [40, p. 6]. People settled in shacks of corrugated metal sheeting on plots of 300 square meters, set out in blocks [44]. Since this intervention assumed that informal settlement was a temporary problem, it has not been sustainable as the growth of informal settlements in Namibia has continued to be more permanent. Namibia's concept of the reception area was based on a further assumption that people would be resettled following the 'Windhoek City Council Squatter Policy' of that time. This was not successful. The city authority could not contain growth within the planned boundaries of the reception areas. "Instead, what ended up happening was that the areas attracted further settlers even before the sites had been laid out or could be provided with rudimentary services" [40, p. 6]. The failure of the reception area approach to the control of informal settlement caused the City of Windhoek to develop new policies and strategies. However, it led to spiralling informal settlements in the city's westside.

Figure 4 comprises 4 photo frames (a–d), representing the *Okahandja Park* and *Ekunde 2* Informal Settlements. Despite the lack of planning (including infrastructure and other essential urban services), residents try to ensure a liveable environment.

Looking at the photos (Frames a–d), one can conclude, from a planning perspective, that informal settlements in Windhoek are where spatially isolated people are confined in constricted spaces far removed from the bubbles of the city. These people live in these isolated spaces because they cannot access their land. Yet, they try to make their environment as liveable as possible. However, when there is no formal planning, it is difficult for residents to adhere to formal rules and regulations. The consequence of this approach is that in Windhoek, 85% of the informal settlers do not own the land they occupy in informal settlements [45]. Existing data indicate that 2004, informal settlers constituted about 29% of Windhoek's population of over 250,000 [46]. This rose to 40% in 2018 and has increased to 45% today. These settlements have limited access to public municipal services and continue to grow exponentially.

3.3 The COVID-19 Scenarios in Informal-Windhoek

The conditions of informal settlements in Windhoek (as shown in the photos) are so unplanned that they are (and will be) vulnerable to any virus-related pandemic disaster. Corburn [47] linked planners' poor roles in urban management to the difficulty in managing COVID-19 in informal settlements. Corburn [47, p. 1] asked: "How can modern city planning, a profession that emerged in the late nineteenth century intending to improve the health of the least well-off urban residents but lost this focus throughout the twentieth century, return to its health and social justice roots?" This question, which directly hit at the heart of planning, raised the issue of "(re)considering urban planning interventions" in managing a viral disaster (such as the COVID-19) in cities [1, p. 115]. In raising this question, Corburn and Sverdlik [48] identified a link between living in informal settlements and human health. Many other scholars agree with this assertion. Raju and Ayeb-Karlsson [5] have noted the health vulnerabilities of informal settlements to diseases. Ezeh et al. [49] and Corburn et al. [50] labelled them unhealthy places. This label has merits because informal settlements are places of high densities, infrastructure deficits, and economic deprivation.

As in every other part of the world facing informal settlement challenges, informal settlements in Windhoek are not liveable (from a planning perspective). They are settlements where people experiencing poverty live. Liveability—a factor of adequate living conditions usually compromised in informal settlements—hinges on providing housing, infrastructure, and livelihood opportunities. A typical depiction of life in

Frame a: Picture showing a distant view of the environmental landscape of the Okahandja Park Informal Settlement. Note that the houses are made of corrugated iron sheets. The settlement is sandwiched between erosion and landslide-prone foothills — ©Photograph by authors.

Frame b: Picture showing the provision of public water supply systems. A communal water tap in the Ekunde 2 informal settlement — ©Photograph by authors.

Frame c: Picture showing a closer view of the typical state of housing in the Ekunde 2 Informal Settlement. Despite its poorly serviced substandard housing made of corrugated iron sheets, residents paint their houses colourfully to enhance architectural aesthetics — ©Photograph by authors.

Frame d: Picture showing the Ekunde 2 informal settlement where, despite their unplanned environment, they use local efforts to ensure a clean environment post-COVID. However, this is not sustainable due to a lack of formal planning. Note its poorly serviced substandard housing made of corrugated iron sheets — ©Photograph by authors.

Fig. 4 Pictorial representations (Frames a–d) of Okahandja Park and Ekunde 2 Informal Settlements in Windhoek

informal settlements in Windhoek is where people live in shacks without electricity or regular tap water. Where these amenities are provided, they are shared as community facilities. For instance, where there is a toilet in informal settlements, they have shared toilets that are available at the rate of 1 bathroom to more than 200 persons (1:100). In an extreme situation, an informal settlement in Windhoek can be described as a place where a household of seven people could be living in a small shack, bushes are used as toilet, and water taps are non-existent. In these places, houses become damaged after each rain, and the informal settlers are barred from renovating their homes due to municipal regulations. The renovation of informal structures in the exact location is allowed only if the structure owner gets authorisation from the leadership of the community and the municipal authority. Such permission is necessary so that the community leadership and the municipal management can inform the police [35]. Without informing the policy through this procedure, the police can arrest anyone engaged in such renovations.

The emergence of the COVID-19 pandemic added more liveability burdens to Windhoek's informal settlements. The first case of COVID-19 was reported on the 13th of March 2020. "This was an imported case of a Romanian couple (a 35-year-old man and a 25-year-old woman) who travelled from Madrid, Spain via Doha, Qatar" [51, p. 25]. The pandemic either caused or exacerbated the following problems.

1. Declaration of the State of Emergency: The national government declared a state of restricted emergency movements and introduced lockdown measures [52].
2. Introduction of strict hygiene measures: During the pandemic, the government initiated stringent social distancing measures and limited mobility interventions.
3. Urban workers retrenched from formal employment: Due to the stay-at-home measures introduced by the government, employing agencies (especially in the private sector) retrenched [4]. This exacerbated the already burgeoning unemployment in the city.
4. Negatively unstable economic growth recorded: Namibia's pre-COVID economic growth rate forecast fluctuated from 1.5% in 2020 to −6.9% in the post-COVID period of 2021 [53].
5. Food Insecurity increased: About 36 per cent of the total population of Namibia (nearly 290,000 people) became food insecure, "21 per cent moderately and 15 per cent severely food insecure" [51, p. 25].
6. Namibia's school feeding schemes stopped: The pandemic led to the halting of Namibia's nationwide government school feeding programme [3].
7. Cemeteries in Windhoek became fully utilised. The municipal authority had to resort to cremation to bury the dead.
8. High crimes, gender-based violence and teenage pregnancies increased [54].

COVID-19, a viral disaster, continued to shake up the urban health terrain of Namibia until a comprehensive vaccine-based solution was implemented. The recovery plan has involved strict funding, policy changes and engagement in coping behaviour. These challenges had (and still have) tremendously negative impacts on an already stressed Namibian economy [2]. The viral disaster had at least four distinct peaks over its occurrence. Its aftershocks, which are still being felt, can be explained from economic, environmental and social perspectives. However, as the focus of this chapter is limited to post-COVID life in informal-Windhoek, the chapter proceeds by providing scenarios of life as influenced by COVID-19.

4 Post-COVID Scenarios in the Informal-Windhoek

The post-COVID challenges of informal-Windhoek can be viewed from different perspectives. This chapter adopts a current analysis of print media depictions of life in informal-Windhoek (newspaper) during the period of ten months (January to October 2022). The scenarios in this chapter are derived from three prominent newspapers in Namibia. These include *The Namibian*[2] and the *New Era*[3] Newspapers. In this regard, Table 1 depicts the background (general) scenarios caused by COVID-19, which affected informal-Windhoek.

Any general adverse effect caused by the COVID-19 pandemic has more significant specific negative urban development consequences in the informal part of Windhoek. An analytical understanding of the COVID-19 scenarios shown in Table 1 leads to one profound inference. These scenarios are the foundation for the recent experience of COVID-19 in informal-Windhoek. They also present a platform for understanding the current (and specifically relevant) lived experiences influenced by COVID-19 in informal-Windhoek, as presented below.

Cuts in government expenditure momentarily halted school feeding programmes: In the Namibian Newspaper (09/06/2022), a 12-year pupil from the informal-Windhoek noted: "*When the pandemic started, we were told to go home, and when we came back, they were no longer giving us food at the kitchen.*" Further consequences, as reported by the pupil, are that: "*Some of the kids who used to get food from there stopped coming to school, and the ones who wanted to share their lunch boxes with others were told they could not because it would spread the virus.*" According to one of the teachers from this primary school, "*The programme does not only help children from the school but also helps the community. Children from outside the school also come to eat.*" The teacher noted, "*When the soft porridge is served, I see students come in huge numbers. Some children will even tell the teachers they did not attend school because there was nothing to eat.*"

Informal economic activities: About 60% of Namibia's employed population is in informal employment—these include informal trading, which is the primary source of livelihood in the informal settlements [57, p. 154]. In terms of business, the informal economy operators in Namibia are Namibians (96.5%), mainly women

---

[2] The Namibian is the largest daily newspaper in Namibia. It is published in English and *Oshiwambo* languages.

[3] The New Era is a daily national newspaper owned by the government of Namibia. It publishes in English and five Indigenous languages: *Otjiherero, Oshiwambo, Damara/Nama, Silozi, and Khwedam.*

Table 1 Media representation of COVID-19 and its effects in informal-Windhoek

Scenarios of COVID-19 were reported in *The Namibian* and *New Era* Newspapers	Direct effect on informal-Windhoek (based on overall content analysis)
4065 Namibian lives lost	More than 3200 lives were lost in the informal settlements. This reflects nearly 80% of COVID-19 deaths in Namibia
More than 45,000 jobs were lost	More than 43,600 reported jobs are lost by people living in informal settlements. This reflects nearly 97% of jobs lost as a result of COVID-19
The COVID-19 pandemic has exposed inadequate health and sanitation facilities at educational establishments	The residents in informal settlements were restricted to poorer health and sanitation facilities. Lockdowns meant pupils who used to access food through school feeding programmes lost such access
800 days of prioritising COVID-19-related regulations to contain the spread of the virus	"Lockdowns and restrictions caused close to 15,000 Namibians to lose their jobs. This figure excludes the informal sector, which was not allowed to operate during the height of the pandemic this year" [55, p. 6]
200,000 more Namibians are being pushed into poverty in urban areas	The informal settlement residents were already in abject poverty before COVID-19. Their chair of this poverty push can only be described as a degeneration from poor to ultra-poor
1,6 million people in Namibia are living in poverty country-wide	Up to 50% of people experiencing poverty in Namibia live in informal settlements
Real gross domestic product contracted by 8.5%. The rebound is expected to be slower than initially expected, with growth projected at 1.2% in 2021 and 2.4% in 2022	Based on the abovementioned indices, the real gross domestic product contractions will decrease in informal settlements (in individual income, production, and residents' sales). This may lead to an upward push in unemployment rates in the informal sector since it relies on formal sector demands for informal services
58% of Namibians are food insecure, translating to about 1.5 million hungry people	According to a separate study conducted by the United Nations Development Programme, the pandemic has pushed the informal sector back into survival mode, relative poverty and unemployment. At least 51.9% of the working population in Namibia earn their livelihoods in the informal economy [56]
As of the 9th of October 2022, a total of 958,674 vaccine doses have been administered. The vaccination rate was less than 36%	Most of those hesitant to take the vaccine live in informal settlements and believe in conspiracy theories surrounding the emergence of COVID-19. This also means that any new wave in the spread of the pandemic will drastically affect those in the informal settlements

(continued)

Table 1 (continued)

Scenarios of COVID-19 were reported in *The Namibian* and *New Era* Newspapers	Direct effect on informal-Windhoek (based on overall content analysis)
To help contain the spread of the virus, the government has spent N$300 million (about US$16.3 million) on vaccines against COVID-19	This could be money that could have been spent on informal settlement upgrading or any other initiatives that would benefit people living in informal settlements in Namibia
Namibia's economic growth outlook is subject to significant uncertainty given the unknown profile of the pandemic and the likelihood of further restrictions in activity if additional infection waves surface	Those living in informal settlements are the most at risk (in terms of facing uncertainty) of any future COVID-19 wave or any new viral disaster
There is no 'policy certainty' on COVID-19	As above

(69.3%) and mostly hold junior secondary education levels (41.1%) [58]. According to a report by *the Namibian* Newspaper (09/06/2022): "*A study conducted by the United Nations Development Programme (UNDP) revealed that revenue for men who own businesses in the informal sector decreased to N$2770* (about USA$152.45) *per month, while women's monthly income dropped to N$1350.53* (about US$74.33) *after COVID-19 restrictions were lifted. Before the pandemic, businesses owned by men made about N$6873.25* (about $378.42) *per month, and women-headed enterprises earned an average revenue of N$3254.71* (about US$179.19)." This scenario indicates a decline in people's entrepreneurial profits in informal settlements. It also points to women being the most affected. This means that the core of the households in the informal settlements has been seriously affected by income generation and food insecurity.

A mix of potentially good and bad post-COVID recovery policies for informal settlements: Namibia's national governments and municipal authorities treat the urban informal economy as undesirable. This emanates from the highly restrictive planning measures that limit the opportunity of informal settlers to exercise their rights and privileges as urban settlers. Various mixes of policies have been introduced post-COVID. The *New Era* Newspaper (01/08/2022) reported, "*The municipality of Windhoek is planning to formulate a new policy to prohibit home shops from operating in residential areas.*" This expected policy has unsettled many people in informal settlements since all use their residential shacks as places of business. While this policy has not been implemented (expected to be abandoned), many in the informal settlements feel more vulnerable now.

Further reports relate to finalising a policy document that will assist in transforming Namibia's informal sector to contribute to the mainstream economy. According to the Ministry of Industrialisation and Trade, this policy is "to be accompanied by an attendant Act of Parliament later on, which provides ample space for ensuring the mainstreaming of incentives and business infrastructure for our MSMEs, both formal and informal." To date, no disaster-preparedness strategies have been designed to raise pandemic alerts in informal-Windhoek.

Health services for informal settlements remain merged with broader health care services: Health services for people in informal settlements were not prioritised during the peak period of the COVID-19 pandemic. This led to overwhelming deaths coming from this section of the urban space. As of 2022, health services for informal settlements are still not distinguished from other forms of urban health—mainstreaming health services for informal settlers in implementing urban plans [1].

A summary of the post-COVID life of people in an informal settlement is clearly stated in the Namibian (23/08/2022), which noted that: "*Lockdown measures implemented by the government aimed at combating the spread of COVID-19 mostly affected people living in poverty, who are working in the informal economy, and who were the hardest hit by such well-meaning interventions.*" Based on the above issues discussed, it is reasonable to conclude that informal settlements in informal-Windhoek are COVID-stressed. COVID stress means they have not entirely (and may never) returned to normalcy. The question that arises is, what does normalcy mean to informal settlers?

In informal-Windhoek, normalcy means having adequate housing, enjoying open space with infrastructural connections that give a consistent supply of water and electricity, and, most of all, enjoying the right to own and exercise land rights securely. This definition of normalcy is something they never had and a situation that got even worse post-COVID. Therefore, a realistic demand should be that, even in their present circumstances, informal settlements should be facilitated to become resilient to future diseases or any wave of COVID-19. This cannot be done without secure access to land. Doing it requires the formalisation of informal settlements, which is an activity that only the government can authorise and facilitate in collaboration with communities.

5 Conclusion: Making a COVID-Stressed Informal-Windhoek Resilient

This chapter has shown that "the COVID-19 pandemic is a disaster that combines a biological threat with various vulnerabilities, such as the organizational and response capacity of health systems, overcrowding, informality, social work practices, and public transport" [59, p. 1]. The difference between the COVID-19 pandemic and other disaster is that "unlike other threats such as earthquakes, hurricanes and floods, which last for minutes, days, or weeks, an epidemic can last for years" [59, p. 1]. It affected all informal settlements in Namibia. Being entirely qualitative—in its structure, approach and output—this study presented a scenario of post-COVID narratives of informal settlers to recommend measures to improve resilience in informal settlements in the city. However, the study is "phenomenological" because it focused on the "contextualisation and reporting of experience" [60, p. 6; 61]. While the study's strength hinges on using extracts depicted in the National Newspapers of Namibia to present the COVID-19 experiences realistically, its weakness lies in its lack of primary empirical data.

To recommend measures that could improve the resilience of informal settlements to a viral disaster, it is essential to grasp further the general circumstances of informal settlements in Windhoek. Claims that informal settlements or the informal economy is a recent development in Namibia are far from reality for one reason: during colonisation, the colonial administrators considered native settlements where they chose not to extend the arm of their laws to be informal. The colonial segregationist planning system was premised on the categorisation of races. Now, in the post-colonial era of Namibia, there is a post-colonial planning system that focuses on cultural differences [19]. While this chapter disagrees with such colonial categorisations or impressions, it does prove that informality (whether of settlements or economies) was not an unknown phenomenon before Namibia's independence. Namibia's post-colonial government must provide an all-embracing leadership that prioritises the development of informal settlements in the country's development.

While these informal settlements did face a basket of risks during the COVID-19 pandemic, these communities also showed resilience. Its resilience may have been motivated by the country's experience with previous pandemic outbreaks, such as HIV/AIDS. For instance, existing community sensitisation structures used for HIV/AIDS helped to reduce the spread of COVID-19 and its associated stigma. For its HIV/AIDS programmes, "Namibia has adopted strategies such as the treat-all approach, also known as the test and treat strategy, whereby all individuals diagnosed HIV positive are immediately initiated on antiretroviral therapy" [62, p. 3]. The country applied similar strategies in encouraging citizens to become COVID-19 aware and undergo testing when ill.

However, since the usual "normal" of informal settlements is not admirable by any standard, transformative resilience is necessary for the future. By transformative resilience, rather than returning to the previous "normal," they need to embrace active engagement with other stakeholders in the urban space rather than exist as recipients of assistance or aid [63]. In this regard, informal settlements within the informal-Windhoek should be facilitated as follows:

1. (Re)Education, capacity development, capacity building and training on the informal settlement upgrading and development will be a starting point for improving or building transformative resilience. This must include health awareness creation.
2. There is a need for all (including urban residents, governments and the generality of urban planners) to change their mindset on the relevance of informal settlements. Accepting that the informal settlements (the home of the informal economy) are a crucial stakeholder (and sector) in the nation-building and development process can lead to a better appreciation of their contribution to a country.
3. The COVID-19 disaster exposed the communication gap between informal settlement organisations (and Civil Society Organisations that promote their welfare)

and policymakers. Building better relationships and bridging the communication gap between local government and the informal settlements is crucial for boosting human dignity and the informal economy.
4. An informal-settlement-tailored housing plan is necessary to reorganise sharing as a survival strategy in informal settlements. Critical is that efforts at improving informal settlements should prioritise sanitation/sanitation and enhance space standards that can save a life during viral disasters, such as the one experienced with COVID-19.
5. Based on the lessons learned from the COVID-19 disaster, people in informal settlements or organisations representing them should be included in the governance and planning processes for urban COVID-19 recovery.

The above resilience measures have been recommended based on the experience of informal settlements in informal-Windhoek, Namibia. However, these measures (even within the frame of urban and rural contexts) are expected to be applicable elsewhere—in Africa, Asia, Latin America or Australia [63–68]. This is because, while informal settlements may differ in their climatic and material landscapes, they share familiar challenges in living conditions. They are areas within cities that consist of "substandard housing" and that are "poorly serviced and/or overcrowded, and therefore unhealthy, unsafe, and socially undesirable" [28, p. 383].

References

1. Chigbu UE, Onyebueke VU (2021) The COVID-19 pandemic in informal settlements: (re)considering urban planning interventions. Town Plann Rev 92(1):115–121. https://doi.org/10.3828/tpr.2020.74
2. Lendelvo SM, Pinto M, Sullivan S (2020) A perfect storm? The impact of COVID-19 on community-based conservation in Namibia. Namibian J Environ 4:1–15. https://researchspace.bathspa.ac.uk/id/eprint/13294
3. Mweuta E (2021) COVID bites school feeding programmes: hungry pupils stay home, sleep in class. The Namibian 35(235):1. https://www.namibian.com.na/covid-bites-school-feeding-programme/
4. Ozili P (2020) COVID-19 in Africa: socio-economic impact, policy response and opportunities. Int J Sociol Soc Policy 42(3/4):177–200. https://doi.org/10.1108/IJSSP-05-2020-0171
5. Raju E, Ayeb-Karlsson S (2020) COVID-19: how do you self-isolate in a refugee camp? Int J Public Health 65:515–517. https://doi.org/10.1007/s00038-020-01381-8
6. Chigbu UE (2020) E-tracking COVID-19 disruptions to the global development agenda on land. Int J Environ Sci Nat Resour 26(1):556176. https://juniperpublishers.com/ijesnr/IJESNR.MS.ID.556176.php
7. Weyer J, Grobbelaar A, Blumberg L (2015) Ebola virus disease: history, epidemiology and outbreaks. Curr Infect Dis Rep 17(5):1–8. https://link.springer.com/article/10.1007/s11908-015-0480-y
8. Ahmad J, Ahmad MM, Su Z, Rana IA, Rehman A, Sadia H (2022) A systematic analysis of worldwide disasters, epidemics and pandemics associated mortality of 210 countries for 15 years (2001–2015). Int J Disaster Risk Reduction 76:103001. https://doi.org/10.1016/j.ijdrr.2022.103001

9. Grossi G, Vakulenko V (2022) New development: accounting for human-made disasters—comparative analysis of the support to Ukraine in times of war. Public Money Manage 42(6):467–471. https://doi.org/10.1080/09540962.2022.2061694
10. Dick OB, San Martín JL, Montoya RH, del Diego J, Zambrano B, Dayan GH (2012) The history of dengue outbreaks in the Americas. Am J Trop Med Hyg 87(4):584. https://www.ncbi.nlm.nih.gov/pmc/articles/PMC3516305/
11. Khan M, Adil SF, Alkhathlan HZ, Tahir MN, Saif S, Khan M, Khan ST (2020) COVID-19: a global challenge with old history, epidemiology and progress so far. Molecules 26(1):39. https://doi.org/10.3390/molecules26010039
12. Mushagalusa CA, Etter E, Penrith ML (2021) Review of African swine fever outbreaks history in South Africa: From 1926 to 2018. Onderstepoort J Vet Res 88(1):1–10. https://www.ncbi.nlm.nih.gov/pmc/articles/PMC8517827/. http://www.scielo.org.za/scielo.php?script=sci_arttext&pid=S0030-24652021000100014
13. Parıldar H (2020) Infectious disease outbreaks in history. J Tepecik Educ Res Hosp 30(Supp: 2):19–26. https://tepecikdergisi.com/jvi.aspx?un=TERH-93764&volume=30&supp=2
14. Tognotti E (2013) Lessons from the history of quarantine, from plague to influenza A. Emerg Infect Dis 19(2):254. https://www.ncbi.nlm.nih.gov/pmc/articles/PMC3559034/
15. United States President's Emergency Plan for AIDS Relief (2022) Namibia country operational plan 2022 strategic direction summary. PEPFAR Namibia, Windhoek. https://www.bing.com/search?form=MOZLBR&pc=MOZI&q=Namibia+Country+Operational+Plan+2022+Strategic+Direction+Summary
16. Douglas KM (2021) COVID-19 conspiracy theories. Group Process Intergroup Relat 24(2):270–275. https://doi.org/10.1177/136843022098
17. Earnshaw VA, Eaton LA, Kalichman SC, Brousseau NM, Hill EC, Fox AB (2020) COVID-19 conspiracy beliefs, health behaviors, and policy support. Trans Behav Med 10(4):850–856. https://doi.org/10.1093/tbm/ibaa090
18. Baker N (2021) The Lab-Leak Hypothesis. New York Magazine, 4.https://doi.org/10.1080/13668803.2014.981506
19. Kohima JM, Chigbu UE, Mazambani ML, Mabakeng MR (2023) (Neo-)segregation, (neo)racism, and one-city two-system planning in Windhoek, Namibia: what can a new national urban policy do? Land Use Policy 125:106480. https://doi.org/10.1016/j.landusepol.2022.106480
20. Friedman F (2001) Deconstructing Windhoek: the urban morphology of a post-apartheid city. Development Planning Unit, University College London, Working Paper No. 111. https://www.semanticscholar.org/paper/DECONSTRUCTING-WINDHOEK%3A-THE-URBAN-MORPHOLOGY-OF-A-Friedman/c5ed11ebfc88b12c7b0f0875f7ea7df3d0cd8189
21. Chigbu UE (2013) Fostering rural sense of place: the missing piece in Uturu, Nigeria. Dev Pract 23(2):264–277. https://doi.org/10.1080/09614524.2013.772120
22. Jauch H (2012) Poverty, unemployment and inequality in Namibia. TEMTI Series of Economic Perspectives on Global Sustainability-EP 02-2013. TEMTI-CEESP/IUCN. https://www.bing.com/search?form=MOZLBR&pc=MOZI&q=Poverty%2C+Unemployment+and+Inequality+in+Namibia.+TEMTI+Series+of+Economic+Perspectives+on+Global+Sustainability-EP+02-2013
23. Mphambukeli TN (2019) Apartheid. In: The Wiley Blackwell encyclopedia of urban and regional studies, pp 1–6. https://doi.org/10.1002/9781118568446.eurs0437
24. City of Windhoek (2021) Interesting facts about Windhoek. Official website of the city of Windhoek. http://www.windhoekcc.org.na/info_facts.php. Accessed 14 Apr 2023
25. Pendleton WC (1996) Katutura: a place where we stay (life in a post-apartheid township in Namibia: Katutura Before and Now). Centre for International Studies, Ohio
26. The Namibian Newspaper (2021) Okahandja unites against COVID-19. July 2. https://www.namibian.com.na/okahandja-unites-against-covid-19/
27. Uhlendahl T, Ziegelmayer D, Wienecke A, Mawisa ML, du Pisani P (2021) Water consumption at household level in Windhoek, Namibia: survey about water consumption at household level in different areas of Windhoek depending on income level and water access in 2010.

Project Report. Institute for Culture Geography, Albert Ludwigs University and University of Freiburg, Germany. https://www.bing.com/search?form=MOZLBR&pc=MOZI&q=Water+consumption+at+household+level+in+Windhoek%2C+Namibia%3A+Survey+about+water+consumption+at+household+level+in+different+areas+of+Windhoek+depending+on+income+level+and+water+access+in+2010
28. Harris R (2009) Slums. In: Kitchin R, Thrift N (eds) International encyclopedia of human geography. Elsevier, Amsterdam, pp 383–388. https://doi.org/10.1016/B978-008044910-4.01079-8
29. Khalifa MA (2015) Evolution of informal settlements upgrading strategies in Egypt: from negligence to participatory development. Ain Shams Eng J 6:1151–1159. https://doi.org/10.1016/j.asej.2015.04.008
30. Chigbu UE (2015) Repositioning culture for development: women and development in a Nigerian rural community. Community Work Fam 18(3):334–350. https://doi.org/10.1080/13668803.2014.981506
31. Chigbu UE (2019) Anatomy of women's landlessness in the patrilineal customary land tenure systems of sub-Saharan Africa and a policy pathway. Land Use Policy 86:126–135. https://doi.org/10.1016/j.landusepol.2019.04.041
32. Renzaho A (2020) The need for the right socio-economic and cultural fit in the COVID-19 response in sub-Saharan Africa: examining demographic, economic political, health, and socio-cultural differentials in COVID-19 morbidity and mortality. Int J Environ Res Public Health 17(10):3445. https://doi.org/10.3390/ijerph17103445
33. Adewunmi YA, Chigbu UE, Kahireke U, Simbanegavi P, Mwando S, Issa AA, Hayford S (2023) A multi-faceted approach to improving public services in low-income housing in Windhoek, Namibia. Sustainability 15(6):4885. https://doi.org/10.3390/su15064885
34. Weber B, Mendelsohn J (2017) Informal settlements in Namibia: their nature and growth—exploring ways to make Namibian urban development more socially just and inclusive. John Meinert Printing, Windhoek. https://www.bing.com/search?form=MOZLBR&pc=MOZI&q=Informal+settlements+in+Namibia%3A+their+nature+and+growth+%E2%80%93+Exploring+ways+to+make+Namibian+urban+development+more+socially+just+and+inclusive
35. Kaminzi M, Karuuombe M (2021) Barely enough room for eight. The Namibian 35(189):5. https://www.namibian.com.na/barely-enough-room-for-eight/
36. Adewunmi YA, Chigbu UE, Mwando S, Kahireke U (2023) Entrepreneurship role in the co-production of public services in informal settlements—a scoping review. Land Use Policy 125:106479. https://doi.org/10.1016/j.landusepol.2022.106479
37. The Namibian Newspaper (2017) The Namibian Newspaper, April 25. https://namibian.com.na/the-toilet-under-the-rock/
38. Gall M, Wiener M, de Oliveira CC, Lang RW, Hansen EG (2020) Building a circular plastics economy with informal waste pickers: recyclate quality, business model, and societal impacts. Resour Conserv Recycl 156:104685. https://doi.org/10.1016/j.resconrec.2020.104685
39. Dachaga W, de Vries WT (2021) Land tenure security and health nexus: a conceptual framework for navigating the connections between land tenure security and health. Land 10(3):257. https://doi.org/10.3390/land10030257
40. Kohima J, Mabakeng R, Alexander L, Ricardo P, Simataa V (2022) Reception areas: analyzing the delivery of land for housing in Namibia. International Federation of Surveyors (FIG) Congress, September 11–15. Warsaw, Poland. https://www.fig.net/resources/proceedings/fig_proceedings/fig2022/ppt/ts06c/TS06C_kohima_mabakeng_et_al_11562_ppt.pdf
41. City of Windhoek (2019) City of Windhoek integrates social media into its communication strategy. Aloe 66:2
42. Nghiulikwa RV (2008) Re-situating and shifting cultural identity in contemporary Namibia: the experience of rural-urban migrants in Katutura (Windhoek). Unpublished Master Thesis. University of the Western Cape, Cape Town. http://hdl.handle.net/11394/2720
43. Republic of Namibia (1992) Local Authority Act No. 23 of 1992. The Government of Namibia: Windhoek. https://www.bing.com/search?form=MOZLBR&pc=MOZI&q=Local+Authority+Act+No.+23+of+1992

44. World Bank (2002) Upgrading of low-income settlements: Namibia Country Assessment Report. Retrieved from https://web.mit.edu/urbanupgrading/upgrading/case-examples/overview-africa/country-assessments/download/NAMIBIA.pdf
45. Karuaihe ST, Wandschneider PR (2018) Limited access to services for the urban poor in Windhoek, Namibia. Dev South Afr 35(4):466–479. https://doi.org/10.1080/0376835X.2018.1475219
46. Karuaihe ST (2019) Namibia's urban poor are stuck in Limbo, without land or services. allAfrica, September 19. https://allafrica.com/stories/201909200106.html. Accessed 6 May 2022
47. Corburn J (2009) Toward the healthy city: people, places, and the politics of urban planning. MIT Press, Cambridge, MA. https://onlinelibrary.wiley.com/doi/full/10.1111/j.1541-1338.2011.00498_2.x
48. Corburn J, Sverdlik A (2019) Informal settlements and human health. In: Nieuwenhuijsen M, Khreis H (eds) Integrating human health into urban and transport planning. Cham, Springer, pp 155–171. https://link.springer.com/chapter/10.1007/978-3-319-74983-9_9
49. Ezeh A, Oyebode O, Satterthwaite D, Chen YF, Ndugwa R, Sartori J, Caiaffa W (2017) The history, geography, and sociology of slums and the health problems of people who live in slums. The Lancet 389:547–558. https://doi.org/10.1016/S0140-6736(16)31650-6
50. Corburn J, Vlahov D, Mberu B, Riley L, Caiaffa, WT, Rashid SF, Jayasinghe S (2020) Slum health: arresting COVID-19 and improving well-being in urban informal settlements. J Urban Health 97:348–957. https://link.springer.com/article/10.1007/s11524-020-00438-6
51. Amesho JN, Ahmadi A, Lucero-Prisno III DE (2020) The calculated responses against COVID-19 in Namibia. Pan Afr Med J 37(1):25. https://www.ncbi.nlm.nih.gov/pmc/articles/PMC7796846/
52. Marenga R, Amupanda JS (2021) The coronavirus and social justice in Namibia. Politikon 48(2):206–225. https://doi.org/10.1080/02589346.2021.1913803
53. Ajambo E (2020) COVID-19: an emerging development challenge, but opportunity for the sustainable development goals (SDGs) in Namibia. Windhoek, Namibia, United Nations. https://namibia.un.org/en/51073-%E2%80%9Ccovid-19-emerging-development-challenge-opportunity-sustainable-development-goals-sdgs
54. Keulder C (2021) The COVID-19 pandemic one year on. The Namibian 35(131):6. https://www.namibian.com.na/the-covid-19-pandemic-one-year-on/
55. Petersen S (2021) COVID-19: the virus that united Namibia. The Namibian Weekend 36(83):6. https://www.namibian.com.na/6216600/archive-read/Covid-19-The-virus-that-united-Namibia-in-2021
56. Petersen S (2022) Counting the cost of COVID: 4000 lives, 15000 jobs lost. The Namibian 36(218):5. https://www.namibian.com.na/6222353/archive-read/Counting-the-cost-of-Covid
57. Endjala T, Botes L (2020) Surviving like a bird. Afr Sociol Rev/Revue Africaine de Sociologie 24(1):151–173. https://www.jstor.org/stable/10.2307/26918069
58. Nakuta JB (2021) Code of conduct for the Namibian Informal Economy: breaking new grounds. Lexnet Publishing House, Windhoek, Namibia. https://www.bing.com/search?form=MOZLBR&pc=MOZI&q=Code+of+Conduct+for+the+Namibian+Informal+Economy%3A+Breaking+new+grounds
59. United Nations (2021) The coronavirus disease (COVID-19) pandemic: an opportunity for a systemic approach to disaster risk for the Caribbean—COVID-19 Report. United Nations Office for Disaster Risk Reduction, Geneva, Switzerland. https://www.cepal.org/en/publications/46732-coronavirus-disease-covid-19-pandemic-opportunity-systemic-approach-disaster-risk
60. Chigbu UE, Atiku SO, Du Plessis CC (2023) The science of literature reviews: Searching, identifying, selecting, and synthesising. Publications 11:2. https://doi.org/10.3390/publications11010002
61. Chigbu UE, Nweke-Eze C (2023) Green Hydrogen Production and Its Land Tenure Consequences in Africa: An Interpretive Review. Land 12(9):1709. https://doi.org/10.3390/land12091709

62. Munyayi FK, van Wyk BE (2023) Determinants and rates of retention in HIV care among adolescents receiving antiretroviral therapy in Windhoek, Namibia: a baseline cohort analysis. BMC Public Health 23(1):1–12. https://bmcpublichealth.biomedcentral.com/articles/10.1186/s12889-023-15356-w
63. Aslam S, Schmidt-Sane M (2022) Evidence review: COVID-19 recovery in South Asian urban informal settlements. Social Science Humanitarian Action Platform, Pakistan. https://www.socialscienceinaction.org/resources/evidence-review-covid-19-recovery-in-south-asian-urban-informal-settlements/#COVID-19_and_urban_inequalities
64. Amoo SK, Harring SL (2012) Property rights and land reform in Namibia. In: Chigara B (ed) Southern African development community land issues: towards a new sustainable land relations policy. Routledge, New York, p 222. https://www.taylorfrancis.com/chapters/edit/10.4324/9780203806555-21/property-rights-land-reform-namibia-sam-amoo-sidney-harring-sam-amoo-sidney-harring
65. Chigbu UE (2015) Ruralisation: a tool for rural transformation. Dev Pract 25(7):1067–1073. https://www.tandfonline.com/doi/full/10.1080/09614524.2015.1071783?src=recsys
66. Chigbu UE, Izugbara CO, de Vries WT (2018) Land, culture, culture loss and community: rural insights from sub-Saharan Africa. In: The Routledge handbook of community development: perspectives from around the globe. Routledge, United Kingdom, pp 98–114. https://www.taylorfrancis.com/chapters/edit/10.4324/9781315674100-8/land-culture-culture-loss-community-uchendu-chigbu-chimaraoke-izugbara-walter-de-vries
67. Chigbu UE, Bendzko T, Mabakeng MR, Kuusaana ED, Osei Tutu D (2021) Fit-for-purpose land administration from theory to practice: three demonstrative case studies of local land administration initiatives in Africa. Land 10:476. https://doi.org/10.3390/land10050476
68. Chitekwe-Biti B (2018) Co-producing Windhoek: the contribution of the Shack Dwellers Federation of Namibia. Environ Urbanisation 30(2):387–406. https://doi.org/10.1177/0956247818785784

Uchendu E. Chigbu is a Professor in Land Administration at the Namibia University of Science and Technology (Namibia). He studied at Abia State University (Nigeria), University of Reading (United Kingdom) and Technical University of Munich (Germany). His fundamental research interest is around the socio-spatial aspects of land, property studies and spatial sciences (including land administration, land management, land policy and governance). He is the lead innovator of the "Tenure Responsive Land Use planning" (TR-LUP) and "Urban-rural Land Linkages" (URLLs) approaches to tackling land use and urban-rural development, respectively. UN-Habitat has endorsed the TR-LUP and the URLLs.

Penehafo Ricardo is a Research Assistant in the Department of land and Property Sciences at the Namibia University of Science and Technology (NUST). She is the co-manager of the Network of Excellence on Land Governance (NELGA) in the Southern African region. Her academic background includes a bachelor's degree in Regional and Rural Development and Bachelor's Honors Degree in Land Administration from NUST. Currently, she is a registered master's candidate at NUST. Her research interests include land tenure, urban sanitation and tenure-related flexible approaches to informal settlements upgrading.

Cathrine Marenga is a Research Assistant in the Department of Land and Spatial Sciences at the Namibia University of Science and Technology (NUST). She is the Namibia Country Director for the Youth Initiative for Land in Africa, a youth organization that advocates for youth participation in land governance. She holds a bachelor's degree and bachelor's Honors Degree in Land Administration from the NUST and is currently pursuing her master's degree in Spatial Science at the same university. Her research interests are on urban development, with specializations in urban informality, bottom-up planning and urban policy development.

Malcon L. Mazambani is a Lecturer in the Department of Land and Spatial Sciences within the Faculty of Engineering and the Built Environment at the Namibia University of Science and Technology (NUST). He holds a Master of Science in Land Management and Land Tenure from the Technical University of Munich (Germany), and a Bachelor of Technology in Land Management (Land Information Systems) from NUST, Namibia. He is well-versed in the application of GIS tools for land administration. His areas of research are land administration, communal land administration, and geographical information systems (GIS).

What Implications Do Transformational Mobility and the Real Estate Market Hold for Cities in the Post-Pandemic Era?

Tiziana Campisi, Talha Tasnim, and Kh Md Nahiduzzaman

Abstract The profound impacts of the COVID-19 pandemic and subsequent global economic recession have reshaped employment, residential choices, and mobility patterns. Practices like remote work, distance learning, and virtual socialization adopted during the pandemic persist in the post-pandemic era. Escalating fuel costs, coupled with the economic downturn, strain traditional daily commutes by cars and public transit too. The pursuit of cost savings has fueled the popularity of alternative modes such as walking, scooters, and electric bikes. Hybrid and remote employment opportunities have emerged as cost-effective and convenient alternatives, influencing residential choices away from city cores and other popular places, leading to significant savings in rent and property ownership costs. This shift has ripple effects on real estate markets, urban growth dynamics, and investment strategies at both individual and aggregate (institutional) levels that have impacts on daily mobility patterns. Understanding these evolving effects on land use and changing growth patterns is critical. Against this backdrop, this paper aims to unravel the dynamics of such transformation and delineate its effects on contemporary cities, offering insights into the implications for future urban landscapes. In the face of this transformative dynamic, a reassessment of current strategies for urban growth and the formulation of sustainable, and affordable mobility plans become imperative.

Keywords Hybrid and remote employment trend · Residential location choice dynamics · Post-pandemic transformational planning · Transformational urban mobility · Urban growth

---

T. Campisi (✉)
Department of Engineering and Architecture, University of Enna Kore, Cittadella Universitaria, 94100 Enna, Italy
e-mail: tiziana.campisi@unikore.it

T. Tasnim
BRAC Urban Development Programme, BRAC Center, 75 Mohakhali, Dhaka, Bangladesh

K. M. Nahiduzzaman
Faculty of Applied Science, The University of British Columbia (UBC) Okanagan, Kelowna, BC V1V 1V7, Canada

K. M. Nahiduzzaman (ed.), *Making Sense of Planning and Development for the Post-Pandemic Cities*, City Development: Issues and Best Practices,
https://doi.org/10.1007/978-981-97-5481-6_8

1 Introduction

The dynamics of cities and mobility systems are undergoing significant transformation, influenced by a range of factors, notably the COVID-19 pandemic as well as the recent energy crisis. Emerging trends such as virtual and hybrid employments and the growing reliance on and the convenience associated with e-services, such as online groceries and daily shopping, are reshaping mobility patterns and locational preferences for residences and their functions within the city core and the surrounding vicinities [1–3]. The pandemic has highlighted gender disparities, exacerbating economic and social inequities. Government-imposed restrictions, implemented globally until the end of the pandemic, have disproportionately impacted women, exposing them to gender-based discrimination while impeding their access to essential urban services [4, 5]. Among others, the pandemic has caused gender impacts that the functional urban areas will need to consider; especially, the increase in gender and economic-social inequity [6, 7]. Metropolitan areas, in particular, face significant challenges in providing access to vital resources such as healthcare, education, and digital infrastructure, disproportionately affecting vulnerable populations.

The COVID-19 pandemic and its aftermath have catalyzed unprecedented transformations in employment dynamics, residential preferences, and mobility behaviors, reshaping the fabric of urban life. Practices such as remote work, distance learning, and virtual socialization, necessitated by the pandemic, have permeated into the post-pandemic era, fundamentally altering traditional modes of work, education, and social interaction [8, 9]. Concurrently, the confluence of escalating fuel costs and economic uncertainties has placed considerable strain on conventional modes of daily commuting, prompting a surge in the popularity of alternative transportation modes such as walking, scooters, and electric bikes in pursuit of cost savings and convenience.

Furthermore, the proliferation of hybrid and remote employment opportunities has prompted a paradigm shift in residential choices, with many individuals opting to migrate away from densely populated city cores and other vicinities. This exodus from urban centers has been fueled by the allure of reduced living expenses, including substantial savings in rent and property ownership costs, as well as the newfound flexibility afforded by remote work arrangements. Consequently, this shift has unleashed ripple effects across various dimensions, impacting real estate markets, urban growth dynamics, and investment strategies at both individual and institutional levels, thereby reshaping daily mobility patterns and altering the spatial configuration of cities [1, 10]. As cities grapple with the far-reaching implications of these transformative trends, it becomes imperative to critically assess their ramifications for urban development and planning strategies. Understanding the interplay between evolving employment patterns, shifting residential preferences, and changing mobility behaviors is essential for fostering sustainable and resilient urban environments in the post-pandemic era.

The post-pandemic landscape has witnessed notable shifts in residential preferences and mobility behaviors, necessitating a re-evaluation of urban planning paradigms. Adapting to these changes is crucial for fostering sustainable and resilient cities. Understanding evolving residential patterns and transportation preferences is vital for developing infrastructure and policies that promote environmental sustainability, social equity, and economic resilience. Insights from studies such as [11] and [12] highlight the importance of addressing these dynamics to ensure cities can effectively respond to future challenges and thrive in the post-pandemic era. This necessitates a multidisciplinary approach that integrates insights from urban planning, transportation engineering, economics, sociology, and environmental science domains to inform policy decisions and shape the future trajectory of urban and territorial landscapes. Therefore, this paper attempts to underscore and respond to the following critical questions:

- In the post-pandemic era, what strategies can cities employ to adjust their urban infrastructure and transportation systems to align with evolving mobility patterns and promote sustainable modes of transportation?
- What approaches can urban planners and policymakers utilize to foster inclusive and equitable urban development amidst evolving residential preferences and spatial dynamics?
- In what ways can cities leverage emerging technologies and data-driven methodologies to streamline land use planning, enhance urban resilience, and cater to the evolving requirements of varied urban communities?

2 Transformational Mobility Behavior and Transportation Infrastructure in the Post-Pandemic Era

Since 2020, the pandemic has instigated significant shifts in people's travel preferences. Particularly in the logistics domain, there has been a surge in the movement of goods and services due to the growing reliance on e-commerce. This trend has led to an increased use of outdoor and public spaces, alongside a shift in demand within urban transportation systems. In the pre-COVID-19 era, these systems were primarily characterized by a well-defined demand for commuting and mobility, often adhering to standardized office, shopping, recreation, school and other trips [13, 14]. During this period, institutions and transportation service providers proposed various measures to enhance user experience and accommodate relatively inflexible demand, thereby devising innovative solutions to adapt to these changes.

Simultaneously, the surge in e-commerce transactions and the diversification of goods available over online have resulted in the closure of many conventional retail stores that had only physical presence in the commercial centers [1]. The objectives outlined in the 2030 Agenda underscore the significance of sustainable urban development, which includes accessible and sustainable public transportation, fostering economic, social, and territorial cohesion, and enhancing citizens' quality of life.

The agenda also aims for climate neutrality by 2050. Implementing a coordinated strategy is essential to address the challenges posed by COVID-19 and navigate through the green and digital transitions, thereby initiating an economic recovery that accelerates sustainable development. Since March 2020, transportation demand has become more intricate, with remarkable varying user behaviors, patterns, and models compared to the pre-pandemic period [15]. Therefore, remote and hybrid jobs have increasingly become a 'norm' in the post-pandemic years, especially in large metropolises.

We are witnessing a radical change in travel related to school and office, and the meaning of "rush hour" itself is changing as well. Whereas before the emergence of COVID-19, demand was inelastic whereby, for example, offices had defined opening and closing hours; in 2020–2022 it has become completely elastic and sometimes unpredictable, as even large companies have established differentiated opening hours and adaptive rules. From 2022, an emphasis is made to the necessity of planning and implementing strategies aimed at integrating and sharing data and information on mobility because of the considerable variation and indeterminateness of demand, demonstrated by the pandemic.

Thus, it stems the need to take greater account of the economic and social conditions of users in the light of the various crises that have occurred and may appear in the future, as these aspects not only strongly influence modal choices and travel motivations, but also improve public transport services and infrastructure. Particular attention will have to be given to both passenger and freight transport, in the light of the growing emergence of e-commerce and last-mile logistics [16]. From a logistics and business perspective, the pandemic of COVID-19 led to the accelerated expansion of the home delivery services sector, resulting in the development of convenient types of working platforms and business models.

The rapid spread of e-commerce and rider deliveries has raised the need for enhanced safety of delivery personnel and further training on the digital tools, such as apps and interactive platforms [17]. Digitalization has helped address some of the immediate challenges arising from the pandemic, particularly during periods of confinement, and that among the many inequalities highlighted by the COVID-19 pandemic—the digital divide appears to be one of the most serious inequalities. As the pandemic has induced an accelerated use of digital platforms and technologies, and they have become essential for teleworking, distance learning, e-commerce, e-health, e-government, digital democracy and digital entertainment [18, 19]. The transport sectors that have undergone the most upheaval have been those related to public and shared as well as private transports [20, 21].

As a result of the COVID-19 pandemic, the local public transportation sector has undergone various experimental measures aimed at containing the virus while ensuring continued urban mobility during the pandemic emergency. The application of quota systems and heightened sanitation protocols has restricted and diminished the demand for public transport, occasionally resulting in service cancellations or suspension. These localized restrictions have led to a reduction in public transportation availability and, in some instances, heightened user anxiety [22, 23].

The pandemic period has highlighted the need to rapidly increase investment in sustainable public transport and its accessibility, as well as in better walking and cycling infrastructure to ensure the safety of road users, especially those with reduced mobility and other disabilities. Therefore, the pandemic highlighted the need to plan and design more walking and cycling infrastructure, and transformational urban planning and mobility solutions to make the urban areas more resilient and adaptive to mobility demand. Moreover, disruptive forces such as pandemic should be seen as an opportunity to address the challenges associated with increasing transport congestion and greenhouse gas emissions. Cohesive urban planning policy can help realize the potential of artificial intelligence and smart city model, for example, by improving administrative capacity and intellectual (digital) skills while encouraging the transition from experimentation to expansion of smart city initiatives through funding and specific support.

Therefore, it will be necessary to:

- Promote best practices in the implementation of a single, multimodal ticketing system that allows people to travel easily and safely [24, 25].
- Rethink complementary forms of transport such as Demand Responsive Transport in order to meet the demands arising from areas with weak transport demand e.g., suburbs, rural areas and/or areas that have found themselves without a local public transport service [26, 27].
- Make the transport system multimodal and complementary, using MaaS digital platforms and including shared and on-demand mobility services [28, 29].
- Promote sustainable urban mobility through environmentally friendly transport systems [30].
- Promote and develop sustainable public transport systems in urban areas and adapting public transport capacity to the growing demand for daily business travel within or outside urban centers (development of complementary services such as DRT for areas with low demand) [31].
- Re-evaluate the investments of urban mobility and prioritize digital infrastructures that will benefit all passengers, including persons with reduced mobility [32].

There is a need on the part of the transport supply side, i.e. on the part of public and private actors providing transport services, for system integration and concrete data sharing. Because, considering the high fragmentation of demand, it is necessary for the user to have real-time access to data, availability of different means of transport and to the possibilities of purchasing relevant tickets (or the single integrated ticket). We must aim to provide travel solutions and integrate the commercial offer as well: the real challenge is to capture in this change in demand an innovation that can bring benefits to all transport actors, is to ensure that during an entire journey the best user experience is put at the forefront.

Considering the above, it will be critical to:

- Support cities in preparing their sustainable urban mobility plans (SUMPs) to the highest standards and promote sustainable mobility at local and regional level, including through SUMPs and public transport projects [33].

- Promote the sharing of best practices among urban areas on sustainable (green) urban planning and infrastructure, clean energy, energy efficiency, zero-emission public transport, active mobility (walking and cycling) infrastructure, efficient water management and sustainable and circular waste management [34].
- Revising the guidelines on the development and implementation of SUMPs, promoting multimodality and addressing some of the repercussions of excessive population concentration in certain urban areas, such as traffic congestion and increased transport costs, through, for example, balanced socio-economic arrangements to ensure that plans are non-discriminatory [35].
- To promote coordination between security and sustainability measures in urban areas; to encourage parking facilities at the entrance to urban areas (such as park-and-ride systems) to provide easy access to different modes of public transport in order to substantially reduce urban congestion, CO_2 emissions and surface and underground parking in urban centers, as well as to restore and improve the attractiveness of urban centers and attract potential customers to support the weakened retail sector in the post-pandemic period and increase value locally [36].
- Improving air quality in urban areas to minimize risks to human health and counteract the increasing levels of environmental noise pollution in urban areas [37].
- Note that personal mobility devices (electric scooters and e-bikes) pose a number of safety problems in urban areas; encourages the development of guidelines for Member States on how to deal with these safety problems [38].

3 Transforming Faces of the Land Use: What Do We Know So Far?

In the wake of the global pandemic, the real estate market has been undergoing a profound transformation, shaped by the surge of hybrid work and thriving e-commerce landscape [39]. This transformative wave is tangible across various aspects of daily life, ushering in a new era of preferences in housing, commuting, and urban development. Examining this shift in the context of cities in the North America (Canada and the USA) provides illuminating examples of such changing dynamics.

One of the most notable effects of the post-pandemic landscape is the diminishing attractiveness of prime urban areas as hubs of residential living. Traditionally priced for their proximity to workplaces, urban cores are witnessing a decline in preference as the rise of hybrid work allows individuals to work from the comfort of their homes that could be located anywhere far from the city [40]. In Toronto, Canada, for instance, the once high-demand downtown condominiums are facing a shift in interest towards suburban and outlying areas. Similarly, in the USA, cities like New York and San Francisco are experiencing a re-evaluation of the centrality of their urban cores in residential choices. The condominium boom in Manhattan has seen a slowdown, and the suburban areas of New Jersey and Connecticut are witnessing increased interest.

The ripple effect of this transformation extends to suburban, rural, and smaller towns, which are now becoming sought-after residential location [41]. Factors such as increased living space and lower land prices are driving this transformation. In Vancouver, Canada, the city's outskirts and nearby towns like Surrey and Langley are witnessing a surge in housing demand as people seek larger properties with spacious surroundings. In the USA, cities like Austin, Texas, Raleigh, and North Carolina are witnessing a population influx, driven by the appeal of suburban living and the affordability it offers [42].

This shift towards decentralized work arrangements has led to prioritizing larger homes or properties in less urbanized areas. The emphasis is now on comfort, tranquility, the availability and affordability of open spaces over the hustle and bustle of city-centric locations [43]. Calgary, Canada, for example, has seen a surge in demand for single-family homes in suburban areas, reflecting a preference for a quieter living. In the USA, cities like Nashville, Tennessee, are experiencing a similar trend, with suburban areas gaining popularity for their tranquility and family-friendly environments.

Furthermore, the transition to hybrid work has significantly altered commuting patterns, resulting in a reduced reliance on public transport and the urge to be live in the centralized locations [44]. This shift not only translates to reduced traffic congestion but also prompts transformations in contemporary land use pattern. Urban and suburban planning is adapting to meet the demand for more spacious homes, green spaces, and community-centric amenities [45].

The flexibility afforded by remote work is prompting a re-evaluation of traditional perspectives on residential areas and urban development [46]. In cities like Montreal, Canada, the demand for properties with home office spaces has risen, reflecting the growing importance of work-from-home arrangements. Similarly, in the USA, cities like Seattle and Denver are witnessing a surge in interest in neighborhoods with a mix of residential and commercial spaces, catering to the evolving work-life dynamic.

The enduring impact of these changes is redefining the structure of cities and reshaping the dynamics of housing and urban landscapes [46]. As individuals continue to prioritize comfort, space, and a blend of urban amenities with suburban tranquility, the evolving real estate market reflects a new paradigm that is likely to persist in the post-pandemic era. This transformation is not just a response to immediate challenges, but a fundamental shift in the way people envision and engage with their living spaces, leaving an indelible mark on the future of urban and suburban life in the global cities [47].

4 The Effects of Evolving Transportation Dynamics and Shifts in the Real Estate Market on Urban Morphology

The COVID-19 pandemic has had a significant impact on all aspects of people's lives. At the economic level, there was an abrupt halt in the first quarter of 2020 partly due to the high degree of uncertainty caused by the situation that people have experienced [48]. This section attempts to offer an in-depth analysis of the impacts of home office and hybrid employment models, active transportation and public transit changes, and changing real estate prices on contemporary land uses and city structures.

I. *Home Office and Hybrid Employment*

The post-pandemic surge in housing development and private investments hints at forthcoming shifts in the real estate market, signaling a growing preference for small-town living and a potential rise in demand for properties in such areas. This trend reflects changes in the local real estate scene, fueled by factors like return migration and the allure of small-town life, spurred or accelerated by pandemic-driven remote work trends. Educational cities have felt the pandemic's impact, with decreased short-term rentals due to student departures and remote work replacing traditional transfers, particularly in Italy and other European markets. Despite initial setbacks, these markets rebounded by late 2021–2022 [49].

Experts foresee significant growth in the logistics and residential sectors, while offices remain a core investment domain. However, retail segment uncertainty persists, with the residential sector, especially multifamily dwellings, viewed as the most resilient amid economic downturns [50, 51]. The rise of remote work has decentralized employment hubs, shifting focus away from city centers to suburban and rural locales that offer affordable accommodations with an ever-aspired opportunity for nature living. This preference for larger living spaces to accommodate home offices reshapes residential preferences and changes the contemporary construct and evidence of urban density and its gradients as the land uses.

Adapting to these shifting dynamics requires reorganizing both theoretical and pragmatic structures to meet new health standards and adjusting to smaller office spaces and home-based work challenges. This demands exploration and further research-based evidence for innovative planning strategies and context-sensitive modeling approaches in guiding urban space reorganization, fostering resilience and inclusivity. Therefore, real estate development must align with broader urban transformation goals, embracing sustainability and integrating green spaces and resilient infrastructure to meet transformational needs.

II. *Active Transportation and Public Transit*

In the post-pandemic period, managing urban crowds has become a critical need along with the rising demand for community crowds and recreational spaces, highlighting the challenge associated with accommodating sudden influxes of crowds safely and efficiently. Given the evolving landscape of demographic dynamics across the urban and peripheral, cities are witnessing a surge in demand for active transportation and sanitized public transit systems. Tracking these shifts, a range of indicators has emerged, leveraging geolocated data from GPS sensors and smartphones. In response to the pandemic's aftermath, Google introduced the Covid-19 Community Mobility report, utilizing aggregated data to gauge crowd density in various locations, aiding in identifying peak hours and ensuring safer mobility. This emphasis on pedestrian-friendly infrastructure underscores the urban space adaptation to evolving mobility patterns, essential for crowd management dynamics and livability.

These mobility trends not only provide crucial intelligence but also inform critical decisions about crowd management to the fight against the pandemic. As cities transition post-pandemic, smart mobility solutions are gaining traction, incorporating data analytics, Demand-Responsive Transport (DRT) services, IoT for real-time traffic monitoring, and contactless technologies. This shift supports resilient urban transportation networks, aligning with sustainability objectives.

Notably, there's a growing preference for active transportation modes like walking and cycling, driven by health-conscious commuting choices and reduced reliance on crowded public transit systems. Amidst changing demand patterns, public transit faces challenges in maintaining service levels, necessitating adaptive strategies to support urban mobility and transit-oriented development (TOD), in line with broader sustainability and public health welfare [52–54].

III. *Changing Real Estate Prices*

At the housing market level, demand for online shopping is increasing and densification of work-residential space is under scrutiny, the ultimate impact caused is not yet clear and will be proportional to the duration and intensity of restrictive measures, by the tools made available in the coming months, and by the consolidation of new behaviours. At the moment, the most immediate consequences have been recorded on rents, which have fallen significantly as a result of the collapse in demand until 2021, particularly in European markets that rely on work and tourism [55, 56].

The residential sector, for instance in Italy, has always been characterized by good volumes of buying and selling between individuals, but few corporate investments, in contrast to the rest of Europe. Over the past few months, however, operators have reported significant movements, despite restrictions, which bodes well that the residential sector can become an asset class of choice for investors institutional investors, given its anti-cyclical nature. In Italy, the lack of adequate product is pushing operators to focus on real estate development or redevelopment projects even if we note some investment on existing income assets. In general, investors (especially foreigners) focus mainly on urban transformation areas in the metropolitan area such as Milan and there remains interest in towards student housing [57, 58].

Real estate markets are experiencing variances in pricing, with some areas seeing increases due to heightened demand for suburban and rural properties, while others, particularly in city centres, face declines [59]. This fluctuation is reshaping investment focuses and prompting a re-evaluation of land use planning and development strategies to meet the evolving demands. These trends signify a pivotal transformation in urban planning and development, emphasizing the need for flexible, resilient, and sustainable approaches to accommodate the evolving dynamics of work, transportation, and living preferences in the post-pandemic era.

5 Conclusion

This paper attempts to provide a comprehensive exploration of the dynamic consequences of COVID-19 pandemic, traversing the realms of mobility, land use, and the real estate sector. It serves as a testament to the transformative disruption witnessed in how cities and their citizens navigate through "new" living and employment paradigms. In particular, the chapter elucidates the burgeoning trend towards suburbanization propelled by the widespread adoption of remote employment, alongside discernible shifts in transportation preferences towards more sustainable and adaptable choices. Moreover, it shed lights on the evolving demands on urban infrastructure and planning practices, necessitating a fundamental re-evaluation of the contemporary planning and development strategies.

Amidst these transformative shifts, cities are compelled to re-conceptualize their approach to development, with an emphasis on accommodating evolving preferences and behaviors while prioritizing sustainability, resilience, and inclusivity in urban design and policymaking. The synthesis of the discussions underscores the intricate interplay between mobility patterns, land use dynamics, and the real estate landscape, accentuating their critical role in shaping the future trajectory of urban environments in the post-pandemic era.

Looking ahead, future research endeavors in the domain of transformational and resilient cities should prioritize the following avenues:

- **Long-Term Impact Assessment:** Comprehensive studies are warranted to assess the long-term impact of the pandemic-induced shifts on urban mobility, land use patterns, and the real estate market. This entails examining how these changes manifest over time and their implications for urban development trajectories.
- **Policy Innovation and Implementation:** Research efforts should focus on identifying innovative policy interventions geared towards fostering sustainable and resilient urban development. Moreover, there is a pressing need to explore mechanisms for effectively implementing these policies, considering the diverse contexts and challenges faced by cities worldwide.
- **Community Engagement and Social Equity:** Future research should delve into the role of community engagement in shaping urban transformation processes and ensuring social equity. This entails understanding community perspectives, needs, and aspirations, and integrating them into urban planning and decision-making processes to foster inclusive and equitable urban development.

References

1. Nahiduzzaman KM, Holland M, Sikder S, Shaw P, Hewage K, Sadiq R (2021) Urban transformation towards a smart city: an e-commerce induced path-dependent analysis. J Urban Plan Dev 147(1):04020060
2. Nahiduzzaman KM, Lai SK (2020) What does the global pandemic covid-19 teach us? Some reflections. J Urban Manage 9(3):35
3. Dingil AE, Esztergár-Kiss D (2021) The influence of the Covid-19 pandemic on mobility patterns: the first Wave's results. Transp Lett 13(5–6):434–446
4. Profeta P (2020) Gender equality and public policy during COVID-19. CESifo Econ Stud 66(4):365–375. https://doi.org/10.1093/cesifo/ifaa018
5. Siriwardhane P, Khan T (2021) The gendered nature of the risk factors of the COVID-19 pandemic and gender equality: a literature review from a vulnerability perspective. Sustainability 13(23):13375. https://doi.org/10.3390/su132313375
6. Olayode IO, Severino AG, Campisi T, Tartibu LK (2022) Comprehensive literature review on the impacts of COVID-19 pandemic on public road transportation system: challenges and solutions. Sustainability 14(15):9586. https://doi.org/10.3390/su14159586
7. Campisi T, Georgiadis G, Basbas S (2022) Developing cities for citizens: supporting gender equity for successful and sustainable urban mobility. In: Gervasi O, Murgante B, Misra S, Rocha AMAC, Garau C (eds) Computational science and its applications—ICCSA 2022 workshops. Springer International, pp 410–422. https://doi.org/10.1007/978-3-031-10592-0_30
8. Saez K (2023) Problematizing workplace learning for the work-from-home and hybrid workforce following COVID-19
9. Florida R, Rodríguez-Pose A, Storper M (2023) Critical commentary: cities in a post-COVID world. Urban Stud 60(8):1509–1531
10. Nahiduzzaman KM (2021) COVID-19 and change dynamics in the transformational cities. J Urban Manag 10(2)
11. Larsson A, Hatzigeorgiou A (2023) Designing smart and resilient cities for a post-pandemic world: metropandemic revolution. Taylor & Francis, p 394
12. Moraci F, Errigo MF, Fazia C, Campisi T, Castelli F (2020) Cities under pressure: strategies and tools to face climate change and pandemic. Sustainability 12(18):7743
13. Tsvetkova A, Kulkov I, Busquet C, Kao P-J, Kamargianni M (2022) Implications of COVID-19 pandemic on the governance of passenger mobility innovations in Europe. Transp Res Interdiscip Perspect 14:100581. https://doi.org/10.1016/j.trip.2022.100581
14. Böcker L, Olsson LE, Priya Uteng T, Friman M (2023) Pandemic impacts on public transport safety and stress perceptions in Nordic cities. Transp Res Part D: Transp Environ 114:103562. https://doi.org/10.1016/j.trd.2022.103562
15. Politis I, Georgiadis G, Papadopoulos E, Fyrogenis I, Nikolaidou A, Kopsacheilis A, Sdoukopoulos A, Verani E (2021) COVID-19 lockdown measures and travel behavior: the case of Thessaloniki, Greece. Transp Res Interdiscip Persp 10:100345. https://doi.org/10.1016/j.trip.2021.100345
16. Campisi T, Russo A, Tesoriere G, Bouhouras E, Basbas S (2021) COVID-19's effects over e-commerce: a preliminary statistical assessment for some European countries. In: Gervasi O, Murgante B, Misra S, Garau C, Blečić I, Taniar D, Apduhan B, Rocha AMAC, Tarantino E, Torre CM (eds) Computational science and its applications—ICCSA 2021. Springer International, pp 370–385. https://doi.org/10.1007/978-3-030-86979-3_27
17. Campisi T, Russo A, Basbas S, Bouhouras E, Tesoriere G (2023) A literature review of the main factors influencing the e-commerce and last-mile delivery projects during COVID-19 pandemic. Transp Res Proc 69:552–559. https://doi.org/10.1016/j.trpro.2023.02.207
18. Simeli I, Tsekouropoulos G, Vasileiou A, Hoxha G (2023) Benefits and challenges of teleworking for a sustainable future: knowledge gained through experience in the era of COVID-19. Sustainability 15(15):11794

19. Hamid S, Azhar M (2023) Behavioral intention to order food and beverage items using e-commerce during COVID-19: an integration of theory of planned behavior (TPB) with trust. British Food Journal 125(1):112–131
20. Abdullah M, Ali N, Dias C, Campisi T, Javid MA (2021) Exploring the traveler's intentions to use public transport during the COVID-19 pandemic while complying with precautionary measures. Appl Sci 11(8):3630
21. Tirachini A, Cats O (2020) COVID-19 and public transportation: current assessment, prospects, and research needs. J Public Transp 22(1):1–21
22. Tully MA, McMaw L, Adlakha D, Blair N, McAneney J, McAneney H, Smith L (2021) The effect of different COVID-19 public health restrictions on mobility: a systematic review. PLoS ONE 16(12):e0260919
23. Campisi T, Basbas S, Trouva M, Papas T, Akgün-Tanbay N, Stamatiadis N, Tesoriere G (2022) Anxiety, fear and stress feelings of road users during daily walking in COVID-19 pandemic: Sicilian cities. Transp Res Proc 62:107–114
24. Duca G, Trincone B, Bagamanova M, Meincke P, Russo R, Sangermano V (2022) Passenger dimensions in sustainable multimodal mobility services. Sustainability 14(19):12254
25. Umrigar N, Pitroda JR (2023) Multimodal urban transportation system for medium size cities. J Inst Eng Ser A 1–17
26. Campisi T, Canale A, Tesoriere G, Ali N, Ignaccolo M, Cocuzza E (2022) An analysis of the integration of DRT services with local public transport in post-pandemic period: some of the preliminary insights in the Italian context. In: International conference on computational science and its applications. Springer International, Cham, pp 496–508
27. Pavanini T, Liimatainen H, Sievers N, Heemsoth JP (2023) The role of DRT in European urban public transport systems—a comparison between Tampere, Braunschweig and Genoa. Future Transp 3(2):584–600
28. Hensher DA (2020) What might Covid-19 mean for mobility as a service (MaaS)? Transp Rev 40(5):551–556
29. Campisi T, Garau C, Acampa G, Maltinti F, Canale A, Coni M (2021) Developing flexible mobility on-demand in the era of mobility as a service: an overview of the Italian context before and after pandemic. In: International conference on computational science and its applications. Springer International, Cham, pp 323–338
30. Rigogiannis N, Bogatsis I, Pechlivanis C, Kyritsis A, Papanikolaou N (2023) Moving towards greener road transportation: a review. Clean Technol 5(2):766–790
31. Campisi T, Georgiadis G, Basbas S, Al Rashid MA (2023) The post-pandemic public transport crisis: a statistical analysis of travel habits in Sicily, Italy. Transp Res Proc 69:576–583
32. Yannis G, Chaziris A (2022) Transport system and infrastructure. Transp Res Proc 60:6–11
33. Annunziata A, Desogus G, Mighela F, Garau C (2022) Health and mobility in the post-pandemic scenario. An analysis of the adaptation of sustainable urban mobility plans in key contexts of Italy. In: International conference on computational science and its applications. Springer International, Cham, pp 439–456
34. Sharifi A, Khavarian-Garmsir AR (2020) The COVID-19 pandemic: impacts on cities and major lessons for urban planning, design, and management. Sci Total Environ 749:142391
35. Charradi I, Campisi T, Tesoriere G, Abdallah KB (2022) A holistic approach to SUMP strategies and actions in the post-pandemic and energy crisis era. In: International conference on computational science and its applications. Springer International, Cham, pp 345–359
36. Ortega J, Tóth J, Péter T (2021) Planning a park and ride system: a literature review. Future Transp 1(1):82–98
37. Jonidi Jafari A, Charkhloo E, Pasalari H (2021) Urban air pollution control policies and strategies: a systematic review. J Environ Health Sci Eng 19:1911–1940
38. Useche SA, O'Hern S, Gonzalez-Marin A, Gene-Morales J, Alonso F, Stephens AN (2022) Unsafety on two wheels, or social prejudice? Proxying behavioral reports on bicycle and e-scooter riding safety—a mixed-methods study. Transp Res F: Traffic Psychol Behav 89:168–182

39. Stephens D (2021) Resurrecting retail: the future of business in a post-pandemic world. Figure 1 Publishing
40. Bedi A, Rahadi RA (2024) Adapting to change: a study of post-pandemic millennial housing preferences. Ekombis Rev J Ilmiah Ekon dan Bis 12(2):1733–1744
41. Turok I, Seeliger L, Visagie J (2021) Restoring the core? Central city decline and transformation in the South. Prog Plan 144:100434
42. Lee H, Airgood-Obrycki W, Frost R (2024) Back to the suburbs? Millennial residential locations from the Great Recession to the pandemic. Urban Stud
43. Bernat S, Trykacz K, Skibiński J (2022) Landscape perception and the importance of recreation areas for students during the pandemic time. Int J Environ Res Public Health 19(16):9837
44. Bond-Smith S, McCann P (2022) The work-from-home revolution and the performance of cities. University of Manchester
45. Boules C, Kato Y (2023) Just transition or just transitioning? potentials and limitations of urban growers' adaptations to the impacts of the COVID-19 pandemic. Sustainability 15(12):9340
46. Ilham MA, Fonzone A, Fountas G, Mora L (2024) To move or not to move: a review of residential relocation trends after COVID-19. Cities 151:105078
47. Nahiduzzaman KM, Ahsan M (2023) Asian cities in and beyond COVID-19. In: Hu R (ed) Routledge handbook of Asian cities. Routledge, London
48. Tamine S, D'Acci L (2022) What impact had the covid-19 on the internal migration and the real estate prices? Politecnico di Torino. https://webthesis.biblio.polito.it/secure/25287/1/tesi.pdf
49. Engle S, Stromme J, Zhou A (2020) Staying at home: mobility effects of COVID-19. SSRN Electron J. https://doi.org/10.2139/ssrn.3565703LNCS Homepage, http://www.springer.com/lncs. Last accessed 25 Oct 2023
50. Warren MS, Skillman SW (2020) Mobility changes in response to COVID-19. http://arxiv.org/abs/2003.14228
51. Wan T, Lu W, Sun P (2023) Equity impacts of the built environment in urban rail transit station areas from a transit-oriented development perspective: a systematic review. Environ Res Commun 5:092001
52. Hrelja R, Rye T (2024) Delivering transit oriented development (TOD) in low to medium density contexts. Actor relationships and market conditions in smaller Swedish cities. Int J Sust Transp 18(3):236–249
53. Farnes K, Hurst N, Wong WW, Wilkinson S (2024) An exploratory study on the benefits of transit orientated development (TOD) to rail infrastructure projects. Smart Sust Built Environ
54. Guglielminetti E, Loberto M, Mistretta A (2023) The impact of COVID-19 on the European short-term rental market. Empirica 50(3):585–623. https://doi.org/10.1007/s10663-023-09576-x
55. Plzáková L, Smeral E (2022) Impact of the COVID-19 crisis on European tourism. Tour Econ 28(1):91–109. https://doi.org/10.1177/13548166211031113
56. Pavesi AS, Cia G, Dott G, Morlacchi A (2022) The relationship between social housing and urban regeneration. Politecnico di Milano
57. Theurillat T, Rérat P, Crevoisier O (2015) The real estate markets: players, institutions and territories. Urban Stud 52(8):1414–1433. https://doi.org/10.1177/0042098014536238
58. Helbich M (2015) Do suburban areas impact house prices? Environ Plan B Plan Des 42(3):431–449. https://doi.org/10.1068/b120023p
59. Li X, Zhang C (2021) Did the COVID-19 pandemic crisis affect housing prices evenly in the US? Sustan, 13(21):12277.

Prof. Tiziana Campisi is serving as an Assistant Professor, Department of Engineering and Architecture, University of Enna Kore, Cittadella Universitaria 94100, Enna (En), Italy. She obtained her degree in Civil and Environmental Engineering at the University of Catania and her specialised degree (cum Laude) in Civil Protection Engineering at the University of Enna

Kore where she also obtained her PhD degree. She is an expert in road quality management and sustainable mobility. She has published more than 130 manuscripts in international and indexing journals and conferences. She is responsible for the local unit of the project (PRIN2017) titled "WEAKI TRANSIT"(CUP Code: F74I19001290001), supported by MIUR (Ministry of Education, University and Research) and funded by the PRIN 2017 programme. She serves as Editor in Chief for the Open Transportation Journal Bentham and as a member of several Editorial Boards and she is a reviewer for several international journals and conferences. Her research interest lies on transport supply/demand, sustainable and shared mobility, road safety, vulnerable road users, and microsimulation of road traffic.

Talha Tasnim is an Officer in Monitoring and Evaluation (M&E) at BRAC's Urban Development Programme, one of the world's largest non-governmental development organizations. He is currently pursuing an MBA at Quantic School of Business and Technology and holds a Bachelor's degree in Urban and Rural Planning from Khulna University. Talha's work focuses on driving innovation and extracting valuable insights from data to make a lasting impact in the development sector. He specializes in child-friendly urban planning and urban informality, contributing to various publications and research projects. Based in Dhaka, Bangladesh, Talha is committed to enhancing urban living conditions through thoughtful planning and evaluation.

Prof. Kh Md Nahiduzzaman is Nordic trained Canadian urban planner who received PhD from the Royal Institute of Technology (KTH), Sweden, and MPhil from Norwegian University of Science and Technology (NTNU), Norway. Prior to joining Mohammed VI Polytechnic University (UM6P), Morocco, Prof. Nahiduzzaman held faculty positions at the University of British Columbia Okanagan (Canada), King Fahd University of Petroleum & Minerals (Saudi Arabia), and the Department of Urban Planning and the Environment in the Royal Institute of Technology (Sweden). His core research interest is urban resilience through the lenses of digital twins, urban transformational planning, smart cities, and climate risks and vulnerability assessment. He is a recognized researcher in the field. He is heavily engaged on collaborative research projects with the key governments and industrial partners, and (indigenous) communities across British Columbia, Canada and Morocco. So far, he has secured reputed research grants worth more than $3 million from the prestigious national and international funding agencies. He has published more than 100 scientific manuscripts in the top ranked journals, books and other reputed proceedings. Many of his works such as non-structural flood mitigation, floodplain maps and land use planning strategies in BC have direct impact on the provincial policies. He is an "Executive" and "Book Review" Editor of the Journal of Urban Management, Editor-in-Chief of "City Development: Issues and Best Practices" journal, and Editorial Board Member of the Journal of Urban Planning and Development. He is also seating on many national and international scientific committees, fora and think-tanks. Prof. Nahiduzzaman is the editor of the book "Making Sense of Planning and Development for the post-pandemic Cities" by Springer. He is also a co-chair of "Smart and Connected Cities" track in the prestigious US-Africa Frontiers program to be held on February 18-20, 2025, in Kigali, Rwanda.

Urban Planning and Development

Development Incentives in the South African Local Government and Administration, a Significant Growth Factor for Municipal Planning, Innovation and Service Delivery During and Post-Covid 19

E. M. Lentsoane and George Onatu

Abstract The world recognises South Africa as one of the African countries that are economically growing and developing. This recognition has been noticeable despite levels of Moody's Investors Services downgrades. The negative rating recorded means that the persistent deterioration in fiscal strength and weak growth, which needs restructuring with associated policies to ensure efficacy are properly addressed. For the government to restore its economic outlook both during the pre and post-Covid-19 pandemic requires systematic and deliberate reform of legislation, policy, and implementation strategies. It further requires proper strategic planning, cost–benefit analysis, and monitoring and evaluation by various ministries when performing their constitutionally mandated functions. Metropolitan municipalities in the Gauteng Province have been struggling to provide much-needed essential services and to attract more investors to boost the ailing economy and create employment. This paper thus assesses the value of development incentives as an innovative measure to overturn the negative economic outlook and render local government financially sustainable in dealing with service delivery and infrastructure provision. The research answered two main questions: whether municipal planning is innovative enough to influence the country's economic growth and social development? and if development incentives can attract, retain, and restore investment in a particular geographical area. To further evaluate whether the benefits of development incentives can have a spill-over benefit to other municipalities and, eventually, the country at large to transform the economy and create viable and sustainable communities. The study further identified sectors where various development incentives are explored with practical results. Existing research and case studies relevant to this topic were

E. M. Lentsoane
University of Limpopo, Polokwane, South Africa

G. Onatu (✉)
University of Johannesburg, Johannesburg, South Africa
e-mail: gonatu@uj.ac.za

K. M. Nahiduzzaman (ed.), *Making Sense of Planning and Development for the Post-Pandemic Cities*, City Development: Issues and Best Practices,
https://doi.org/10.1007/978-981-97-5481-6_9

evaluated to validate and support the findings and recommendations made. The variable used is the development principles as the guiding factor for municipal planning, infrastructure provision, and service delivery.

Keywords Municipal planning · Development incentives · Covid-19 pandemic · Service delivery · Innovation and technology · Compliance

1 Introduction and Background

Growth and development tend to reject any form of policy development and execution that promotes compartmentalisation. Chapter 7 of the South African Constitution explicitly stipulates the principal objectives of local government, which encompasses a broader spectrum of responsibilities beyond service delivery. In psychology for example, compartmentalisation is used to denote the defence mechanism to suppress thoughts and emotions [1]. Such suppression levels can make the individual disconnect from the world's key activities and realities. The same analogy is borrowed to reflect on governance when compartmentalising municipal planning and service delivery as the sole responsibility of local government. Section 40 (1) of the Constitution provides three spheres of government which are distinctive, interdependent, and interrelated. Accordingly, Lentsoane [2] recognition is given to the importance of collective legislative participation supporting the three spheres of government to tackle issues of high economic interdependencies. As such, Judge Jafta in June 18 2010, rendering his judgement in the matter between the City of Johannesburg Metropolitan Municipality v Gauteng Development Tribunal, expanded on Sect. 40 of the Constitution by stating that [3]:

> The functional areas allocated to the various spheres of government are not contained in the hermetically sealed compartments, but that notwithstanding, they remain distinct from one another. The distinctiveness lies in the level at which a particular power is exercised.

Local government through various legislation, operate within multifaceted arrangements that juggle multiple systems, structures, institutions, procedures, processes, and frameworks to meet the local communities' basic needs. Additionally, the local government must facilitate integrated infrastructure projects planning and economic development programmes that attract investments and promote market access and global competitiveness. Furthermore, through good governance, the local government must be able to connect and synergise several provisions of legislation and policies to successfully yield tangible municipal planning outcomes [2].

However, the intelligence needed to synergise and satisfy the government's competing expectations on growth and development depends not only on municipal planning attributes but also on the municipalities' ability to adapt and adopt an entrepreneurial mindset. The entrepreneurial mindset thus requires an in-depth understanding of the market trends and various development incentives appropriate to attract the strategic investment necessary to elevate the effectiveness of a land use management system. How this can be carried out effectively and efficiently is still a very big debate in the literature.

2 Approach and Methodology

The study is exploratory in nature as it indicates how municipalities understand and use development incentives to change their economic growth trajectory and outlook [4, 5]. In this book chapter, development incentives from different countries were studied to understand associated practices, successes and failures. The selected research case countries were randomly sampled. As such, each participant had an equal chance of being selected as purposive sampling was seen as appropriate for choosing exceptional [6, 7] also support this approach. The United Arab Emirates, China, the USA-California and South Africa- the City of Tshwane municipality comprise the total number of cases that were investigated and studied. The research population as indicated is accepted in the literature as all constitute the collection of objects, events, or individuals having some common characteristics that the researcher is interested in studying [8, 9]. Documents relating to development incentives from selected cases were examined and interpreted to gain the insight necessary to develop and expand on the empirical knowledge [10, 11]. Submitting that document analysis when employed in a research study provides scholars and interested parties' details that can be seen as factual to be shared and used in formal and informal settings [12, 13]. Thereby, enabling freedom in active development and capturing what is intended to be demonstrated.

As mentioned above, this research employs the case study approach and policy content analysis/review of relevant literature. Noor [14] saw case studies as concerned with how and why things happen, allowing the investigation into contextual realities and the differences between what was planned and what actually occurred [14, 15]. The emphasis remains on how and why it is essential to use development incentives and the consequences thereof based on evidence from best practices. Pan [17] in support of this assertion refers to a case study as a method, strategy, research design, or methodology [16, 17]. To call a case study a method necessitates that a case study is a technique, procedure, or mean for gathering evidence or collecting data. A case study is also looked upon as a research design that guides research right from the questions stage to the conclusions and includes steps for collecting, analyzing, and interpreting evidence according to pre-established propositions, units of analyses, a logic for linking the data to the propositions, and application of set criteria for interpreting the findings [18]. When a case study is used as a methodology, it is interpreted as a theory and analysis of how research should proceed.

According to Gerring [19], it all depends on what one is arguing. According to Eckstein et al. [20], a case study could be defined and explained as a phenomenon for which we report and interpret only a single measure on any pertinent variable. The singular variable we investigated is the application of development incentives. The documents reviewed includes various pieces of legislation and policy statements that were instrumental in formulating the development inventive evidence to ensure the credibility, reliability and validity of documents analysed and in support of qualitative research undertaken [21–23]. There were limitations in this study as only document and content policy analysis were used, and no interviews or questionnaires were part of the study. Thus, the limitations of this study concerned generalizability and extrapolation of the research results. Generalizability refers to the researcher's ability to generalise the results from the sample to the population from which it was drawn [24]. However, the researcher remains ethical in the chosen methodology as provided by Drew et al. [25, p. 56], who noted that bringing ethical awareness is very important as ethics has become an essential cornerstone for conducting effective and meaningful research. A case study is also very relevant here because it enabled us to provide more rigorous details of the phenomenon as compared to other research designs.

3 Definition of Key Concepts: Review of Literature

3.1 Development Incentives

Liu et al. [26] defines economic development incentives as direct financial benefits that firms provide to incentivise their opening, expansion, or retention. This paper describes development incentives as a package of monetary and non-monetary benefits from the government dedicated to attracting, retaining, and increasing the number of businesses in specific geographic areas or sectors considered economic growth magnets. Development incentives are vital to levelling the business playing field and reducing barriers hindering entrepreneurship and emerging, small, and medium businesses from participating in the economy [27]. Central governments are noted to use incentives not only to promote economic growth but to guide innovation and sustainability at different sectors [28]. Development incentives used in this chapter requires a high level of innovation and an entrepreneurial mindset to be impactful.

3.2 Municipal Planning

Municipal planning is founded on a legal framework since it is a place-making tool to regulating land use and management strategies, thereby promoting ownership, people's livelihoods safety, development, and maintenance of places [29–31]. Municipal planning is listed in Schedule B as a matter for which only the local government is responsible. However, the effectiveness of municipal planning requires both the provincial and national governments to work together with the private sector as de-risking agents in order to manage public and private investments effectively [32–34]. Structured development incentives and approaches that are flexible due to change of circumstances can discourage monopolies in the market and provide measures that facilitate the growth of small and medium businesses [35].

Municipal planning is dependent on a functional land use management system. Thus, the land use management system is an all-encompassing apparatus and a technique of regulating and managing land uses and conferring land use rights through the application of land use schemes and land development procedures (Spatial Planning and Land Use Management Act, 2013) [36]. The land use management system provides a consistent value chain associated with the land development processes, translating land-use activities into coherent and structured land development projects. This is noted to promoting economic growth and physical and social development. It further refers to a mechanism that enables the commissioning of a strategic and integrated institutional arrangement that deals with land development, management, and regulatory activities.

The private sector is the key contributor to the country's Gross Domestic Product (GDP). Therefore, the more businesses flourish in designated sectors, the higher the likelihood of economic growth and municipal financial viability. This is to submit that the ease of doing business and conducive environments for the private sector are critical factors to the actual reduction of deprivation traps and an increase in service delivery [37, 38]. Therefore, local governments can consider the application of development incentives as a system to advance technology and innovation [39–41].

3.3 Innovation in Local Government

Innovation is the task of endowing human and material resources with new and greater wealth-producing capacity. It is a specific tool of entrepreneurs, how they exploit change as an opportunity for a different business or service [42]. Innovation, entrepreneurship, and commercialisation are mutually inclusive [43, 44]. Thus, the government must be innovative, adopt and adapt the entrepreneurial mindset, and understand commercialisation. These three concepts are crucial in providing development incentives to the private sector and essential to the economic growth and service delivery of any country [45].

3.4 Entrepreneurship in Local Government

Entrepreneurship is often understood as a driver of innovation and productivity and an engine for sustainable economic growth [46, 47]. Entrepreneurs are those persons (business owners) who seek to generate value through the creation or expansion of economic activity by identifying and exploring new products, processes, or markets. Dimov [48] states that entrepreneurs act purposefully to exploit opportunities in the face of uncertainty. Entrepreneurship has three elements: product or service, customer or a user, and infrastructure that sources, creates, and distributes the product and service. Considering these three elements, municipalities are meant to operate in a system consistent with land use management, land development, and economic rules and arrangements. Therefore, entrepreneurship becomes possible through the six domains of the entrepreneurial ecosystem: policy; finance; markets; human capital; support; and culture when integrated within municipal planning [45, 46].

The Organisation for Economic Cooperation and Development Report (2012 and 2014) provides that policy, markets, human capital, support, culture, and finance are the six domains of the entrepreneurial ecosystem [49]. Therefore, the 6 domain of the entrepreneurial ecosystem suggests that the imbalance between income generation, primary services demand, job creation, and economic growth lies on the municipal intelligence to lean on commercialisation as a strategy to reduce gaps in finance while managing 'citizens' expectations.

3.5 Commercialisation of Assets in Local Government

Municipalities have various assets that should be captured in an asset register to be evaluated in line with the Property Valuation Act, 2014 and other applicable laws. Assets in this context are defined as resources with economic value owned and managed by municipalities or other entities on the 'municipality's behalf. The Municipal Financial Management Act defines assets as resources controlled by an entity because of past events and from which future economic benefits or service potential are expected to flow to the entity. Municipal assets and asset management systems are vital in keeping a credible investment portfolio that renders the municipality credit worthy. The benefit of municipal assets includes increased revenue, attracting investment, improved land valuation, and generally improving the 'citizens' quality of life [50]. Therefore, the infrastructural assets in the municipality can be used for commercialisation purposes and as one of the revenue streams to generate revenue.

3.6 Entrepreneurial Mindset

Dweck [51] define mindset as the cognitive belief consisting of interrelated beliefs, assumptions, and knowledge to process information, inform decision-making, and guide behaviour. Therefore, an entrepreneurial mindset is a specific set of beliefs and thought processes that drive entrepreneurial behaviours. The entrepreneurial mind rejects failure as an alternative. A mindset of an entrepreneur is:

1. Highly engaged, solution-oriented, resourceful
2. Opportunity driven.
3. Curious, creative, critical, and explorative
4. Understands the power of brand.

5. Take ownership.
6. Value the unique selling points that set them apart
7. Want to be the first to the market or to reach the customers first
8. Spot opportunities amid challenges
9. Search and find new markets for a new and existing product
10. Network with a clear intent
11. Maintain quality control and
12. Resilient

Considering the definition of the entrepreneurial mindset, the development, provision, and effective management of the development incentives requires municipalities to think and behave like entrepreneurs [52, 53]. The municipality must be positioned strategically and to take a posture of a winning team. The municipality is brand, and employees are brand ambassadors. Thus, their conduct must be of exceptional excellence in making the municipality worthy of investors.

Scholars also argue that managers micromanage and often encourage predictability, which can stifle the growth prospects of an organisation. Therefore, when considering technological advancement with 4IR, where artificial intelligence can replace managers, municipalities need to transform, appoint, and groom managers to have entrepreneurial ambitions in order to respond to various developmental, political and economic trends in alignment with the five competitive forces [54].

4 Case Studies—Development Incentives in Different Countries

4.1 United Arab Emirates (UAE)

With the government eliminating barriers that hinder the flow of investment, the UAE makes investments an enabler of the national economy by offering various incentives to investors to increase the number of small and medium enterprises and start-ups in the country. The incentives range from 100% company owners available to all nationalities; more than 40 free zones allow tax exemptions, 100% ownership for foreign investors, and grant investors flexibility in selecting the most appropriate zones for establishing businesses. The development incentives played a crucial role during the COVID 19-pandemic as UAE government in managing the economic growth provided for the postponement of rent payments by a period of 6 months; Facilitating instalments for payments; Refunding security deposits and guarantees [55]; Cancelling fines for both companies and individuals; permitting temporary contracts that allow the free movement of labour between companies operating in the free zones to continue for the rest of the year.

4.2 China

The Survey of Global Investment and Innovation Incentives (2020) states that the local government in China has introduced investment incentives projects that substantially contribute to local employment, economic development, and technological innovation to lower the investment cost. These incentives are case-by-case negotiated, including tax refunds, tax credits, and free office space leasing. Approximately 18 stimuli in China are designed to encourage a particular economic activity or disadvantaged businesses or owners. About 80% of the fees are set aside for Research and Development Activities. According to [56] they noted that Chinese government have successively promulgated relevant Research and Development (R&D) policies to promote structural adjustment for example, Photovoltaic power generation industry and the improvement of technological progress. The proportional increase in R&D investment in the photovoltaic power generation industry through several incentives have greatly promoted China's photovoltaic power generation tremendously which now ranked among the world's largest in terms of installed capacity [56]. They noted that the Golden Sun Demonstration Project of 2009 resulted to two-third of its R&D spent on renewal energy since 2010. Employment incentives are also vital for China as enterprises are provided with VAT refunds by employing a certain percentage of persons with disability. Additionally, there are various preferential incentives related to sustainable development whereby energy-serving technologies and products are incentivised accordingly [57]. These moves and strategies are very significant when promoting an inclusive economy in any country.

4.3 USA-California

The recognition of the benefit of development incentives made California's economy grow remarkably, from over 12.4% of unemployment rate and a $2.7 billion deficit in 2010 to a 4.2% unemployment rate, clearing the deficit in 2019. Since 2014, the government has allocated over $829 million in income tax credits to 974 companies that are projected to create over 95,000 jobs and make $17.9 billion in investments. The government used legislation to directly allocate incentives, and as a result, California in 2017 became the 5th largest economy in the world. In a separate research, it was also noted that in the state of California, policy makers are noted to allocating billions of dollars each year for tax incentives and relief for creative industries [58]. This approach to targeted incentives and economic development in the United States is noted to take root in the Great Depression of the 1929. During that time, policy makers in the Southern States enacted bond programmes that subsidized factories and other facilities thereby lowering firms' effective capital costs [59]. This trend is noted to be continued in some states still despite academic research that questions whether targeting incentives are an effective strategy to address unemployment and to increase investment per se.

4.4 South Africa—The City of Tshwane Municipality Case Study Area

On 15 March 2020, President Cyril Ramaphosa declared the coronavirus pandemic outbreak in the country a" National Disaster". From this day, various travel restrictions, business activities, gatherings and other social activities were prohibited. The City of Tshwane Metropolitan Municipality is one of the municipalities affected. It is one of the Metropolitan Municipalities in South Africa with a land area of 6,298 km^2 and a population of 2,921,488 which form part of Northern Gauteng and make up the capital city of South Africa. The City of Tshwane adopted a strategic Incentive Framework as adopted by Council resolution of 27 November 2014. This is one of the outcomes of its Vision 2055, specifically targeted at creating a conducive place to live, work and Play. The City of Tshwane's development incentives are meant to catalyse economic growth and development [5]. The development incentives aim to create an investor-friendly environment that will encourage developers and investors to consider the municipality the first point of call when exploring investment opportunities.

They provide various incentives ranging from 100% rebate from payment of rates and taxes on development meeting asset criterion based on spatial policies and the National Building Regulation and Standard Act, 1977 (Council resolution: 2015). The City of Tshwane 2021 presented to the Treasury Managers Forum in its Economic Development and Recovery plan to build the economy and support small businesses so severely impacted by the Covid-19 pandemic. According to the report and survey, 47.9% of companies were temporarily closed, 33.4% of the respondents indicated a decrease in their ability to access financial resources, 32.9% indicated that the prices of materials, goods and services increased more than normal, 36.4% of businesses have laid off staff in the short term, and 89.6% businesses had a turnover that was lower than their normal range. Faced with these challenges and crisis, the city developed five pillars of economic recovery [61]. These include financial relief that supports and facilitates access to the Social Relief Distress Grant (SRDG) released by the Presidency that results in a cash transfer of R350 ($20) for unemployed adults 18–59 not supported by any other social security schemes before (CGD, 2022) [60]. Job creation and re-skilling; Enterprise Development & support; Facilitating access to existing Covid-19 support and Investment in Aftercare and business retention services (Fig. 1).

To this end, the City of Tshwane also adopted a Disaster Management Plan using Covid-19 as a case for the first time. There were also the property rate payment deferment to all tose who require assistance and the halting of interest arrears on all those owing municipal accounts for 2020/21 (property/utilities), as well as waiving development applications/bulk contribution fees until 2020/21.

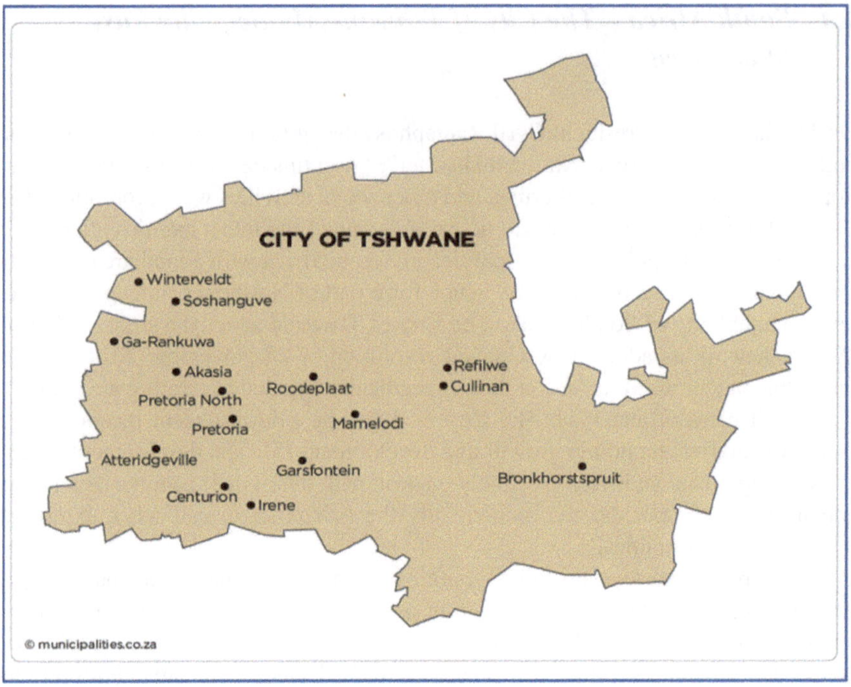

Fig. 1 City of Tshwane administrative area [61]

5 Conceptual Framework for Development Incentives

5.1 Strategic Compass

Development incentives is noted to hold mutual benefits; hence when it is provided by the municipalities it should be expected to be of interest, attractive, and viable to capture the attention of prospective investors and stimulate more entrepreneurs [62, 63]. In addition, they must offer unique sets of investment opportunities. With the strategic compass, the municipality can refine its ideas of development incentives to align this with each sector's unique selling points and qualities. Strategic aspects of the spatial planning and land use management must provide an in-depth understanding of the customers and an analysis of the existing competition for development incentive which can create value and improve the municipality's spatial, social, and economic outlook.

5.2 *The Relevance of Spatial Planning and Land Use Management to Development Incentives*

Analysis of Spatial Planning shows that it is a process that requires a strategic thinking ability whereby the vision/goals are properly set and understood. The principal purposes are clearly defined to enable the formulation and prioritisation of programmes and projects. Strategic thinking advocates that proper planning prevents poor performance. Programme priorities need to be at a strategic and operational level to yield desired outcomes. In addition, programme prioritisation also depends on the established and definite purpose that can allow the alignment of related variables and focused decision-making [64]. Strategic planning determines hierarchies of importance in goal setting, budgeting, implementation, management, and reporting. Spatial planning thus becomes outcome based.

Development incentives are often linked with the strategic implementation plan of the municipality, looking at the Spatial Development Framework and other spatial policies or strategies [65]. In the South African context, development incentives will be managed at the municipal level. Still, they will require collaboration and coordination efforts from the three spheres of government, especially in aligning development and economic aspiration and related programmes of the National, Provincial and Local government. Municipal planning is highly regulated. There is a myriad of applicable spatial policies influencing economic growth and service delivery.

Figure 2 contains the constitutional development duties of local government. In addition, policies are used to control the development agenda of local government and this also influences local government programmes. Therefore, municipalities must take cognisance of these multiple policy desired outcomes and reflect it in their development incentives. These policy objectives are used to measure and evaluate the impact of development incentives on economic growth and service delivery.

Given the nature of municipal development duties, myriad of legislation that regulates municipal planning. Table 1 indicates the number of legislations to be consider when implementing an effective land use management system. Each piece of legislation further guides the type of development incentive packages that can bring value to the customer and the municipality.

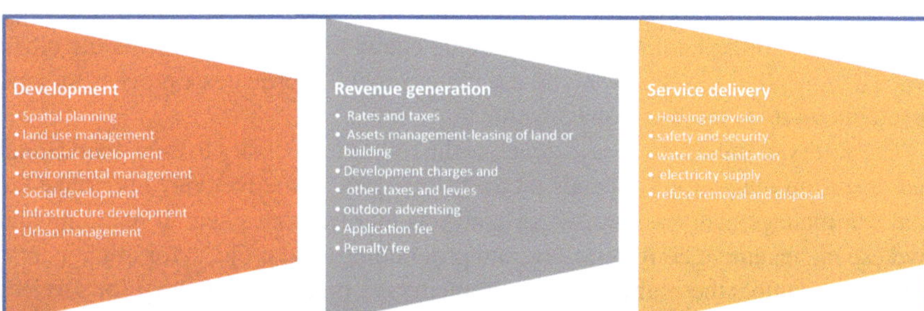

Fig. 2 Summary of municipal developmental duties

Table 1 Policies and strategic legislation that influence & guide development incentives

Legislation developed pre-1994 (directly used by municipalities as part of their municipal planning function)	Legislation developed post-1994 (directly used by municipalities as part of their municipal planning function)	National and provincial legislation (to be considered by municipalities in performing their municipal planning function)
Deeds Registries Act 47 of 1937 State Land Disposal Act 48 of 1961 The Subdivision of the Agricultural Land Act 70 of 1970; National Building Regulations and Building Standards 103 of 1977 Gauteng Town Planning and Townships Ordinance No. 15 of 1986 Tobacco Product Control Act, 1993	Gauteng Removal of Restriction Act 3 of 1996 Interim Protection of Informal Rights Act 31 of 1996; Local Government: Municipal Structures Act 117 of 1998 Prevention of Illegal Eviction from Unlawful Occupation of Land Act 1998 Land Survey Act The White Paper on Local Government 1998 The Rationalization of Local Government Affairs Act 10 of 1998 Municipal Systems Act 32 of 2000 Municipal Finance Management Act 56 od 2003 Land Reform (Labour Tenants) Act 4 of 2011 Spatial Planning and Land Use Management Act 16 of 2013	Housing Act 107 of 1997 National Environmental Management Act, 107 of 1998, include regulations and activities National Heritage Resources Act 25 of 1999 Gauteng Transport Infrastructure Act 8 of 2001 Mineral and Petroleum Resources Development Act 28 of 2002 National Environment Management: Protected Areas Act 57 of 2003 National Health Act 61 of 2003 National Environmental Management: Biodiversity Act 10 of 2004 Water Services Act 108 of 1997 National Environment Management: Waste Act 59 of 2008

Source Lentsoane [2]

Figure 3 describes how the land use management system connects and translates the municipal development duties and various pieces of legislation into small components that should work systematically. The multiple components of the land use management system are unique yet complementary and interdependent; therefore, each part needs to be fully operational to be effective. Each component plays a critical role in the land development value chain. One feature informs the other, hence being called a system for land use management. It provides a process for land development and land use management.

The ability of municipalities to generate the revenue necessary to fund their capital and operation obligations depends on their competence to effectively undertake the land use management system, which connects and translates spatial planning and land use management considerations into programmes and projects for implementation. This requires the municipality to implement a mechanism to levy for services rendered. Furthermore, the viability of the municipality depends on its capacity to

Fig. 3 Components of a land use management system

secure maximum return on investment. Therefore, the municipality must do everything in its power to generate revenue not only for capital projects but also to meet the economic and social needs of the community and stakeholders. Economic conditions include attracting, retaining and growing investment from both the locals and others outside the municipality's boundary. It was shown that compliance with legislation, policies, strategies, processes, and procedures in place could contribute significantly to increasing the return on investment made by the municipality. Such returns may have a spill-over effect on other municipalities influencing the province's Gross Domestic Product and, ultimately, the country. This will then require the municipality to be financially sustainable to remain economically and socially competitive with other striving municipalities (Fig. 4).

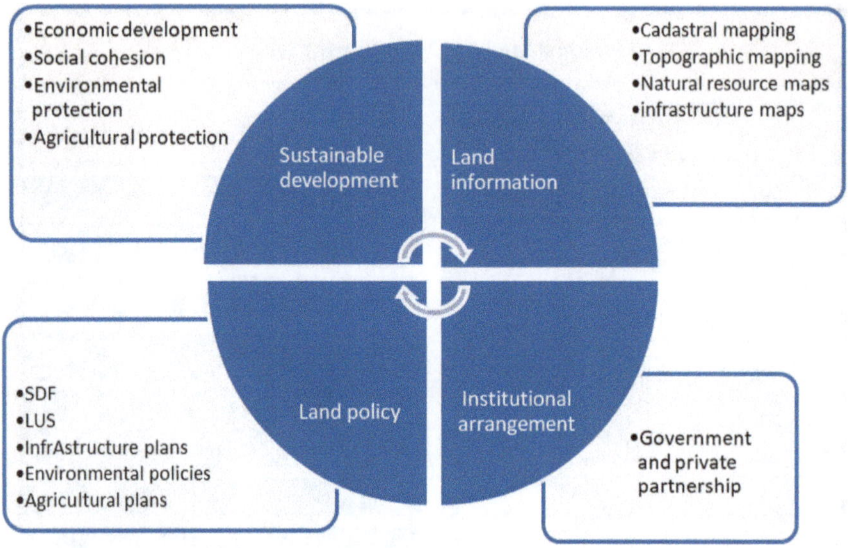

Fig. 4 The land management paradigm [71]

5.3 Development Trends and Assumption Mapping Strategy Compass for Development Incentives

Trend Analysis, Ideation, and Assumption Mapping process play a significant role in producing credible development incentives. The Economic Development in Africa Report (2021) states that South Africa between 2001 and 2015 experienced poverty reduction but inequality and unemployment increased [66]. The World Bank Report [67] states that South Africa was projected to rebound from 6.4% in 2020. This follows two consecutive years of recession. The Statistics South Africa Report (2021) states that the country has recorded its fourth consecutive quarter of growth, expanding by 1.2%; however, 1.4% smaller than before the Covid-19 pandemic. Municipalities must be competent in understanding the global economic outlook before implementing development incentives.

Blank [68] mentions that behind every new product and business idea, hide leap of faith assumptions, and if proven false, these critical and unproven assumptions can make or break you. Assumptions Mapping is about interrogating the hypothesis relating to the effectiveness of development incentives by looking at desirability (does the market want development incentives?), feasibility (can we deliver at a scale?), and viability (can development incentive create value?) of development incentives can be made clear and linked with all applicable policies and budget. Assumption mapping help with risk management. Therefore, municipalities need to undertake these exercises to be able to establish how the development incentives actually affect and impact on their functions and operations.

5.4 *Understanding the Players in the Economy*

Economic growth respect no administrative boundaries. The MDG Gap Taskforce Report (2009) provides that challenges and mutual interests are better tackled through partnership for better developmental and economic outcomes [69]. Comprehension of the rules of engagement pertaining to role players in the economy demands a deeper understanding of geopolitics, including political uncertainties, intergovernmental relations, technological imperatives, environmental considerations, social and cultural (social attitude, social development, and change population), and legal imperatives and legislation reform. This will ensure that the municipality is better positioned to deal with development and growth outcomes as a result of development incentives. Few South African financial partners are crucial to the country's economic growth. The European Union and South Africa, together with SADC countries, signed the Southern African Economic Partnership Agreement (SADC EPA) that regulates trade in goods between the two regions: the Southern African Development Community, Africa Union, and part of the BRICS (Brazil, Russia, India, and South Africa) [70]. Thus, when developing development incentives, municipalities must look beyond intergovernmental relations and Public–Private partnerships and recognise and appreciate all existing partnerships, regional treaties and their associated arrangements.

5.5 *Land Administration System*

Data management plays a crucial role in municipal planning. Thus, managing and monitoring the impact of the development incentives will require a land administration system that will apply property rights and related values. The land administrative system refers to recording and disseminating information about land ownership, value, and use. Therefore, Cadastre is another component that must be included in the land administration system.

An accurate impact assessment of development incentives involves land administration, land use management, and cadastral systems [72]. The land management paradigm is indicative of a functional and effective land use management system whereby the Spatial Development Framework, Land Use Scheme, and other relevant infrastructure and environmental laws determine the developmental capability and associated possible return on investment in the short- and long-term planning, The development incentives to yield tangible results must be connected to the land administration system.

6 Discussions

Development incentives, as can be seen, should be area or sector-specific. The municipality can maximise value and gain a Return on Investment by focusing on development incentives. Furthermore, there must be clear criteria stipulating the rules of engagement as to when and how Development Incentives will be granted. The municipality must, before granting incentives, consider the following:

1. The total private capital investment in the project includes site acquisition, site improvement, building development and equipment purchases.
2. Number, type, and quality of full-time jobs to be created for residents and jobs retained.
3. Skills development to be offered.
4. Wage level and associated benefits for a job created by the project.
5. Potential future expansion and associated revenue generation relating to the business.
6. Potential increased job creation, development, and integration of small and emerging businesses to the main large established business.

Service delivery, economic growth, and land development are the municipality's responsibilities through municipal planning processes. Development Incentives can be either monetary or non-monetary, ranging from exemption or reduction in development charges, utility rate (for example, water and electricity), fee waivers, and relaxing the land development controls as demonstrated by the City of Tshwane post Covid-19. Monetary Development Incentives are benefits/rewards associated with financial relief but do not take the form of direct cashback. Instead, the development incentive is an indirect cash relief system, where through a land development application or building plan, a municipality may determine which incentives are appropriate to be offered. The development incentives may be granted by looking at the following:

1. **Easy access to government-owned strategic land**: municipalities must profile, rezone, and service their land to have proper developmental rights. The land must be made available to investors or start-ups who want to participate in the economy but have no land. The municipality may enter into a partnership agreement to ensure there is mutual benefit. Land can be made available to individuals or groups if their business meets the set development incentive criteria. This approach reduces the cost of low-income housing development.
2. **Density bonus**: may be granted based on the Spatial Development Framework densification strategy where high densities are supported along major public transportation routes. This is very prevalent in developed countries such as United States and Canada. Furthermore, strategic location accessibility and proximity to development opportunities will be used as a determining factor density bonus.
3. **Development charges, rates, and taxes**: short-term waivers may be used as an incentive for qualifying development. The development charges may

be exempted, and other rebates can be applied on merit. Sector/ Infrastructure Departments must compile a list/matrix indicating methods of incentives applicable and percentages thereof.

4. **Urban management/revitalisation**: incentives may consider reducing payable rates and taxes to development that support urban management and using adaptive reuse techniques and technologies. Such methods may be certified by considering energy efficiency and building control laws. The developer must prove why the development qualifies for rates and taxes short-term waver. For example, the waiver may be between 3 to 8 months. The reduction may be between 5 and 20% of monthly payable rates and taxes. The municipality must be able to provide a criterion and work out the affordability plan for the waiver. Where the adaptive reuse project is for social housing purposes, the reduction must be higher or more extended period or both.
5. **Energy-efficient**: tax rebate-alternative energy such as solar should not be seen as a luxury but a necessity in line with climate change protocols. Therefore, municipalities must exempt or waive utility-related fees to businesses using alternative energy sources. Climate change and consequences must be acknowledged and noted.

Municipalities should provide basic infrastructure to its citizen. Thus, incentives on infrastructural assets such as roads, energy, and water must be done with due diligence. In addition, monetary stimulus requires the budgeting system of the municipality to change to be able to cater for the financial relief offered. The financial relief propels the municipality to establish a Development Incentive Fund. The National Treasury and the Department of Economic Department must support and guide municipalities to ensure inclusive growth and service delivery with municipal planning.

As the literature suggests, development incentives are provided to stimulate economic growth, not to deal with incompetency. Development incentives if not done for proper reasons and with reliable research results can result to series of unexpected resources challenges to the municipality and entity concerned. The literature review and case study shared have demonstrated that monetary incentives can yield positive results in increasing economic growth, increasing employment, reducing inequality, and dismantling multiple deprivation traps associated with poverty that is now exacerbated by the Covid-19 pandemic. Furthermore, it is found that effective Development Incentive requires a municipality to identify sectors with a maximum economic impact to participate in the program. Moreover, to ensure that emerging and small businesses benefit fairly from these incentives, and not overshadowed by well-established companies. There is need for an economic and social partnership agreements to be entered into by all stakeholders participating in the programmes so that all understands the rule of engagement. Development Incentives must be used for social and economic relief and must be specific and deliberate to uplift the local communities so that distressed communities become viable. This propels that for well-established businesses to qualify for the development incentives they must demonstrate how they are going to support the emerging and small business.

They must further indicate the number of quality jobs to be created and produce a skills development programme catering for local people especially youth, women, and people with disability.

There must be thorough further research on the following trend so that an informed decision is taken on the nature of development incentives to be introduced and financed:

- Development trends
- Spatial development trends
- Growth management trends
- Migration tends,
- Socio-economic trends and
- Market trends

Understanding the implication of these 5 listed trends may have a direct and indirect impact on the municipal's ability to effectively perform municipal planning and render services. These trends must be understood in relation to 3 entrepreneurial assumptions that seek to gain insight into the investor's interest and the municipal capacity and capabilities in the offering, management and sustaining the development incentives long term.

7 Conclusion, Summary and Recommendations

Municipal planning-related policies, systems budgeting, and strategies must evolve, moving away from traditional planning to adopt design thinking and entrepreneurial mindset. Municipalities need to revisit their strategic plan and review its performance in dealing with all kinds of deprivation traps found in their jurisdiction, especially now after the Covid-19 pandemic and its effects. Furthermore, to analyse what the ripple effects of the deprivation traps have on the country's economic growth and service delivery.

Providing Development Incentives should be a last resort where economic growth is affected by global events just like we have seen how the COVID-19 pandemic became destruction and disruption at the same time. Silo mentality in planning, budgeting, execution has been long identified as the primary challenge in government. To date, the silo mentality persists. This persistence can be attributed amongst others to the government posture on all legislative ad operational matters.

Municipalities lacking the entrepreneurial mindset have failed to position themselves as a brand and to truly behave like a worthy band, this includes identifying and capitalising on their unique selling points and what value can be generated. Municipal planning and service delivery can be seen as the primary value proposition of the municipality, if this is true what are the key activities to ensure that there are various revenue streams that are exceptionally generating revenue for the municipality.

Therefore, the municipality should relook at its financial model, investment, and fundraising strategies before resorting to development incentives. Often municipalities provide monetary and non-monetary incentives. Non-monetary development incentives can include exceptional professional services rendered; coordination and availability of accurate information; education awareness targeting developers; Cleaner and safe environment; Safe and secure neighbourhood (visible policing); Expedited and quality decision making; Reliable, responsive, and accountable officials; Welcoming and conducive environment; and Providing recognition certificates certified by the Mayor and Municipal Manager to all complying business as per the business performance standard approved by the municipality.

Exceptional customer service is another culture that can be learned from the business communities. It creates credibility. Municipalities can borrow the Ritz-Carlton customer service model. The Ritz-Carlton takes customers and employees as unmatched assets. Understanding that without customers (citizens) and investors the municipality ceases to exist. The Retz-Carlton has this practice:

- Employees first—arguing that customers are important, but it starts with employees doing the job they are supposed to do.
- Hiring takes patience—must have a great team in place. This means hiring the right people
- Your market is your customer—for a business to succeed, you must understand your market. Knowing what they want and need before you can begin to meet their expectations.

The above three Carlton practices can be adopted by municipalities to attract, retain, and increase investment milage. Exceptional customer service boosts investor confidence and citizens/customer satisfaction, thus increasing economic growth. De Berg and Wilderon [73] in support of this approach define organisational culture as a shared perception of organisational work practices within organisational units that may differ from other organisation units to another. It is not adequate for municipalities to be known as simply implementing agents of the government. They need to position themselves as brands and do more as a lot is expected from them. A brand is the most asset government can possess. It is how a municipality is perceived based on reputation, an identity created, and voice heard. It is our submission that when municipalities develop an entrepreneurial mindset, the non-monetary incentives should be embedded in the day-to-day operations of the municipality. These are what distinguishes the municipal character from the rest. When a municipality excels in this so-called monetary development incentive, the municipality will move away from being an incentive provider and become an environment where businesses and investors fight amongst themselves to do business with the municipality. Innovative spirit and entrepreneurial mindset in municipalities will be the growth factor for municipal planning and service delivery. Hence, a win–win situation for all stakeholders including the private sectors, investors and the government at large.

References

1. Ashkenas R (2012) Managing yourself: breakthrough your mental bureaucracy. Harvard Bus School Rev.
2. Lentsoane EM (2020) Inconsistencies in national and provincial legislation and their impact on municipal planning processes and service delivery, Doctoral dissertation, North-West University
3. Van Wyk J (2012) Planning Law. Juta and Company Ltd
4. Rogerson CM (2009) Strategic review of local economic development in South Africa. Final report submitted to Minister S Shiceka of Department of Development Planning and Local Government (DPLG). Commissioned by DPLG and GTZ.
5. Nkhabu MF (2021) The role of local government in the implementation of Sustainable Development Goals: A case of Tshwane Municipality, Doctoral dissertation, University of Pretoria
6. Guest G (2014) Sampling and selecting participants in field research. In: Handbook of methods in cultural anthropology, vol 2, pp 215–250
7. Ishak NM, Abu Bakar AY (2014) Developing sampling frame for case study: challenges and conditions. World Journal of Education 4(3):29–35
8. Bogopane LP (2013) Evaluation of black economic empowerment (BEE) policy implementation in the Ngaka Modiri Molema District, North West Province, South Africa. J Soc Sci 34(3):277–288
9. Bogopane LP (2013) A critical review of pertinent qualitative research processes, approaches, and tools in social sciences. J Soc Sci 35(3):217–229
10. May C, Finch T, Mair F, Ballini L, Dowrick C, Eccles M, Heaven B (2007) Understanding the implementation of complex interventions in health care: the normalization process model. BMC Health Serv Res 7(1):1–7
11. Corbin JM, Strauss A (1990) Grounded theory research: procedures, canons, and evaluative criteria. Qual Sociol 13(1):3–21
12. Atkinson P, Coffey A (2004) Analysing documentary realities. Qual Res 3:77–92
13. Coffey A, Atkinson P, Delamont S, Lofland J, Lofland L (2004) The handbook of ethnography. Anthropologica 46(2):298
14. Noor KBM (2008) Case study: A strategic research methodology. Am J Appl Sci 5(11):1602–1604
15. Slusher MP, Anderson CA (1987) When reality monitoring fails: the role of imagination in stereotype maintenance. J Pers Soc Psychol 52(4):653
16. Pan SL, Tan B (2011) Demystifying case research: a structured–pragmatic–situational (SPS) approach to conducting case studies. Inf Organ 21(3):161–176
17. Zhang Z, Lee MK, Huang P, Zhang L, Huang X (2005) A framework of ERP systems implementation success in China: an empirical study. Int J Prod Econ 98(1):56–80
18. Yin RK (2003) Designing case studies. Qual Res Methods 5(14):359–386
19. Gerring J (2004) What is a case study and what is it good for? Am Pol Sci Rev 98(2):341–354
20. Eckstein F, Wirth W, Nevitt MC (2012) Recent advances in osteoarthritis imaging—the osteoarthritis initiative. Nat Rev Rheumatol 8(10):622–630
21. Golafshani N (2003) Understanding reliability and validity in qualitative research. Qual Rep 8(4):597–607
22. Patton MQ (2002) Two decades of developments in qualitative inquiry: a personal, experiential perspective. Qual Soc Work 1(3):261–283
23. Patton MQ (2014) Qualitative research & evaluation methods: integrating theory and practice. Sage
24. Polit DF, Beck CT (2010) Generalization in quantitative and qualitative research: myths and strategies. Int J Nurs Stud 47(11):1451–1458
25. Drew CJ, Hardman ML, Hosp JL (2007) Designing and conducting research in education. Sage
26. Liu Q, Pan X, Tian GG (2018) To what extent did the economic stimulus package influence bank lending and corporate investment decisions? evidence from China. J Bank Finance 86:177–193

27. Acs ZJ, Szerb L (2007) Entrepreneurship, economic growth and public policy. Small Bus Econ 28:109–122
28. Wu YA, Ng AW, Yu Z, Huang J, Meng K, Dong ZY (2021) A review of evolutionary policy incentives for sustainable development of electric vehicles in China: strategic implications. Energy Policy 148:111983
29. Nel V (2016) A better zoning system for South Africa? Land Use Policy 55:257–264
30. Kumar-Nair S (2022) The conceptualisation, institutionalisation, delivery and management of public space in the Buffalo City Metropolitan Municipality: towards integrated placemaking and greening practices, Doctoral dissertation, University of Pretoria
31. Nsele, T. (2016). *Examining the effectiveness of the land use management framework in Ntambanana Municipality* (Doctoral dissertation).
32. Phillips SD, Johnson B (2021) Inching to impact: the demand side of social impact investing. J Bus Ethics 168(3):615–629
33. Phalatse S (2022) The role of private finance in infrastructure development in South Africa—a critical assessment. J Infrastruct Policy Dev 6(1):1391
34. Mazzucato M, Qobo M, Kattel R (2022) Building state capacities and dynamic capabilities to drive social and economic development: the case of South Africa. The future of the South African political economy post-COVID 19. Springer International Publishing, Cham, pp 43–74
35. Choi JK, Thangamani D, Kissock K (2019) A systematic methodology for improving resource efficiency in small and medium-sized enterprises. Resour Conserv Recycl 147:19–27
36. Maluleke MN (2017) Implications of Spatial Planning and Land Use Management Act (16 of 2013) (SPLUMA) on land allocation in areas under traditional authorities, Doctoral dissertation
37. Dymitrow M, Biegańska J, Grzelak-Kostulska E (2018) Deprivation and the rural-urban trap. Tijdschr Econ Soc Geogr 109(1):87–108
38. Marutlulle NK (2021) A critical analysis of housing inadequacy in South Africa and its ramifications. Africa's Public Service Delivery & Performance Review 9(1):16
39. Betz MR, Partridge MD, Kraybill DS, Lobao L (2012) Why do localities provide economic development incentives? Geographic competition, political constituencies, and government capacity. Growth Chang 43(3):361–391
40. Department for Business, Innovation & Skills, Great Britain (2010) Local growth: realising every place's potential, vol 7961. The Stationery Office
41. Zheng C, Deng F, Zhuo C, Sun W (2022) Green credit policy, institution supply and enterprise green innovation
42. Baron RA (2006) Opportunity recognition as pattern recognition: how entrepreneurs "connect the dots" to identify new business opportunities. Acad Manag Perspect 20(1):104–119
43. Botha M, Taljaard A (2021) Exploring the entrepreneurial intention-competency model for Nascent Entrepreneurs: insights from a developing country context. Front Psychol 12:516120
44. Botha M, Taljaard A (2019) The bidirectional relationship between entrepreneurial intention and entrepreneurial competencies for nascent and existing entrepreneurs. S Afr J Econ Manag Sci 22(1):1–12
45. Zvavahera P, Chigora F, Tandi R (2018) Entrepreneurship: an engine for economic growth. Int J Acad Res Bus Soc Sci 8(11):55–66
46. Kreuzer T, Lindenthal AK, Oberländer AM, Röglinger M (2022) The effects of digital technology on opportunity recognition. Bus Inf Syst Eng 64(1):47–67
47. Wiramihardja K, N'dary V, Al Mamun A, Munikrishnan UT, Yang Q, Salamah AA, Hayat N (2022) Sustainable economic development through entrepreneurship: a study on attitude, opportunity recognition, and entrepreneurial intention among university students in Malaysia. Front Psychol 13:866753
48. Dimov D (2017) Towards a qualitative understanding of human capital in entrepreneurship research. Int J Entrep Behav Res 23(2):210–227
49. Morgan C, Volante L (2016) A review of the Organisation for Economic Cooperation and Development's international education surveys: governance, human capital discourses, and policy debates. Policy Futures Educ 14(6):775–792

50. Fernholz RM (2010) Infrastructure and inclusive development through "free, prior, and informed consent" of indigenous peoples. Physical infrastructure development: balancing the growth, equity, and environmental imperatives. Palgrave Macmillan, New York, pp 225–258
51. Dweck CS (2012) Mindsets and human nature: promoting change in the Middle East, the schoolyard, the racial divide, and willpower. Am Psychol 67(8):614
52. Kuratko DF, Fisher G, Audretsch DB (2021) Unraveling the entrepreneurial mindset. Small Bus Econ 57:1681–1691
53. Chen MJ, Miller D (2011) The relational perspective as a business mindset: managerial implications for East and West. Acad Manag Perspect 25(3):6–18
54. Porter ME (2008) The five competitive forces that shape strategy. Harv Bus Rev 86(1):78
55. Yasmin T, Refae GAE, Eletter S (2023) Examining the impact of effective government stimulus packages amid COVID-19 pandemic in UAE: an input-output analysis. Int J Trade Glob Mark 18(1):63–81
56. Xin-gang Z, Wei W, Ling W (2021) A dynamic analysis of research and development incentive on China's photovoltaic industry based on system dynamics model. Energy 233:121141
57. Xu Z, Xie N, Wu L (2023) Evaluation on sustainable development of 11 regions in Hebei province. Environ Dev Sustain 1–15
58. Slattery C, Zidar O (2020) Evaluating state and local business incentives. J Econ Perspect 34(2):90–118
59. National Research Council (2012) Rising to the challenge: US innovation policy for the global economy
60. Bassier I, Budlender J, Goldman M (2022). Social distress and (some) relief: estimating the impact of pandemic job loss on poverty in South Africa (No. 2022/80). WIDER working paper.
61. "City of Tshwane Metropolitan Municipality—Map." [Online]. https://municipalities.co.za/map/3/city-of-tshwane-metropolitan-municipality. Accessed 23 Feb 2024
62. Bagautdinova N, Gafurov I, Kalenskaya N, Novenkova A (2012) The regional development strategy based on territorial marketing (the case of Russia). World Appl Sci J 18(18):179–184
63. McMullen JS, Bergman BJ Jr (2017) Social entrepreneurship and the development paradox of prosocial motivation: a cautionary tale. Strateg Entrep J 11(3):243–270
64. Fawcett SE, Smith SR, Bixby Cooper M (1997) Strategic intent, measurement capability, and operational success: making the connection. Int J Phys Distrib Logist Manag 27(7):410–421
65. Demazière C (2018) Strategic spatial planning in a situation of fragmented local government: the case of France. disP Plan Rev 54(2):58–76
66. Gumede V (2021) Revisiting poverty, human development and inequality in democratic South Africa. Indian J Hum Dev 15(2):183–199
67. World Bank (2021) World development report 2021: data for better lives
68. Blank S (2020) The four steps to the epiphany: successful strategies for products that win. John Wiley & Sons
69. Hulme D (2009) The Millennium Development Goals (MDGs): a short history of the world's biggest promise
70. Lumumba-Kasongo T (2015) Brazil, Russia, India, China, and South Africa (BRICS) and Africa: new projected developmental paradigms. Afr Dev 40(3):77–95
71. Enemark S, Fig P (2010). From Cadastre to land governance in support of the global agenda—the role of land professionals and FIG. International Federation of Surveyors. https://www.fig.net/resources/monthly_articles/2010/december_2010/december_2010_enemark.pdf
72. Enemark S (2010) Spatially enabled land administration: addressing societal needs and responding to the global agenda. In: Realizing spatially enabled societies GSDI-12 world conference Singapore, pp 19–22
73. De Hoogh A, den Hartog D, Koopman P, Thierry H, van den Berg P, van der Weide J, Wilderom C (2004) Charismatic leadership, environmental dynamism, and performance. Eur J Work Organ Psy 13(4):447–471

Dr. E. M. Lentsoane is a Chief Executive Officer for the South African Council for Planners (SACPLAN), a regulatory body for the urban and regional planning profession. She has 21 years' worth of experience in the built environment. Her experience cut across governments and the private sector dealing with spatial planning, land use management, economic planning and other activities affecting the built environment. She holds a PhD in Urban and Regional Planning, North-West University; Master in Town and Regional Planning, Pretoria University; Master Entrepreneurship, Management, and Innovation with the University of Bath England. She advocates for urban and regional planning that transcends beyond traditional spatial and land use theories and techniques. Dr Tshepiso's research interest lies in value creation whereby entrepreneurship coexists with spatial planning and land use management balancing the power of the formal and informal economy in creating viable communities and sustainable jobs. She believes that development incentives and a circular economy are innovative approaches for the government to adopt to remain globally competitive, credible and relevant.

Dr. George Onatu is a Senior Lecturer and former Head of Department of Urban and Regional Planning, Faculty of Engineering & Built Environment, University of Johannesburg, South Africa. He is an exceptional academic leader and experienced researcher with several supervisions and publications on case study research, housing development, land use management, spatial planning and consultancy both locally and internationally. He is a board member of several organizations, including the Association of African Planning Schools (AAPS), Member of Asian Pacific Network of Housing Research (APNHR), Member of International Society of City and Regional Planning (ISOCARP) and South African Planning Institute (SAPI), Editorial Board Member of Town and Regional Planning Journal, South Africa and part of the policy and legislative advisory Panel to the former Minister of Human Settlements South Africa. He has visited more than 15 countries on research collaboration, teaching and learning as well as guest speaker. He has DPhil in Engineering Management, Masters in Town and Regional Planning and BSc in Geography and Regional Planning.

Shortcomings of Urban India in Dealing With the Pandemic: A Study on the Response From the Rural Regions in India

Arun Baby M. Wilson and Jayaprakash Chadchan

Abstract Cities as the growth grounds of the Indian Economy play a vital role in upholding the overall development of the country throughout all the sectors. During the pandemic period of COVID-19 waves the country witnessed the downfall of urban life in several aspects. The rural population was less affected by the pandemic than the urban population in India. Since having better infrastructure facilities and service providers the urban life uses ICT in almost all ways where the IOT is less addressed. The e-governance was less appreciated and used for tackling the hits and punches of the pandemic in urban areas wherein the rural area the governance is acting at a different scale like with the National Rural Health Mission (NRHM) who had done a commendable job in developing and maintaining medical records of people in rural areas using the network of anganwadi teachers and ASHA workers. The urban regions were really suffering more from the pandemic situations since the IOT of medical care is less addressed or encouraged compared to the rural areas. The government's involvement in public health is becoming lesser as the private health sector is growing quickly and developing high end infrastructure. The network of medical infrastructure, which is primarily with the private sector in urban agglomerations with the governing authorities, needs to be strengthened as it is done in the rural regions, according to an analysis of how governing bodies, especially in rural regions, managed to tackle the pandemic hit. During the lockdown period, rural India had survived while Urban India was struggling to pass their days. These scenarios clearly point out the need for revisiting the planning and infrastructure deployment strategies in urban regions and bringing in local/regional concepts both physical and cultural than global/international concepts. This paper discusses the shortcomings of urban infrastructure in tackling pandemic situations when compared to the rural regions of India.

A. B. M. Wilson (✉) · J. Chadchan
School of Architecture, Christ University, Bangalore, India
e-mail: arunbaby.wilson@christuniversity.in

J. Chadchan
e-mail: jayprakash.chadchan@christuniversity.in

K. M. Nahiduzzaman (ed.), *Making Sense of Planning and Development for the Post-Pandemic Cities*, City Development: Issues and Best Practices,
https://doi.org/10.1007/978-981-97-5481-6_10

1 Introduction

The world had witnessed the massive pandemic strike during the last few years which turned out to be a very scary and destructive situation and the entire world suffered. Human life was literally shattered as the pandemic situation had adversely affected the social and economic activities and hit on the spine of our daily life. Even though the governments and authorities had implemented preventive measures and alternative means for continuing livelihood activities and controlling the spread of disease there was an evident decline in the economic growth as well as in social wellbeing of the people everywhere across the globe.

On comparing the intensity of impact due to the pandemic strike, the rural regions in India possessed a better command over controlling and containing the spread of disease than the urban regions and the adversity on human life was very less compared to that in urban regions. This paper is studying and discussing the local and regional model of governance and the network of authorities which was the main reason in tackling the pandemic situation and analyzing how the successful rural model can be adapted for urban regions. For the same, this study is intended to analyze the model of rural Governance in Kerala in collaboration with the NRHM and its policies.

2 Methodology

This study aimed at getting into the details of pandemic strike over different regions and analyzed the impact of the same to identify the factors related to it in the first phase with spatial and temporal data collected. With the inference drawn from the initial analysis the study had gone into the details of the local governance and IOT allied with rural regions considered. The data collected are validated through discussions with the forefront warriors in the public health sector taking part in the fight against pandemic. With the help of maps created a network analysis was done identifying the service extent to correlate the factors identified in the first phase and the effectiveness of the system in performing at the needs. It is then compared with the urban governance model and IOT associated with to identify the issues or gaps which need to be addressed. As a result of the study, this paper put forward the very crucial governance amendments and policy inclusions.

3 Kerala Model—The Public Participatory Model

During pandemic days and after the lockdown period as well, the term Kerala model was widely discussed even in international platforms. The way the state of Kerala and the system had contained the pandemic spread was a hot topic for discussion and other governments had looked into it to draw insights. This study is dealing with

Fig. 1 No. of doses of COVID vaccine per 10 persons

the Kerala model which is a deep rooted public participatory model, in detail with relevant studies and data and analyzing the gap between the Urban Governance to the Rural governance Kerala has to validate the strength of public participation which helped in overcoming the pandemic strike.

Kerala in association with the National Rural Health Mission scheme had utilized the IOT associated with the department of health and governing bodies to an excellent level which helped them in containing the pandemic. Even after the central government had issued the ArogyaSetu app and asked everyone to install it to keep the record of disease spread and details of contact with the infected, the efficiency of the app in updating the data was questionable as it completely relied upon ICT but not IOT. wherein the health and governing department in Kerala, especially the rural regions had very well used the IOT which includes Governing bodies (LSG), Community Health Centers, Primary Health Centers and Sub health Centers with Anganwadi teachers and ASHA workers collectively to keep track of the pandemic spread and cluster formation. This action of real time response was further enhanced the rate of controlling the pandemic and even keeping human life in less disturbed situations (Fig. 1).

3.1 *Internet of Things*

The major component in this successful network was people rather than technological interventions which is very rare in urban governance as everything is getting digitized and the direct social involvement is negligible due to the characteristics of urban life. In the rural governance model, the technical domain is being used mostly to store, transfer and analyze data where the collection and updating of data is being mostly carried out by people who are within the network which includes common man to the head of governing departments. Since the data regarding pandemic is purely personal and dynamic, the people had performed way better than machines in handling the data and drawing inferences.

Consideration being given to IOT is very less compared to ICT where in actual ICT is just a component of IOT. In this advanced world, due to the availability of less time and increased density of urban living, the people tend to rely more on online applications and services for every need where the governance is not acting as E-Governance. Enabling all the services in online mode is not what E-Governance is all about but the governing authority has to act in all the formats ensuring everything is falling in place.

3.2 *The Not So Urban Model of Kerala*

Over the years the state of Kerala had a greater rise in the quality of life of the citizens. Irrespective of the fact that Kerala is having a low per capita income compared to other states the parameters/indicators of social and cultural development which are human development index (HDI = 0.84), infant mortality rate (IMR = 12/1,000 live births), male to female ratio (sex ratio = 1,084 females to 1,000 males) and literacy rate of female population (92.07%) are comparable to those of many developed countries and with a population of about 35 million resulting a high population density of about 860 people/sq. km.[1]

The primary healthcare services in Kerala are organized systematically in accordance with National Health Mission, NHM (Previously known as National Rural Health Mission, NRHM) with the network rooted to the village level. The network consists of 230 community health centers and 845 primary health centers which is at the ratio of approximately one per 30,000 people in the State. The system is equipped with 5,320 female junior public health nurses (PHN) and 4,728 junior public health inspectors (JHI) of which around 50% are women working as multipurpose health workers (MPHW), as in a way two MPHW serving every 5,000 people. The network again strengthened with Accredited Social Health Activists (ASHAs), who are community health volunteers, approximately one for every 1,000 to 1,500 people.[2]

The Local self-government Department (LSGD), enabling the bottom to top approach is very much active and organizes the local and regional actions very effectively in Kerala. Along with the health network the LSG as the governing authority had worked seamlessly in containing the Pandemic.

[1] Ministry of Health and Family Welfare [7].

[2] Covid-19 Jagratha Government of Kerala [2].

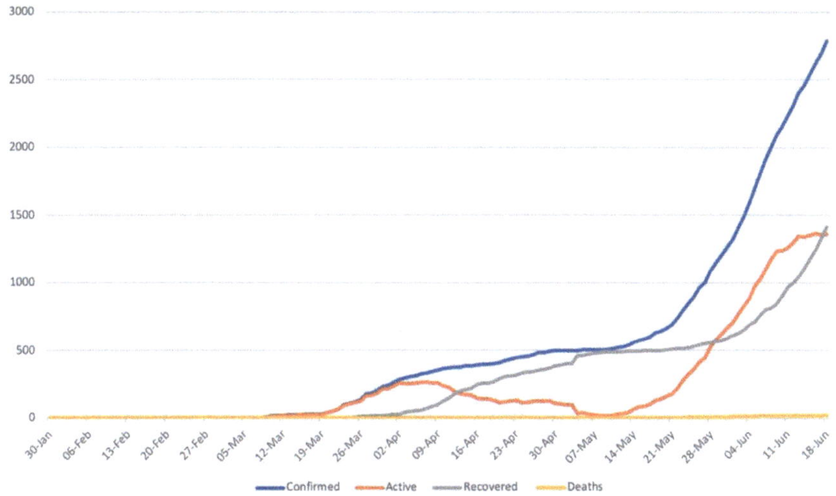

Fig. 2 No. of COVID cases in Kerala

The majority of the COVID-19 cases reported (85%) in Kerala are transferred from international borders and other states, and only the rest are locally acquired. The authorities had confirmed that there is no trace of community transmission with a very rigorous and regular sentinel surveillance activity among the non-COVID-19 suspects. The samples that were collected from the state for a nationwide zero-surveillance had also indicated a very low positivity rate in the state of Kerala compared to other states in India.

3.3 *The Fight With the Pandemic*

There was immense commitment and timely actions were evident in the state by the local and regional network as far as the pandemic control is concerned (Fig. 2). The state had a scary situation before as the NIPAH epidemic in 2018 and 2019 which was addressed by the state health system through being alerted even before the instance of the very first case (Fig. 3). The rapid and strong success with containing and isolating the epidemic has proven the facts that the state has a strong primary health system in place. With the active participation from the community and integration of the data into a system of management which is decentralized at the triple levels of LSG enabled a centralized epidemic control network aligned with the state health network.[3]

[3] Heera [3].

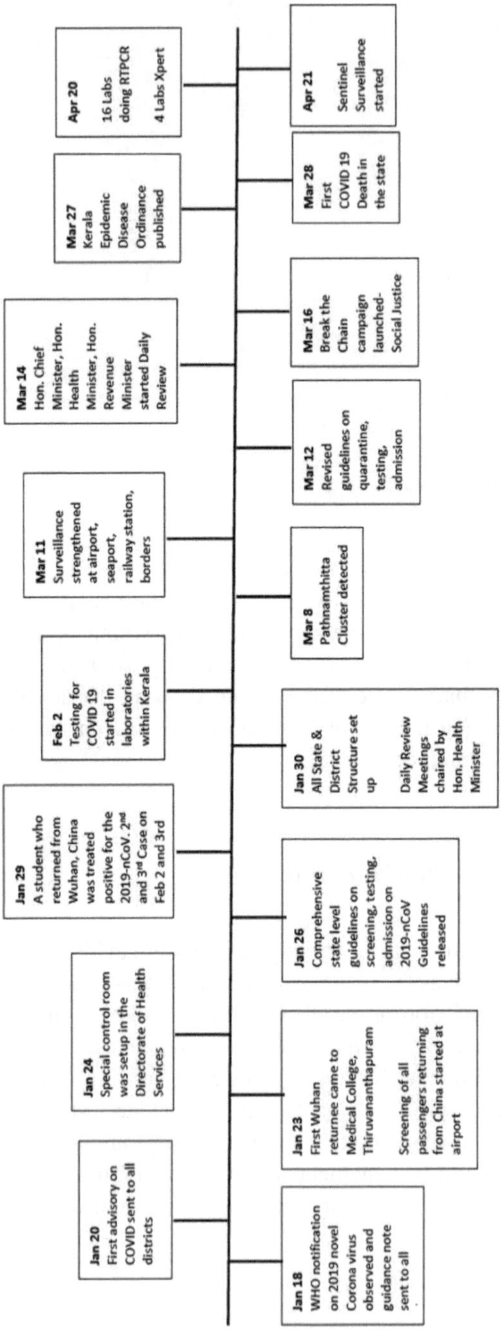

Fig. 3 Milestones in Kerala's pandemic fight

3.4 The System in Action; Internet of Things (IOT)

While all the urban regions were giving priority to digital components and ICT in managing things and capturing data, the network in the state of Kerala had given importance to the most important component but least utilized in urban regions, the local network of people to be in action in the fight against COVID-19.

The state had put up a network for screening and follow-up of every person who is arriving in the state by any mode of transportation (air, sea, rail or road) from abroad or other states of India. Those who are found with symptoms would be transferred to the dedicated hospitals or isolation centers and given with utmost care by the network of people established. There was a portal created by the government for the same called the COVID-19 Jagratha Portal to facilitate the same.[4]

Even Though the GOI had advised everyone to install the Arogya Setu application to monitor the pandemic data, the reliability was less since the data updating had to be done by each individual and the underprivileged those who cannot have a smartphone were not being monitored or mapped. Meanwhile in Kerala, the network of PHN, JHI and ASHA workers along with Anganwadi teachers were constantly updating the data for every individual and thus making sure everyone is being monitored.

Because of the homestead pattern and local setting of amenities in the neighborhood, the people of Kerala had a very effective and comfortable means of quarantine at their own homes. The cost incurred in the authority and inconvenience for the public were very less due to the facilitation of home-based quarantine and local supply materials at doorstep. There were volunteer groups that also came in action to help the health network system during the peak surge of pandemic strengthening the human resource in fighting COVID-19. The intensive care and support extended by the local and regional health network towards the people allowed the authorities to make the public participate in mass testing campaigns conducted which helped in continuous updating of data upon the spread of the disease.

3.5 Dedicated Health Network

The state had set up twenty-seven dedicated COVID-19 hospitals and all confirmed cases were shifted without any delay and taken under care at hospitals until they became negative. The state had ensured uninterrupted treatment facilities to the public who were not yet detected with the disease, especially those with chronic health issues through the network of health department officials (Fig. 4). As part of this in line with the e-sanjeevani initiative by GOI, primary health centers had put up telemonitoring of local people with the help of PHN and ASHA workers and delivered medicines at their doorsteps.[5]

---

[4] Chathukulam [1].

[5] Ministry of Health and Family Welfare [7].

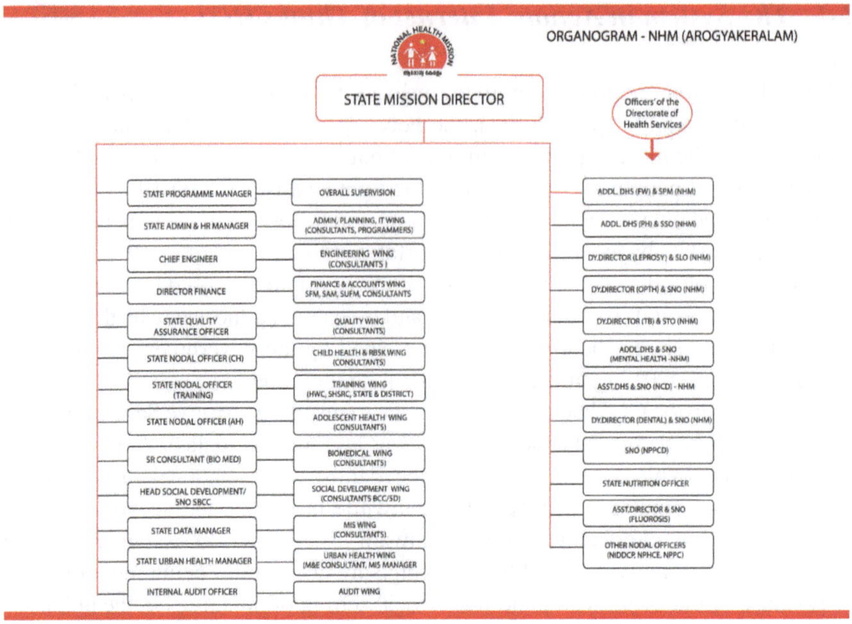

Fig. 4 Organizational chart of NHM—Kerala (*Source* https://arogyakeralam.gov.in/2020/03/25/organogram/)

3.6 Public Participation—A Deep Rooted Society

The major criteria which helped in all the actions by the health department and LSGD was the immense participation by the public in handling all the situations. Right from the lockdown period to the surge happened in festive days, the public volunteers along with the NHM team had taken up the duty in ensuring timely identification, prompt isolation and quarantine, routine testing and the youth had even gone to the extreme in delivering daily needed groceries and goods at the doorsteps (Fig. 5). The state had entrusted the female population in handling the situation where they were part of ASHA working group and Kudumbashree units and managed to maintain the environment in harmony with their dedicated hard work.[6]

It is nevertheless necessary to explain how and under what circumstances the underprivileged classes in a society, including a sizable portion of the rural poor, are able to influence state policy and force it to take into account their demands, even though the state is such a crucial institution in the provision of such public goods. High HDI measures, which are statistical averages, highlight how development in this region has been more inclusive and widespread than in other parts of India. This prompts the crucial political question of how these classes are organized, assimilated into comparatively solid organizations with universalistic ideologies and

[6] Chathukulam [1].

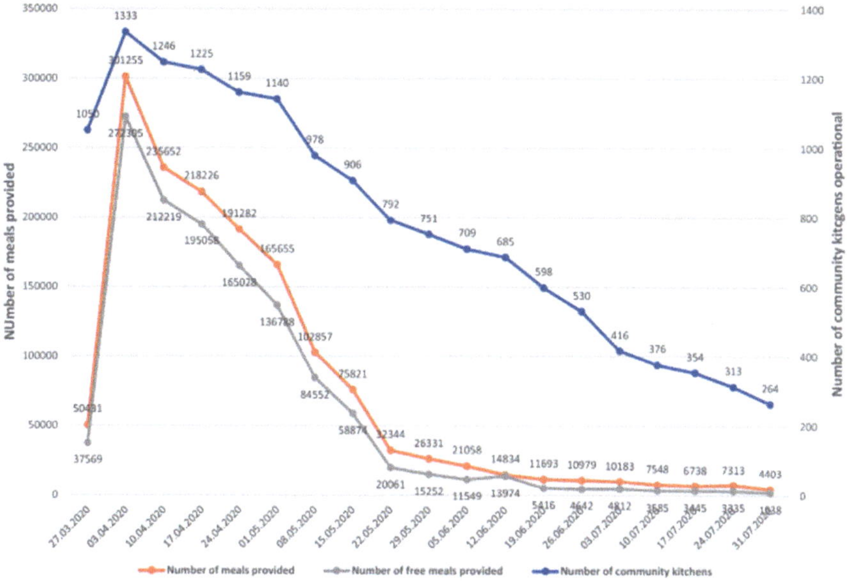

Fig. 5 Statistics on community and public participation in Kerala

programmes, and included in the political system. Kerala provides significant insight into this issue and demonstrates how and why these classes' mobilisations—the character of their movements—are necessary. The first two waves of Covid 19 infections in the state were successfully contained by an orchestrated decentralized response system made up of the state government, local governments, Kudumbashree, the public health system, and the general populace, according to policy experts and development economists.[7]

The two essential components of the Kerala model, as we have argued above, are a pro-poor interventionist state and a mobilised society that interacts with the state through effective mass organisations and parties. In this section, we go through how these two components have worked together to develop and preserve a particular synergy, or "virtuous" relationship. We propose that this may be crucial in figuring out why Kerala has been successful while other parts of the country have found it difficult.[8]

[7] Karan [5].

[8] Israelsen and Malji [4].

3.7 Conclusion

The state of Kerala has managed and contained the pandemic strike not by incorporating any advanced technology systems or hired professionals but the key point is considering the people as the major criteria in the network of everything, the IOT and enabling public participation as it always maintained the deep-rooted bequest in care for others. The level of impact was reduced in the rural regions due to the timely actions by the NHM and LSGD and the local setting pattern where the availability of amenities is in close proximity. Further study and research on the role of active public participation and people as the key role players in IOT have to be encouraged to formulate policy frameworks to enhance the social bonding and well being. The education system can be reformulated including activity oriented courses or credit systems which involves public participation and social interaction to upbring the morality and societal skills of the future generations.

4 The Public Health Network in India

The distribution of Public Health Network under NHM (NRHM and NUHM) is clearly described in the following maps. There is a clear difference visible in the service area distribution of public health networks in rural regions compared to the urban regions as the rural region has better services with the public health network (Fig. 6).

The demographic profile of urban regions in India certainly contain the component of lower class and underprivileged population and they are more serviced with private healthcare networks which are inaccessible for the common due to their lower level of affluence (Fig. 7).

The isolation and quarantine process put the urban population down as the activity and connectivity were completely restricted due to the increased density and high risk of disease spread. Since the handling of data regarding such a huge population, the urban region statistics was difficult to record and maintain as the health network is diverse and not under a common umbrella of governance.[9]

The advantage of low population density and close proximity to basic needs along with the voluntary public participation were very much helped in fighting the pandemic in rural areas especially in Kerala and the key point was the service coverage of NHM network and the public health department acted as the governing authority even by bringing private health facilities under the common health network. Since the expenditure in consulting private hospitals was higher, the common rushed into public health care and with the lower rate of serviceable population per each public health facility the governing authority was able to manage the situation. With the insights drawn from the NIPAH outbreak, the government and the healthcare

[9] Menon [6].

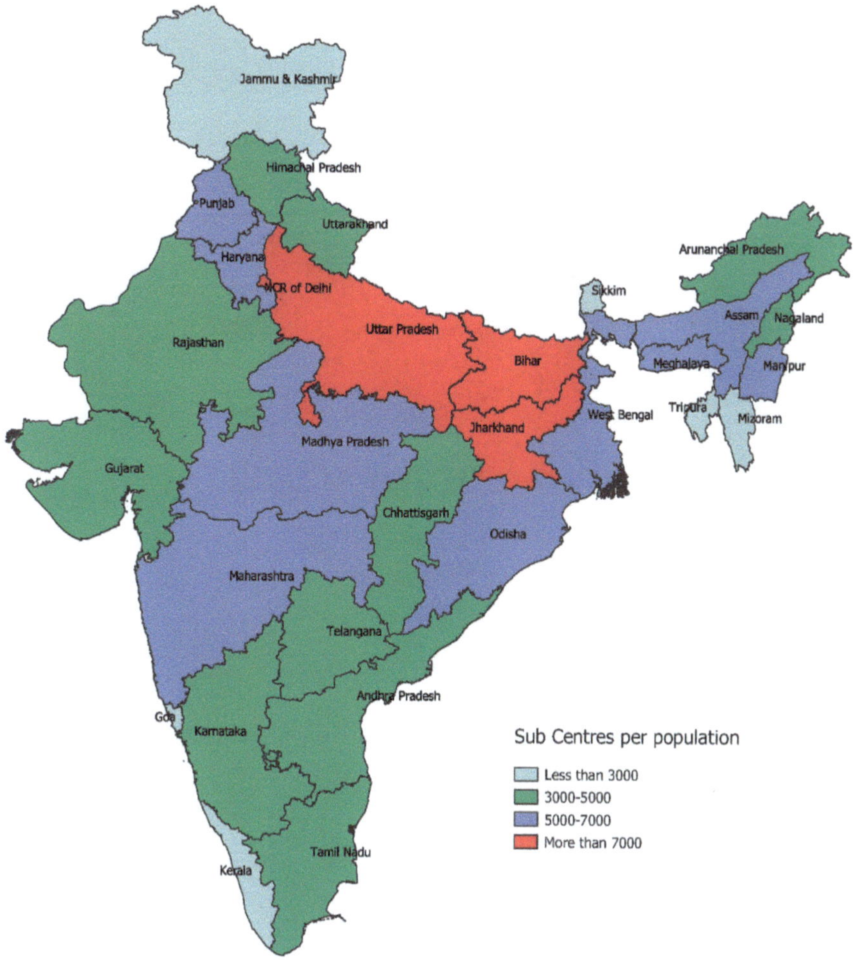

Fig. 6 Map—distribution of sub centers for the population (*Source* National Health Statistics 2021–2022, MoHFW)

network in Kerala were more prepared for any sort of such medical emergencies which also enabled the system to act fast.

There is a considerable shortfall in the human resource needed for the public health network in India. Since the requirement is specifically for professionally certified people, it is hard to meet on short notice since the appointment has to be done by the authorities (Fig. 8). Understanding this situation, the NHM network has taken in the anganwadi teachers and ASHA workers to manage the human resource requirement and in collaboration with the LSG the duties and responsibilities are mapped and managed (Fig. 9).

It is inferred from several studies conducted by the WHO and recorded in the national health accounts global health expenditure database that the importance given

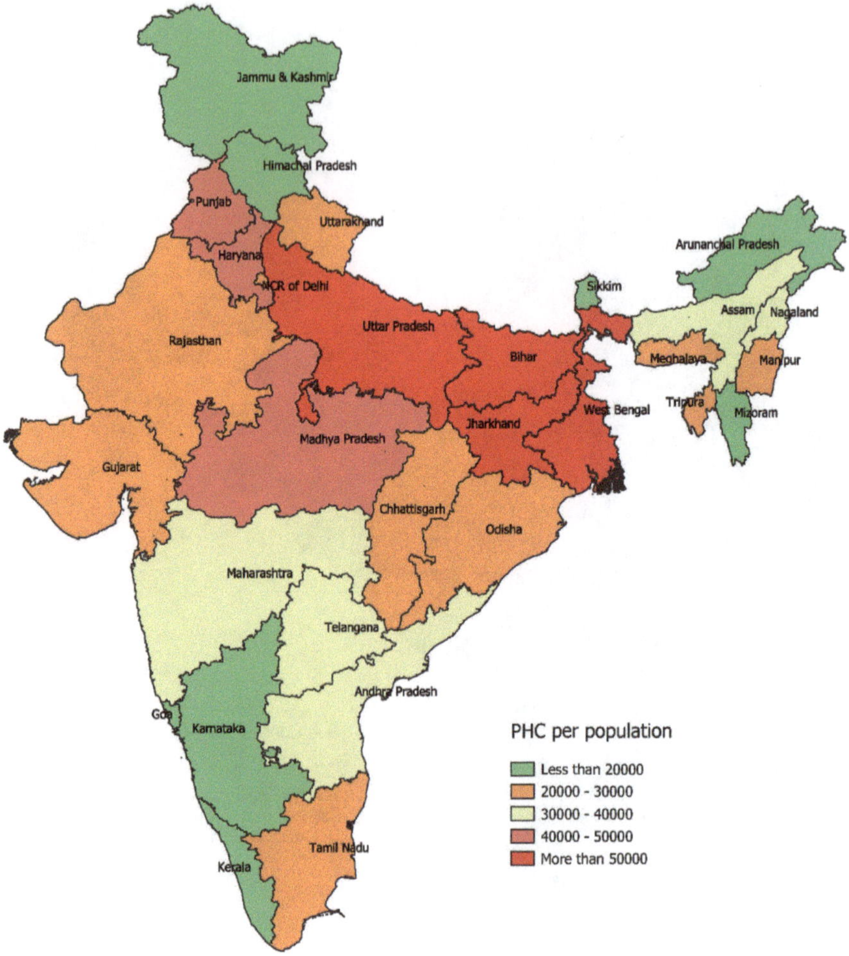

Fig. 7 Map—distribution of PHC for the population (*Source* National Health Statistics 2021–2022, MoHFW)

to the development and enhancement of public health network in India is very less compared to other world countries and the expenditure in providing the required health facilities and professionals is below 5% of the union budget allocation and even less than 2% of the GDP (Table 1).

National Sample Survey Organization, NSSO had identified several criteria and stated that the public health sector needs a huge push at certain levels to meet up the minimum requirements and to be ready for any future medical emergencies (Fig. 10). Comparing the statistics of distribution of medical professional in various states in India, the higher values for the state Kerala defines how the state had stood high in the Nipah outbreak as well as in COVID-19 strike (Fig. 11).

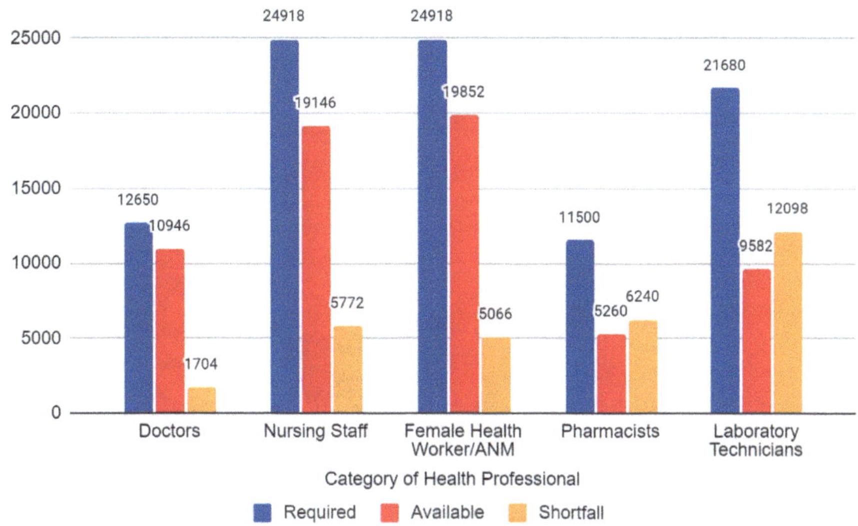

Fig. 8 Shortfall in public health professionals

There are public health facilities in the country still functioning without having minimum human resource/medical professionals which is absolutely causing the downfall of the public health system. This causes the public to lose faith in public health systems and move into private health care facilities for medical needs which again brings down the public health system (Table 2).

Based on the statistical data and analysis, a projection was done to identify the gap in meeting the required workforce for NHM in the coming years and it shows a very alarming situation would occur if the governing authority will not address this issue with utmost seriousness (Table 3).

The dominant share of private players over the health sector had critically affected the common as admission to the outpatient department or treatment in inpatient sessions in private hospitals were way costlier than that in public or government hospitals. The conversion of general hospitals into COVID special hospitals and isolation centers was difficult in the areas where the major player in health care was private parties.

In Kerala, even though the private sector is growing day by day at a higher rate than some of the neighboring states and metro regions, the disparity in per capita income has been reduced which normalized the cost of medication and was made affordable to the common to an extent (Fig. 12). The public health sector is very strong and widely spread as they were able to reach out to everyone and provide medical services (Table 4).

The fatality rate due to the pandemic strike in Kerala was lesser than most of the other states and all India average as well. The progression of fatality was steady and not rising rapidly at any point as the control measures were established effectively (Fig. 13). This study had evolved around the local and rural factors that helped in

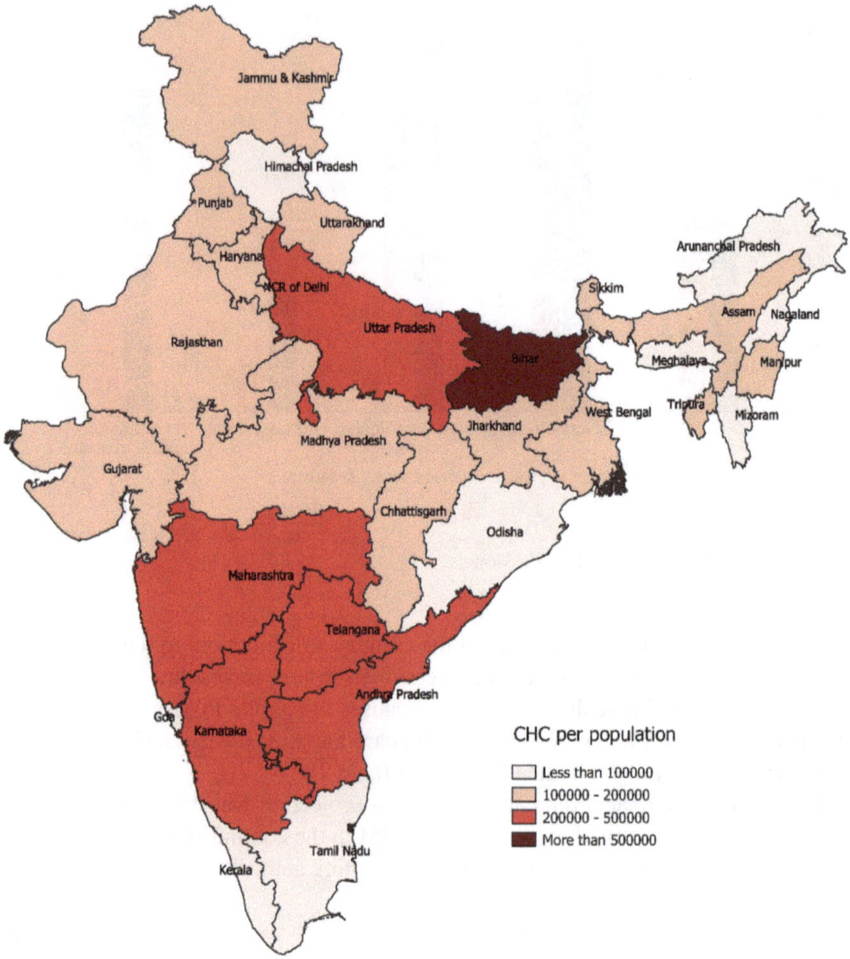

Fig. 9 Map—distribution of CHC for the population (*Source* National Health Statistics 2021–2022, MoHFW)

Table 1 Health expenditure status of countries

Country	Health expenditure as a percentage of budget	Health expenditure as a percentage of GDP
South Africa	14.02	4.33
Brazil	6.93	4.66
Russia	8.42	3.15
Bangladesh	7.84	1.31
India	4.55	1.28
Indonesia	6.63	1.2
China	12.63	3.11

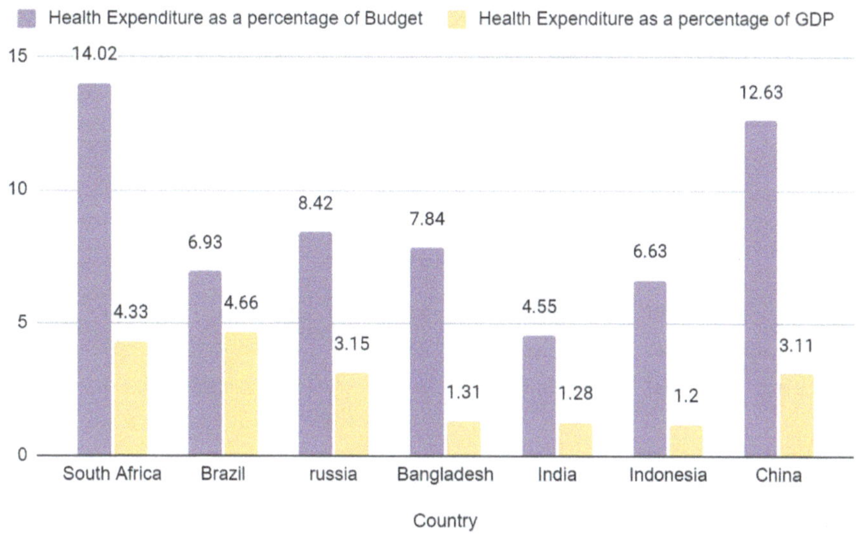

Fig. 10 Health expenditure of different countries

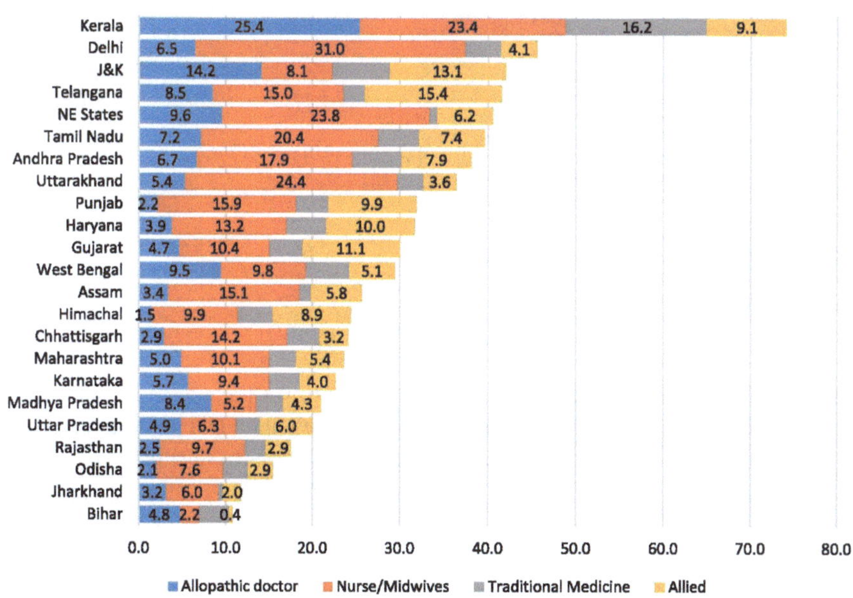

Fig. 11 NHM facility distribution in each state

Table 2 Critical shortfall in medical staff

CHC without specialist doctor		PHC without doctor		PHC without ANM	
State	% of CHC without specialist	State	% of PHC without doctor	State	% of PHC without ANM
Sikkim	100.00	Sikkim	43.00	Sikkim	72.10
Madhya Pradesh	95.40	Madhya Pradesh	37.70	Madhya Pradesh	48.00
Bihar	87.30	Bihar	34.30	Bihar	44.40
Chhattisgarh	82.80	Chhattisgarh	32.80	Chhattisgarh	33.80
West Bengal	79.60	West Bengal	30.70	West Bengal	32.00

Table 3 Forecasted health workforce shortfall/deficiency

Year/ forecast point	Population in billion (India)	Doctors (in million)	AYUSH (in million)	Nurses (in million)	Projected skilled health workforce (in million)	Skilled health workforce needed to reach 25/ 10,000 (in million)	Gap (in million)
2019/ baseline*	1.369	0.65	0.32	0.80	1.77	**3.42**	**1.65**
2025/ forecast mid-point	1.452	0.76	0.42	1.04	2.23	**3.62**	**1.40**
2030/ forecast end-point	1.513	0.93	0.50	1.22	2.65	**3.78**	**1.13**

fighting the pandemic strike better compared to the urban regions and as the result it is identified that there are multiple factors which are public and political in nature which enabled rural India to stand taller during the pandemic era (Fig. 14).

5 The Way Forward

Considering all the criteria mentioned in this paper, human resource is the most important factor that played the vital role in handling the pandemic situation in rural regions which urban India is less considered with. People as the key component of IOT in handling the situation had enabled the rural regions to stay alive and rise above the disastrous hit of COVID-19. At the same time the urban regions were more reliant upon the digital network for everything as the human social network is

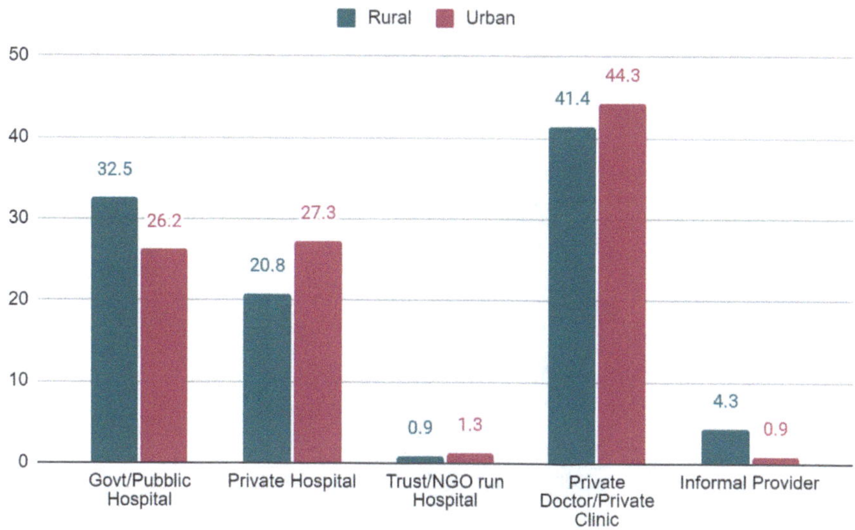

Fig. 12 Distribution of health facilities—public v/s private

Table 4 Distribution of health facilities—public v/s private

	Rural	Urban
Govt/Public Hospital	32.5	26.2
Private Hospital	20.8	27.3
Trust/NGO run Hospital	0.9	1.3
Private Doctor/Private Clinic	41.4	44.3
Informal Provider	4.3	0.9

weaker and not revived. This is the most important fact that the urban regions should learn from the local network that how a physical network of people performed way better than the virtual network of ICT. The human relation values and empathy for the people around was the driving forces for the rural population to stand with the NHM and LSG in dealing with the pandemic spread. As the affluence level got higher the urban had chosen to be under the care of the private health care sector which made the recording and mapping of the pandemic data difficult and thus the handling.

While comparing the strength and strategies that enabled the states like Kerala to fight the pandemic and contain it securely, the study had inferred the following which will help other states and urban regions in future to be prepared for any kind of medical emergencies.

1. Governance and Policies

    a. The pivotal player in the health sector has to be the governing authority which connects the public health system to the people.

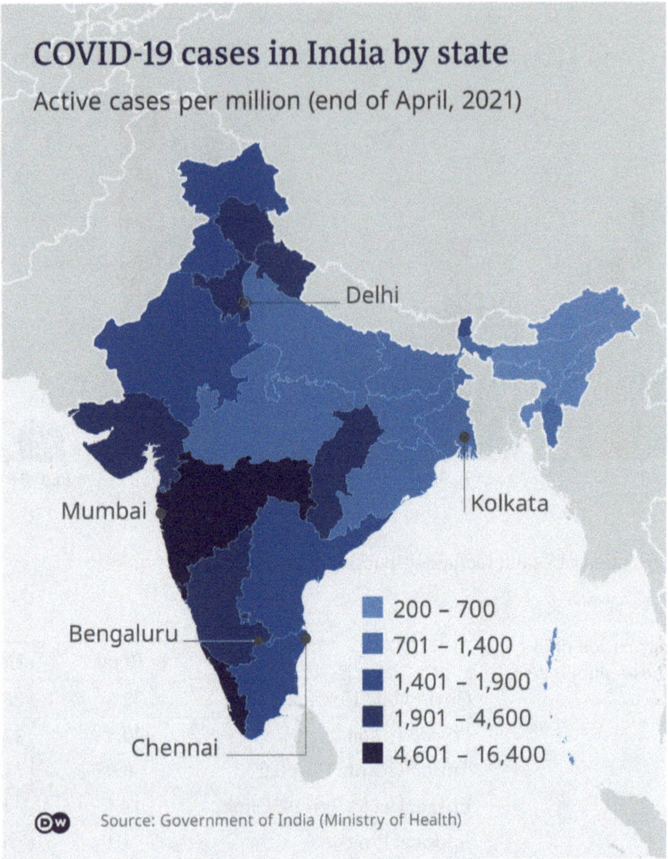

Fig. 13 COVID fatality rate in Indian States—As on April 2021 (*Source* GOI Ministry of Health)

   b. Public health facilities should be equipped with required amount of medical personnel to serve the communities.
   c. Both private and public health systems should be brought under a common governance.to make them accessible for the common.
   d. Immediate actions to be made in addressing the shortfall of public healthcare facilities in urban regions based on the demographic distribution.

2. Local and Cultural aspects: in this era of rapid urbanization, the people who living in or migrating to the urban regions are tend to lead an isolated life irrespective of the social and communal context but the control the state of Kerala had put over the pandemic showed the entire world in holding the social and community values which helped a lot more than the systemic involvement of healthcare and government.

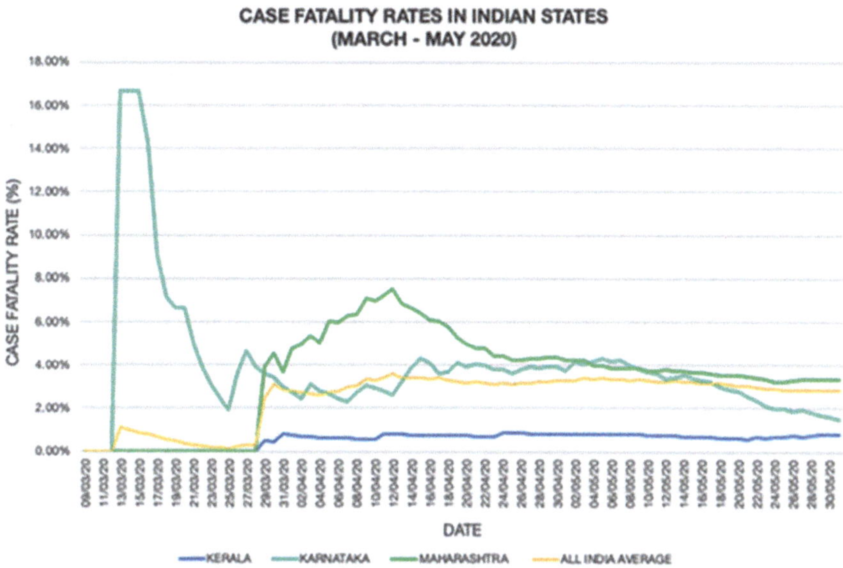

Fig. 14 COVID fatality rate in Indian States (*Source* GOI Ministry of Health)

a. The coordination of the sub centers, anganwadis and ASHA workers with the local community around should be strong and kept in good terms.
b. The human resource should be considered as the prime component of the network rather than ICT where the emotional quotient is one factor that made the people to voluntarily participate in the activities fighting the pandemic.
c. The concern about the neighbors and native people eases isolation, quarantine and containment activities.
d. Social activity groups and other cooperative groups created a platform for the youth to gather and act at the needs of the people during the lockdown period.

The Pandemic era was utilized by the public health sector to build their wealth by overpricing the needy but somehow the government was able to put up standard rates and procedures to safeguard the people from the higher expenditure levels. Still the private health sector is getting bigger as it seems to be more acceptable for the urban population. The public health facilities have to be upgraded to meet a minimum quality and standard for the urban population to accept and use them. It is high time the government should act on this otherwise India will not survive a massive pandemic strike like the COVID-19 outbreak. It is not possible for the governing authority to provide additional doctors day by day as the medical emergency gets worse. To avoid a situation where there is no doctor available for being in charge of a public health facility the government has to post eligible doctors and supporting staff and even constitute a team consisting of NHM officials, Anganwadi teachers and ASHA workers. As the old proverb says "*Prevention is Better than Cure*".

References

1. Chathukulam, JTJ (2021) The Kerala model in the time of COVID19: rethinking state, society and democracy. World Dev 105207
2. Covid-19 Jagratha Government of Kerala (2023, April 12). https://covid19jagratha.kerala.nic.in/
3. Heera R, Rajeev R (2020) Curbing COVID-19: Kerala model. J Oral Maxillofac Pathol 24:222–226
4. Israelsen S, Malji A (2021) COVID-19 in India: a comparative analysis of the Kerala and Gujarat Development Models' initial responses. Prog Dev Stud 21:397–418
5. Karan AN (2020) Size, composition and distribution of health workforce in India: why, and where to invest? Human Resources for Health
6. Menon JC, Rakesh PS, John D, Thachathodiyl R, Banerjee A (2020) What was right about Kerala's response to the COVID-19 pandemic? BMJ Glob Health 5:e003212
7. Ministry of Health and Family Welfare (2022, November 26). https://www.mohfw.gov.in

Arun Baby M. Wilson who is currently working as an Assistant Professor in School of Architecture, Christ University, Bangalore in the state of Karnataka in India. He has completed his Bachelor of Architecture and Masters in Urban and Regional Planning from the prestigious National Institute of Technology, Calicut in India and has experience in project scaling from small-scale residential projects to large-scale Planning projects. He had presented research works on various international platforms and published research papers in the field of Urban, Transport and environmental planning domains. The major areas of interest in research are; Geospatial Analysis for Advocacy Planning, Simulation and Generative AI for decision-making, Materiality and Building services in Architecture, Community planning and Social Development, Transportation Planning and Environmental Sustainability.

Dr. Jayprakash Chadchan who is currently working as a Professor in School of Architecture, Christ University, Bangalore in the state of Karnataka in India. He has completed his Bachelor of Architecture from Karnataka University Dharwar, Karnataka state, India, Masters in Urban and Regional Planning from Institute of Environmental Design, Sardar Patel University, Vallabh Vidyanagar, Gujarat and awarded with PhD from Department of Architecture and Planning, Indian Institute of Technology Roorkee (IIT), India. He is experienced in Urban planning projects and involved in research programs. He had presented research works on various international platforms and published research papers in the field of Urban, Transport and environmental planning domains. The major areas of interest in research are; Architecture, Sustainable Urban Development, Urban Design, Urban Affairs and Policy Studies, Urban Environmental Management, Climate Change Studies, Resilient Cities, Urban Management, Smart City Governance, Urban Transport Planning and Management, Urban Infrastructure Planning and Management, Urban Growth Management, Urban Land Use Planning, Urban Planning and Public Health, Urban Housing and Real Estate, Application of Contemporary Urban Growth, Theories with special focus on (Smart Growth, Smart Cities, New Urbanism, Compact City, Transit-Oriented Development).

Older Adults' Experiences of Social Exclusion in Pakistan: A Qualitative Study

Muhammad Ahmad Al-Rashid, Tiziana Campisi, Muhammad Nadeem, and Bilal Saghir

Abstract Social inclusion is a crucial enabler for older adults' well-being. It has been widely claimed that older adults have been mainly susceptible to social exclusion experiences. Notably, the social exclusion issue has been more profound during the covid-19 pandemic. Hence, the current study investigates older adults' experiences of social exclusion within the less-discussed context of Pakistan. In-depth semi-structured interviews were carried out within the selected Union Councils of Lahore metropolitan Pakistan, including 11 older adults. Utilizing NVivo 12, the thematic analysis technique was mainly adopted to analyze the data. The analysis identified three main themes: restricted access to essential resources, lack of participation, and poor social relations and resources. The themes were interrelated, necessitating engagement across similar lines of professional responsibility and research that looked at many dimensions of social exclusion. This research contributes to understanding older adults' social exclusion experiences, including new outcomes on the significance of essential resources and active engagement and recommendations for overcoming exclusion. Adequate access and active participation in the

M. A. Al-Rashid (✉)
Department of Urban and Regional Planning, Faculty of Built Environment, Universiti Malaya, 50603 Kuala Lumpur, Malaysia
e-mail: ahmad.uetian@hotmail.com

T. Campisi
Department of Engineering and Architecture, University of Enna Kore, Cittadella Universitaria, 94100 Enna, Italy

M. Nadeem
Graduate School of Urban Innovation, Yokohama National University, Yokohama 240-8501, Japan

B. Saghir
Architecture and City Design Department, King Fahd University of Petroleum and Minerals, Dhahran 31261, Saudi Arabia

K. M. Nahiduzzaman (ed.), *Making Sense of Planning and Development for the Post-Pandemic Cities*, City Development: Issues and Best Practices,
https://doi.org/10.1007/978-981-97-5481-6_11

necessary socio-political resources are proposed to confront social exclusion challenges among older adults. The concerned policymakers and social planners could use these findings to develop socially inclusive ageing communities.

Keywords Social exclusion · Resources · Participation · Older adults

1 Introduction

The COVID-19 virus was initially discovered in Wuhan, China, in December 2019 [1] before quickly spreading over the rest of the country and the world. In response, huge measures were taken to protect the public from verified and reported cases. Notably, the lockdown strategy was frequently used and has since been established to control infection successfully [2]. While these initiatives aimed to prevent the spread of disease, they have also contributed to increased social exclusion and discrimination as the number of recorded cases keeps rising. Hence, social exclusion is notable among marginalized groups, such as older adults, which must be addressed [3].

There are two concurrent shifts in social policy regarding older adults living in society. Firstly, the goal is for individuals to 'age in place,' that is, to stay in their familiar residence and neighbourhood even as their vulnerability increases [4]. This perspective is built on the premise that older adults can stay more active and socially integrated by getting old in familiar environments [5] and that neighbourhoods play a crucial role in the well-being and feeling of belongingness of older adults [6, 7]. Age-friendly neighbourhoods that facilitate social inclusion are also essential for older adults to age in place [8]. However, changing sociocultural environments and the widespread COVID pandemic could make community cohesion and sense of belonging complicated, thus increasing the possibility of social exclusion among older adults [7, 9].

Secondly, social exclusion is a matter of concern. Preventing social exclusion is a top priority for the European Union (EU) and international policy [10, 11]. The United Nations and the World Health Organization have equally endorsed the approach [12, 11]. Moreover, numerous studies [13, 14, 15] have indicated the negative consequences of social exclusion in later life, notably poor health, lack of well-being, as well as a higher mortality risk. Given its considerable prominence, there is a need to analyze the nature and level of social exclusion among older adults, particularly in less discussed contexts, to encourage a healthy aging and socially sustainable communities during the post-pandemic time.

The following section briefly overviews the existing literature while identifying the importance and need for social exclusion research. Section 3 details the study settings, design, participants, research technique, and data analysis methods. Section 4 explains the results, while these findings are critically discussed in Sect. 5. Finally, the later sections offer practical implications, research limitations, and conclusions.

2 Literature Review

2.1 *What Is Social Exclusion?*

The term social exclusion was first introduced by [16]. Social exclusion is considered a process rather than a constant state, as well as prescriptive and subjective. The Council of European Union [17] defined social exclusion as a "process whereby certain individuals are pushed to the edge of society and prevented from participating fully." Social exclusion started with acknowledging non-poverty concerns, such as rising job instability, deteriorating family relationships, and the inadequacy of current institutional structures to address them [18]. It was considered that improving their financial status alone would not be enough to keep them up to par with others. The exclusion was increasingly seen as a failure of the social welfare system. Apart from the conventional financial approaches of assisting people experiencing poverty, people sought the 'social integration' of the excluded. Over time, the idea has become increasingly important in political discussions, official documents, social policy, and research agendas in many countries worldwide [19].

Social exclusion has increasingly been recognized as a societal issue deeply connected to a lack of resources, resulting in physical and personal prejudice [20]. It has several interpretations worldwide, which is understandable given the broadness and diversity of this concept. Social capital, material resources, citizen engagement, and access to public services are prominent themes in social exclusion research among older adults [13, 21, 22]. Another important trait is the multidimensionality of social exclusion, which means that people might be restricted from different aspects of their lives [23]. Social exclusion can exist in one or concurrently in either of the (a) social; (b) economic; (c) political, and (d) cultural arenas [24, 20]. It is recognized that the exclusion in one domain can contribute to exclusion in another, either at the same time or across the lifespan, and this accumulated disadvantage could result in a challenging recurrence [25].

2.2 *Existing Research on Older Adults' Social Exclusion*

Old age is one such feature that may put older individuals at risk of exclusion due to functional decline, with mental and physical capacities viewed as major mobility concerns [26]. In addition, other age-related challenges, including a higher risk of illness, physical constraints [27], lack of independence [28], loss of a spouse, relatives, and friends [29], or age-related prejudice and ageism [30], further exaggerates such exclusion process. Similarly, Barnes et al. [31] found that the oldest-old, those living alone and having no children, who have poor health and lack private and public transportation, are more likely to be excluded in different dimensions of social inclusion.

Old age is associated with more significant functional limitations, loneliness, and social exclusion. A decent standard of living requires cohesion and social inclusion [32]. However, research has indicated that social isolation hurts older adults' mental and physical health [31]. Furthermore, an ageing population with a high level of exclusion will also have considerable socio-economic consequences [33]. For instance Nikitas et al. [34] suggest that mobility issues can make it difficult for older adults to participate in most social engagement, resulting in poor confidence, anxiety, and isolation. Furthermore, it could also negatively affect caretakers, social care, healthcare providers, and possibly friends and relatives of older adults [35].

Abundant research has examined older adults' social exclusion following quantitative and qualitative approaches. The available quantitative research includes correlational studies of transport disadvantage and social exclusion [36], gender studies of older adults that highlighted gender and temporal differences in their social exclusion levels [37], prevalence and assessment of social exclusion [38], trajectories and predictors of social exclusion [39]. Some studies, e.g. [40], utilized the GIS techniques and identified areas and spatial differences in social exclusion. In addition, numerous studies have focused on how social exclusion impacts health [41], obesity [42], well-being [13, 43], and life satisfaction [44]. Moreover, Rahman and Sing [45] investigated how older adults' disabilities could promote social exclusion. McCausland et al. [46] recently revealed how the COVID-19 pandemic negatively affects older adults' social inclusion.

Likewise, numerous qualitative research studies have investigated the social exclusion phenomenon. For instance, Fields et al. [36] analyzed the transport disadvantage impact on social exclusion. Dahlberg [47] explored older adults' encounters with neighbourhood exclusion within the context of England. Ekoh et al. [48] discovered the exclusion of older rural women in southeast Nigeria and its impact on their life satisfaction. Mikulioniene and Rapolienė [49] provided essential insights into how older adults experience limited social participation in an Eastern European country. Burns et al. [50] considered the unique experiences of homeless older adults and identified that these participants possess identity and institutional exclusion due to discrimination. Zhang and Yamamoto [51] reveal how cultural attitudes and social norms influence mobility-related social exclusion among Japanese older adults. Restrepo Pineda et al. [52] revealed the social stigmatization experience which ultimately influences the older adults' vulnerability and risk of social exclusion. Barbosa Neves et al. [53] tried to understand the meanings that older adults ascribe to social isolation in the Australian context. Similarly, a recent study [54] explored the challenges faced by homeless older rural–urban migrants and identified social exclusion as a significant challenge.

Existing research He et al. [55] also recognize that the more aged people are more likely to be prone to social inequality, comparatively restricted participation in daily activities, and worsening physical conditions in urban environments. It implies that the increased physical vulnerability and peers' dependence in later life could prompt exclusion among older adults. Similarly, there are differences between men's and women's associated risks of social exclusion [56]. Thus, gender profoundly impacts how older adults experience their life. In addition to gender, the lower economic status of older adults significantly puts them at a more significant disadvantage and more prone to social exclusion [57]. Moreover, older adults' family type and relationship with their children are among the critical factors for social exclusion among older adults [58, 59, 60].

The above review clearly states that numerous factors influence social exclusion among older adults. However, most research on this topic has been conducted in developed countries. Even though the necessity to analyze social exclusion from the experiences of older Asian adults has been highlighted [61, 22], there is limited evidence from the South Asian context. Moreover, few notable research studies [62, 6, 3, 63] have adopted quantitative methods to analyze social exclusion in the Pakistani context. However, to our information, there have been no studies to date that have investigated the experiences of socially excluded Asian older adults using qualitative data. Hence, it is crucial to reflect the aspirations of socially excluded older adults from the Asian context, customize policies to their needs and avoid frustrations of late-life challenges. Thus, the study aims to address the research question, i.e., what is the nature and level of social exclusion among older adults in Pakistan? In this regard, the present research details outcomes on older adults' experiences of their social exclusion based on in-depth semi-structured interviews with older adults living in the selected union councils of Lahore metropolitan Pakistan.

3 Methods

3.1 Study Setting

The present research was undertaken in Lahore metropolitan Pakistan. It is a city with 11 million inhabitants, with 6% being 60 or older [64]. This percentage is about the same as that of the rest of Pakistan. The Lahore metropolitan incorporates several zones their respective municipal corporations govern [65]. The union council is recognized as the smallest administrative unit per the local government system and can be referred to as the neighbourhood in the international context. Geographically, the metropolitan is located in the northeast part of the Punjab province. It is the hub of many industrial and commercial activities and attracts many migrants from various parts of Punjab and surrounding districts [66].

3.2 Study Design, Participants, and Technique

Moustakas [67] suggested that open-ended remarks and questions be utilized to obtain information on the topic and issue using the lengthy interview approach. This study used one-on-one in-depth semi-structured interviews among many qualitative data collection techniques. Numerous researchers have recognized the significance of using semi-structured interviews in qualitative studies [68, 69]. These studies have revealed that semi-structured interviews can help understand the motivations behind public perceptions, beliefs, and judgments. Furthermore, the versatility of a semi-structured interviewing technique allows for a reciprocal investigative process in which the investigator and respondent cooperate to discover meanings [70].

Given the geographical scope of the Lahore metropolitan, it was challenging to cover the whole study area. Hence, this study used a multi-stage sampling approach to collect the data to maintain the probability. Multi-stage sampling involves obtaining a sample from a population by splitting a study population into smaller and smaller groups and then performing a simple random selection to determine the samples [71]. Hence, two out of nine Municipal Administrative Zones within Lahore were selected. Three Union Councils (UCs) were chosen randomly among these chosen zones. First, we selected 'Gulberg Zone' and 'Data Gunj Buksh Zone'. Then, three UCs within the Gulberg Zone were selected, including Model Town, Garden Town, and Gulberg III. At the same time, Mozang, Shadman, and Anarkali were chosen from Data Gunj Buksh Zone. Therefore, this study eventually covered six UCs in Lahore. Figure 1 shows the location map of the study area.

The target participants were older adults (aged 60 years and above) residing in the selected UCs of the Lahore metropolitan. Given the essential typology of older adults identified in the existing literature, an effort was made to select participants of various

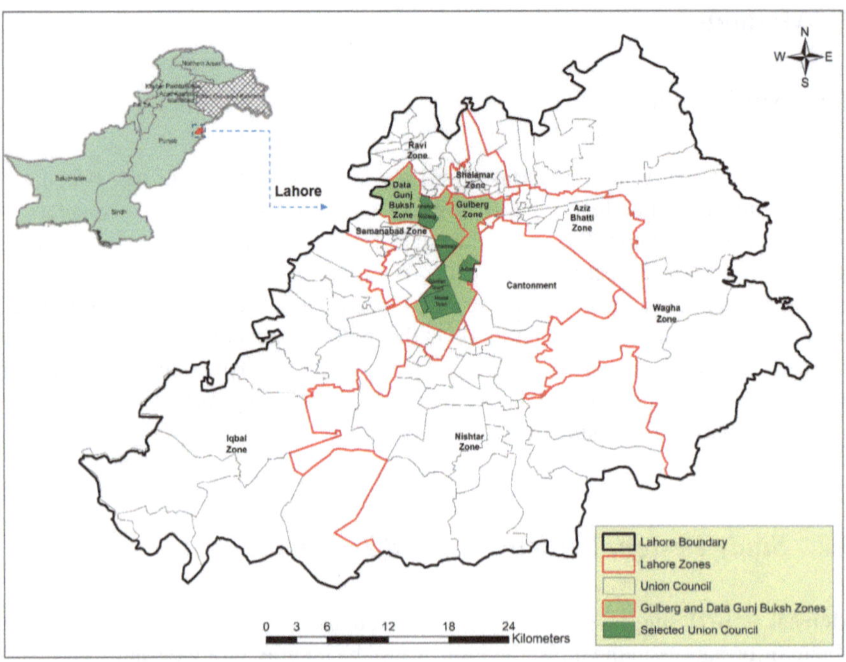

Fig. 1 Location of selected union councils (UCs) in Lahore metropolitan (Adapted from Al-Rashid et al. 72)

aged-groups, gender, socio-economic characteristics, family type and relationships, etc., to investigate the complex social exclusion phenomenon. The interviews were conducted during October-December in 2021, each taking an average of 50 min. We followed a set protocol to portray the experiences of older adults. The interview protocol was pre-tested through recording, and two interviews were initially conducted with older adults in the study area. The protocol was also directed with the help of allied academic scholars and relevant research guidelines. Thus, academic experts examined these interviews, and the research protocol was modified to improve flow.

Numerous researchers have pointed out the justifications for selecting qualitative research samples. For instance Creswell [73, p. 189] states, "*From my review of many qualitative research studies, I have found have found ... phenomenology to typically range from three to ten*". Morse [74] agrees with Creswell [73], indicating that good phenomenological research requires six participants. In line with this, Morse [74] discussed the significance of recognizing data saturation while carrying out qualitative research. Finally, Malterud et al. [75] offered the concept termed 'information power' to describe the discrepancies in applying saturation to diverse qualitative research techniques. Given the significance of information power, we made concerted efforts to consider these characteristics when deciding the data saturation and required sample size. Hence, in total, 11 participants were included in this study.

Before starting the discussion, all invited participants were asked to fill out the informed consent form before starting the discussion. All interviews were conducted face-to-face and in a quiet, distraction-free environment which helped individuals disclose their real-life experiences without worrying about their confidentiality being breached. The recorded responses allowed us to repeat the transcription and further data analysis. Considering the data saturation, the data from eleven interviews were finalized for further analysis. Table 1 summarizes the key characteristics of the participants.

Regarding gender, most informants were females: seven informants were female (63.6%), and four informants were male respondents (36.3%). The participants' ages were evenly distributed (mainly from the 70s to 80s), where the maximum age was 83 and a minimum of 60. Participants from a variety of socio-economic statuses participated in this study. Moreover, there was also an equal representation of older adults from the six selected union councils.

The sample data shown above reflect that the perspectives of a wide range of participants were incorporated into this study. Similarly, we verified the validity and reliability of this research by employing participant validation, emphasizing negative experiences in the data, presenting high-quality quotations as support, and cross-checking derived codes among the research team [76]. Moreover, it is critical to mention that the lead researcher belongs to the case study area and is quite familiar with the local context of older adults. Therefore, he recognized some participants and understood the context before initiating this research.

Table 1 Summary of the participants' profiles

Participant code	Gender	Age	Level of education	Income level	Union Council
P-01	Male	69	Masters	High	Model Town
P-02	Male	62	Illiterate	Low	Anarkali
P-03	Female	67	Illiterate	High	Garden Town
P-04	Male	60	Primary	Low	Shadman
P-05	Female	64	Post-graduation	High	Model Town
P-06	Female	72	Matriculation	High	Gulberg
P-07	Female	83	Bachelors	Low	Mozang
P-08	Male	68	Primary	Middle	Garden Town
P-09	Female	73	Matric	Middle	Shadman
P-10	Female	76	Intermediate	Low	Anarkali
P-11	Female	81	Primary	Middle	Mozang

3.3 Data Analysis

QSR NVivo 12 was employed as the primary instrument to conduct the coding and analysis of qualitative data. All transcripts derived from the recordings were saved as text files and loaded into the NVivo software. Furthermore, the notes taken during the interview sessions were also included in these text documents. These memos served as a reminder of any crucial ideas that came to the researcher's mind during the interview sessions. While examining the transcripts, NVivo enabled the researcher to highlight certain sections or text phrases and assign a code name for the preliminary development of codes. These initial codes served as a foundation for thematic analysis. Moreover, the word cloud analysis was also done using NVivo to reveal the most frequent words used by the participants. Thus, the word cloud and coding approach enabled the investigator to develop themes and sub-themes that might be used for in-depth analysis.

4 Results

This section describes the themes discovered through the face-to-face in-depth semi-structured interviews that depicted the core of the social exclusion behaviour. The interview findings were combined and utilized to explain social exclusion behaviour among older adults. The themes were generated based on the coding and word cloud analysis using NVIVO. Firstly, the overall word cloud of the statements related to social exclusion was developed. The word cloud helped reveal the keywords used by older adults concerning their social exclusion issues. The word cloud further supported the development of the themes and sub-themes for explaining social exclusion.

Fig. 2 Word-Cloud of most-frequently used terms related to social exclusion

Figure 1 shows the overall word cloud of the most frequently used terms related to social exclusion. The top fifty words that older adults frequently use are shown. The selected minimum word length was four, and the words of similar meanings were grouped to enhance the sense and identify the critical words or key concepts.

Figure 2 clearly shows the choice of terms explaining older adults' social exclusion. The focus is on other keywords such as activities, resources, neighbourhood, involve, interact, isolated, relations, etc. The overall cloud analysis shows that older adults' access to essential resources and participation in everyday activities is vital for social inclusion. The word 'public transport' and 'private travel' are also noticeable in the word cloud, suggesting that excellent access and mobility are crucial to their social exclusion phenomenon.

Secondly, the themes and sub-themes were generated from the coding process. With quotations from the interviews supporting the results, each theme and its related sub-themes are debated in the proceeding discussion. The researcher was able to identify a total of seven concepts from the various responses that were collected from the informants. These seven factors were separated into three theme categories: (1) restricted access to essential resources, (2) lack of participation, and (3) poor social relations and resources. Figure 3 visually depicts the themes and sub-themes for social exclusion. The theme related to restricted access to essential resources includes health and shopping/leisure. The lack of participation consists of three sub-themes: socio-economic activities, community or cultural activities, and political activities. At the same time, the poor social relations and resources include two sub-themes: lack of neighbourhood social connectedness and loneliness. All these concepts are related to social exclusion.

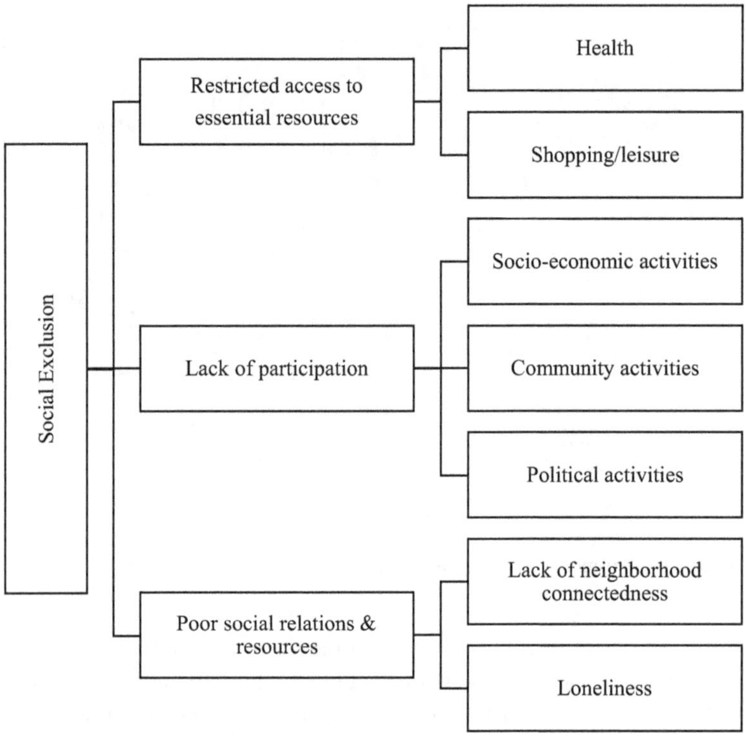

Fig. 3 Themes and sub-themes for explaining the social exclusion

4.1 Restricted Access to Essential Resources

The first theme (i.e., restricted access to essential resources) comprises health and shopping/leisure. Smooth access to health services is necessary for the ageing population. However, the problem is that health-related services are not equally accessible among older adults. The interview results suggested the uneven distribution across the gender and socio-economic level of the participants. In other words, the high-income group participants could easily access health services in their private vehicles (especially cars) without delay. For example, a high-income participant (P-05) indicated the abundant availability of the hospitals in her vicinity, and she usually accesses these facilities in her private cars without any restrictions. She said:

> *"The availability of essential services is within my vicinity. All high-class facilities are available nearby, so I do not need to go outside there... for the medical facility, there are three main hospitals near my home... I use my own transport" (P-05)*

Similarly, another high-income respondent (P-06) revealed the ease of access to health facilities. However, access among lower-income and a few middle-income adults is constrained and time-bound due to the unavailability of private modes. Thus, they rely mainly on their peers, who usually escort them on the Qingqi rickshaw or

Uber. For example, a low-income female participant (P-10) mentioned the availability of health services within and near his residence. However, she is dependent on her brother due to the unavailability of public bus routes. She said:

> *"There are numerous hospitals / private clinics in Mozang, present within the 15-20 mins ride. But there are no specific public bus routes to these health facilities. Thus, I need to wait for my brother to escort me to the hospital on his motorbike" (P-10)*

The second sub-theme in the restricted access to the essential service's theme was shopping or leisure. The nature of access to shopping places was also crucial to the participants when examining their social exclusion behaviour. Leisure and shopping resources could provide prospects for social interaction and improve life quality. Thus, access to these leisure resources is particularly essential for older adults. Like health care accessibility, the participants' feedback provided varying socio-demographic attributes. The accessibility of high-income adults to the shopping areas was far more convenient than the low-income respondents. However, the middle-income participants possessed moderate access to these leisure/shopping areas. For example, the high-income participants (P-05 and P-06) mentioned their residence as adjacent to the central commercial district of Lahore, where all leisure and world-class shopping facilities are available, and all these facilities are easily accessible by their private modes.

> *"….and if want to go for some shopping and leisure, all these things/shops are near to my home. Mostly we go to Liberty." (P-05)*
>
> *"Gulberg III is adjacent to the main commercial hub and business district of Lahore. Thus, all shopping facilities are available within the vicinity…and usually prefer own car to access these services." (P-06)*

Some middle-income participants preferred online taxi services to avoid these issues to access the shopping facilities. For example, P-09 said: "… *and prefer to use online taxis such as Uber and Cream. These are convenient and easy to go for shopping and other leisure activities. It avoids fatigue and public transport hustle.*" In contrast, a low-income male (P-04) mentioned the barriers to accessing the shopping facilities. According to him, there is a need to take multiple transport modes, including metro bus and Qingqi. He said: "… *and need to take a combination of public transport modes to reach the desired shopping places. Hired rickshaw is not affordable. So, commonly, take the metro bus and Qingqi. However, there are numerous accessibility issues with these modes….*"

4.2 Lack of Participation

There was an agreement that the lack of participation has negative consequences among older adults, as discussed in three sub-themes: (1) socio-economic activities, (2) cultural or community activities, and (3) political activities. Effective participation in socio-economic activities is essential to counter ageing challenges. The interview results revealed that compared to high-income participants, the involvement of low-income older adults in economic activities was relatively challenging, with some financial burdens and responsibilities. For example, a low-income participant (P-04) mentioned the issues of getting daily employment. He said: "*Mainly involved in day labour… and sometimes, I wait for many days to get some new task or work… It also increases my financial burdens.*"

However, a middle-income male participant (P-08) discussed his engagement with the real-estate business. He added: "*I am involved with the real-estate business, and I assist my friend… and contribute to my personal as well as my family income. So, at the end of the month, I have some money to spend on my personal needs.*"

Moreover, a high-income male respondent (P-01) offered his part-time services to a local university that helped him financially. He mentioned: "*… and partly offer my teaching services to a local university here in Lahore. It always feels good to be involved with your profession.*" Notably, most older females (except P-05) did not participate in economic activities regardless of income level. Most female participants did not work after retirement and somewhat assisted their families in homemaking.

Another concept in this theme group was the lack of participation in community or cultural activities. The interview results provided exciting insights into the older adults' community activities. Most female participants believed they hardly participated in social events in their locality. One of the critical reasons mentioned during the interviews was the social pressure and limited time owing to their engagement in helping household work with other family members. For example, the participant (P-10) mentioned her reliance on her brother for everyday activities.

> *"… always rely on my brother for help in any activity. I am not used to it and feel reluctant. I hardly get myself involved in any community or social activity." (P-10)*

Another female participant (P-07) discussed the caretaking engagement of her granddaughters as one of the reasons for limited community participation. She added: "*the full-time engagement with my granddaughters keeps me busy. I love to spend time with them…I only engage outside in any cultural activity while accompanying with other family members, which they do hardly*".

Low-income male older adults also mentioned a lack of involvement in local community activities. However, such activity participation is relatively higher among high-income older adults. Similar results regarding political participation among older adults were reported. The findings from the qualitative interviews informed that political involvement was commonly seen among high-income adults. It was observed that some high-income participants were members of the various management committees in their neighbourhoods despite their busy routines. Thus, they tried to attend meetings and social activities arranged by their supported political party. In this regard, P-01 reported: "*… strongly supported the candidate of Pakistan Muslim League-N political party during the recent elections… and often organize and actively participate in the political events mainly conducted at the Model Town Community centre" (P-01).*

In contrast, no such community groups were in unplanned locations such as Mozang or Anarkali. Moreover, some financial and social issues also affected the participation of low-income participants (such as P-02 and P-04) in political activities. They mentioned:

> *"I remain too involved in daily waging. And if I do not earn, I cannot feed my family. So, there is a lack of time and financial resources… and hardly involved in any political event." (P-04)*
>
> *"I have don't have interest in any political party or their relevant activities. I am only concerned for my daily earning." (P-02)*

4.3 Poor Social Relations and Resources

The last theme in the group of social exclusion factors was poor social relations and resources. This group or theme possesses two sub-themes: the lack of neighbourhood social connectedness and loneliness. Neighbourhood social connectedness is an essential dimension of a healthy ageing society. The interview results found that the connectedness among the neighbours is entirely satisfactory in low-income neighbourhoods. A few low-income participants (such as P-10) mentioned that everyone in their area knows each other. He added: "… *and we know everyone in our street.*" Similarly, another participant (P-07) disclosed that she often visits their neighbouring friends and sits together in the evenings. "… *have good relations with our immediate and nearby fellows… and in the afternoon, I often visit them… Especially during summer, we mutually walk and go to some nearby park or public place in the afternoon" (P-07).*

In contrast, the social connectedness among neighbours in high-income areas (i.e., Gulberg, Garden Town, and Model Town) is very low. For example, the participant (P-06) stated she hardly recognized her neighbours. She said: "*I hardly recognize anyone in my street. People rarely come out to walk; they prefer their private car, and to enjoy the natural environment, they use their indoor lawn.*" Moreover, P-05 reported that she generally visits their relatives rather than their neighbours. She added: *"Actually, almost my whole family reside near to my house and within my vicinity. So, I prefer to visit relatives and don't like to interact with neighbours" (P-05).*

In addition to the lack of neighbourhood social connectedness, loneliness was also revealed as one factor of social exclusion. The interview findings also provided varying income levels, family type, and functioning. Low- and middle-income older adults received less loneliness, especially those living in joint families. The interview participants, for example, P-10, disclosed that the level of care does not make them feel lonely. She mentioned: "… *even though I got divorced during my early life. I do not have any son or daughter with whom I can live in this phase of my life. But the level of care I receive from my brother and his children is priceless. They take of me and assist me in meeting my needs" (P-10).*

Moreover, the joint family of the participant (P-07) kept her involved in home-making activities, thus reducing her feelings of loneliness. She said: "… *and most of my time is spent in assisting and helping my granddaughters in homemaking. They also do care of me and keep myself busy... and likes to assist" (P-07).*

Contrarily, high-income older adults personally feel lonely while staying at home. The prime reason behind this is the neglected attitude of their children, i.e., the busy routines and other priorities of their children and grandchildren. For example, regardless of the availability of the servants, she (P-06) feels lonely in her house. She said: "*...and the busy routine of my husband and son. Their full-time engagement in the catering business keeps them busy, and they spend most of their time outside the home. So, it makes me feel lonely during most time of the day.*" Moreover, the participant (P-03) mentioned the lack of care from her sons due to their unavailability during the day. She disclosed: "*...but my sons remain busy in their routine jobs... engagement of my grandsons with their social activities or their life priorities. So, they spend less time at home, which makes me feel isolated.*"

5 Discussion

It has been widely established that pleasant social and physical environments facilitate older adults to live independently and socially integrated [5]. However, this idealised notion does not always exist, generating negative experiences, particularly during the pandemic. In this study, we have discovered the critical aspects of the socio-physical environments that drive social exclusion behaviour among older adults. Based on the in-depth semi-structured interviews with older adults in the Lahore metropolitan, this study aimed to explore the older adults' social exclusion experiences. Given the qualitative data and the above findings, the following critical insights can be summarized to analyze the level and nature of social exclusion among older adults in Pakistan.

The findings revealed that access to essential resources and participation in everyday activities is crucially linked with transport availability and accessibility. It suggests the relationship between older adults' social exclusion and their level of access to different transport modes, i.e., the less transport access, the more social exclusion older adults face or vice versa. A similar result is also indicated in previous research such as [77, 78, 79], who found that the individual who perceives the transport system as accessible encounters less exclusion in daily socio-economic activities.

In line with the existing literature [36, 80], the interview results further suggested that most high-income older adults had no problem accessing essential health and leisure resources. Moreover, they were actively engaged in daily socio-economic and community activities. This is because the higher-income older adults had a readily available private car and could easily access and participate in their desired activities [81, 82, 83]. In contrast, the unavailability of a personal car and the numerous accessibility issues with the public transport services enhanced the social exclusion problem, particularly among low-income older adults. One potential pathway is to concentrate on the positive aspect where an older adult with good perceived accessibility was likely to report less social exclusion experience [3]. Another pathway is to look at the ease of travel to participate in necessary activities to prevent exclusion. The availability of private or public transport reduces social exclusion among older adults. Therefore, the person who experiences high exclusions and cannot travel would have a reduced possibility of coping with social exclusion due to the unavailability or lack of access to public and private transport modes.

Lack of affection with children or grandchildren was among the sources cited for causing loneliness among older adults. Thus, good family functioning was essential, and low functioning appeared as a cause for loneliness, as depicted in the previous literature [58, 59, 60]. It was revealed, for example, that good family functioning, caring attitudes, and close household ties could make older adults feel respectable and vital for their families, eliminating loneliness. Moreover, family members lifestyles and employment status were also related to loneliness among older adults. It appears that the children's employment status and busy routines might cause the problem of a poor social bond among the family members, especially with older adults. Such peers may likely spend more time outside their homes to perform their economic and other preferred social activities (e.g., hanging out with friends or travelling). Thus, evidence suggests that the relationships between older adults and their families or caretakers can help reduce the social exclusion experience.

In addition to family functioning, family type significantly influences older adults' social exclusion. The interview results revealed that the older adults who lived in joint families reported lesser social exclusion than those from the nucleus families, as reported in existing studies [84, 60]. The findings revealed that older adults from joint families had enough quality time with their family members. Some female participants mentioned their participation in homemaking activities during their free time, which reduced their loneliness. However, the older adults from the nuclear families noted their loneliness experiences when their peers were not home. In line with the issues discussed in this research, the existing literature [85, 86] also mentioned that the centuries-old strong joint and extended family system in Pakistani society is changing rapidly and is being replaced by the nuclear family system. Such nuclear family systems are expected to be comparatively weak economically and politically and exaggerate daily problems for older adults [86]. Thus, this is an evident concern for social scientists and researchers to develop appropriate social sustainability policies for older adults.

Neighbourhood social connectedness is notable in reducing loneliness among Pakistani older adults. It allows them to mingle with them during their free time without depending on any transport modes. The findings further suggested that neighbourhood social connectedness was significant among low-income older adults. It might be due to the difference in the neighbourhood characteristics. There are apparent discrepancies between the high-income and low-income neighbourhoods in the local context of Pakistan, where people in high-income urban settings are mainly concerned with their own lives and hardly know their neighbours. Similarly, the older adults from high-income areas had lower social connectedness because they hardly see and recognize neighbours. This was not the case for low-income zones with greater social cohesion among neighbours. Sampson et al. [87] suggested that the lack of identification with their immediate surroundings and perceptions of safety and trust can substantially influence a community's capacity to develop strong bonds and collective resilience. Thus, the lack of contact and recognition with neighbours among older adults affects social connectedness, especially when they are alone at their residence.

5.1 Practical Implications

There was a widespread consensus among participants that the neighbourhood's social character had weakened significantly over time. Older adults become increasingly reliant on others in their community as they age, leading to an inactive lifestyle. However, neighbourhood cohesion, community trust and integration could promote safety and independent mobility. Thus, to enhance the possibilities for a prolonged and healthy life, one critical focus for policymakers and practitioners is to help neighbourhood organizations and local governments strengthen cohesion among older adults and across generations and recognise vigorous dilemmas of anti-social behaviour in programs advocating neighbourhood regeneration. A recent review provides similar recommendations, suggesting the significance of reciprocity and recognition of personal differences in addressing social needs [88].

The focus must also be on the neighbourhood's political environment and community activities. This is vital concerning both neighbourhood social cohesion and community identity. It is also likely that the activity participation would eventually reduce if resource accessibility issues remained unaddressed. Therefore, this dimension turns out to be noticeable in this research. The findings emphasize how older adults with insufficient income cannot participate in the everyday sociocultural and political arenas. However, this situation is especially imperative for older women regardless of financial status. It demands immediate measures from social scientists, policymakers, and local government to focus on trust-building relationships, analyzing older adults' crucial needs, advancing social character in the neighbourhood, and providing opportunities to participate in essential community activities. Moreover, neighbourhood-level communal facilities need to be expanded, i.e., communal services are intended to be used and shared by individuals of different interests and ages, leading to harmonious relations between other groups. In conclusion, age-friendly cities encourage multigenerational interaction and interpersonal enrichment, which can help older adults overcome social exclusion.

Removing socio-physical barriers, including inadequate lighting, substandard road surfaces, poor social support, etc., impeding people with functional disabilities from getting around would reduce social exclusion. It might be claimed that enhancing older adults' ability to remain connected to society prevents social exclusion. Moreover, it could also minimize the fall risks among older adults and substantially influence their level of activity participation. It has been stated that introducing minor infrastructural improvements during the design stage is significantly less expensive than incorporating changes later and that 'inclusive' environments assist all inhabitants. In addition, appropriate and effective public transport services are essential for older adults when surviving a socially active and independent life. Hence, the built environment and transport accessibility are crucial for activity participation and social inclusion among older adults.

The identified overlapping themes and policy areas indicate that practices and policies must interact across professions and organisations and work across typical responsibilities when confronting social exclusion among older adults. Walsh et al. [89] also advocate that the informal practice, where individuals cooperate beyond the system boundaries of the public, private, cooperative, and family/friend domains, could reinforce societies' capability to overcome social exclusion among older adults.

6 Conclusion

The present research discovered the complexity of the social exclusion phenomenon among older adults in Pakistan and identified its vital aspects: restricted access to essential resources, lack of participation, and poor social relations and resources. These different dimensions of social exclusion were inextricably linked. This necessitates cross-system collaboration and research that includes socio-physical attributes of social exclusion. More specifically than other studies on social exclusion, this research emphasizes the significance of socio-political assistance for older adults' inclusion and social cohesion within the neighbourhood. Resource accessibility and family support are crucial attributes of how older adults might be socially included and integrated within the neighbourhood, which must be carefully considered in potential social exclusion research and practice and policy interventions. Hence, understanding older adults' perspectives and their active participation in programs is critical for formulating successful policies to eliminate social exclusion and facilitate healthy ageing in the post-pandemic era.

References

1. Huang C, Wang Y, Li X, Ren L, Zhao J, Hu Y, Zhang L, Fan G, Xu J, Gu X, Cheng Z, Yu T, Xia J, Wei Y, Wu W, Xie X, Yin W, Li H, Liu M, Xiao Y, Gao H, Guo L, Xie J, Wang G, Jiang R, Gao Z, Jin Q, Wang J, Cao B (2020) Clinical features of patients infected with 2019 novel coronavirus in Wuhan, China. Lancet 395(10223):497–506. https://doi.org/10.1016/S0140-6736(20)30183-5
2. Tian H, Liu Y, Li Y, Wu C-H, Chen B, Kraemer MUG, Li B,Cai J, Xu B, Yang Q, Wang B, Yang P, Cui Y, Song Y, Zheng P, Wang Q, V Bjornstad ON, Yang R, Grenfell BT, Pybus OG, Dye C (2020) An investigation of transmission control measures during the first 50 days of the COVID-19 epidemic in China. Science 368(6491):638–642. https://doi.org/10.1126/science.abb6105

3. Al-Rashid MA, Harumain YAS, Goh HC, Ahmad Z (2021) Psychosocial factors of public transport users and social inclusion implications among older women in Pakistan. J Urban Plan Dev 147(4):04021046. https://doi.org/10.1061/(ASCE)UP.1943-5444.0000727
4. Genet N, Boerma WG, Kringos DS, Bouman A, Francke AL, Fagerström C, Melchiorre MG, Greco C, Devillé W (2011) Home care in Europe: a systematic literature review. BMC Health Serv Res 11(1):207. https://doi.org/10.1186/1472-6963-11-207
5. Wiles JL, Leibing A, Guberman N, Reeve J, Allen RES (2011) The meaning of "aging in place" to older people. Gerontologist 52(3):357–366. https://doi.org/10.1093/geront/gnr098
6. Al-Rashid MA, Goh HC, Harumain YAS, Ali Z, Campisi T, Mahmood T (2021) Psychosocial barriers of public transport use and social exclusion among older adults: empirical evidence from Lahore, Pakistan. Int J Environ Res Public Health 18(1):185. https://doi.org/10.3390/ijerph18010185
7. Lager D, Van Hoven B, Huigen PPP (2013) Dealing with change in old age: negotiating working-class belonging in a neighbourhood in the process of urban renewal in the Netherlands. Geoforum 50:54–61. https://doi.org/10.1016/j.geoforum.2013.07.012
8. World Health Organization (2007) Global age-friendly cities: a guide. World Health Organization
9. Seifert A, Cotten SR, Xie B (2020) A double burden of exclusion? Digital and social exclusion of older adults in times of COVID-19. J Gerontol Ser B 76(3):e99–e103. https://doi.org/10.1093/geronb/gbaa098
10. Commission (2017) European pillar of social rights. https://ec.europa.eu/commission/sites/beta-political/files/social-summit-european-pillarsocial-rights-booklet_en.pdf
11. Popay J, Escorel S, Hernández M, Johnston H, Mathieson J, Rispel L (2008) Social exclusion. Final report to the WHO commission on social determinants of health from the social exclusion knowledge network
12. Atkinson AB, Marlier E (2010) Analysing and measuring social inclusion in a global context. United Nations Publications
13. Dahlberg L, McKee KJ (2018) Social exclusion and well-being among older adults in rural and urban areas. Arch Gerontol Geriatr 79:176–184. https://doi.org/10.1016/j.archger.2018.08.007
14. Sacker A, Ross A, MacLeod CA, Netuveli G, Windle G (2017) Health and social exclusion in older age: evidence from understanding society, the UK household longitudinal study. J Epidemiol Community Health 71(7):681–690
15. Yur'yev A, Värnik P, Sisask M, Leppik L, Lumiste K, Värnik A (2013) Some aspects of social exclusion: do they influence suicide mortality? Int J Soc Psychiatry 59(3):232–238. https://doi.org/10.1177/0020764011431792
16. Lenoir R (1989) Les exclus: un français sur dix. FeniXX
17. European Union (2004) Joint report by the commission and the council on social inclusion. In: Bruselas
18. Poggi A (2004) Measuring social exclusion. Universitat Autonoma de Barcelona
19. Levitas R (2005) The inclusive society? Social exclusion and new labour. Springer
20. Sen A (2000) Social exclusion: concept, application, and scrutiny
21. Van Regenmortel S, De Donder L, Dury S, Smetcoren A-S, De Witte N, Verté D (2016) Social exclusion in later life: a systematic review of the literature. J Popul Ageing 9(4):315–344. https://doi.org/10.1007/s12062-016-9145-3
22. Walsh K, Scharf T, Keating N (2017) Social exclusion of older persons: a scoping review and conceptual framework. Eur J Ageing 14(1):81–98. https://doi.org/10.1007/s10433-016-0398-8
23. Burchardt T, Le Grand J, Piachaud D, Hills J, Grand L (2002) Understanding social exclusion. Oxford University Press, Oxford
24. Levitas R, Pantazis C, Fahmy E, Gordon D, Lloyd-Reichling E, Patsios D (2007) The multi-dimensional analysis of social exclusion
25. Mayes DG, Berghman J, Salais R (2001) Social exclusion and European policy. Edward Elgar Pub
26. Lucas K (2006) Providing transport for social inclusion within a framework for environmental justice in the UK. Transp Res Part A Policy Pract 40(10):801–809. https://doi.org/10.1016/j.tra.2005.12.005

27. Jokela M, Batty G, Kivimäki M (2013) Ageing and the prevalence and treatment of mental health problems. Psychol Med 43(10):2037–2045. https://doi.org/10.1017/S0033291712003042
28. Kneale D (2012) Is social exclusion still important for older people?
29. Rook KS (2009) Gaps in social support resources in later life: an adaptational challenge in need of further research. J Soc Pers Relat 26(1):103–112. https://doi.org/10.1177/0265407509105525
30. Allan LJ, Johnson JA, Emerson SD (2014) The role of individual difference variables in ageism. Personality Individ Differ 59:32–37. https://doi.org/10.1016/j.paid.2013.10.027
31. Barnes M, Blom AG, Cox K, Lessof C, Walker A (2006) The social exclusion of older people: evidence from the first wave of the English Longitudinal Study of Ageing (ELSA), Final report
32. Victor C, Scambler S, Bond J, Bowling A (2000) Being alone in later life: loneliness, social isolation and living alone. Rev Clin Gerontol 10(4):407–417. https://doi.org/10.1017/S0959259800104101
33. Balamurugan J (2014) Social exclusion of older people: a case study of rural South India. J Exclusion Stud 4(2):130–145. https://doi.org/10.5958/2231-4555.2014.00003.5
34. Nikitas A, Avineri E, Parkhurst G (2018) Understanding the public acceptability of road pricing and the roles of older age, social norms, pro-social values and trust for urban policy-making: the case of Bristol. Cities 79:78–91. https://doi.org/10.1016/j.cities.2018.02.024
35. DfT (2015) Road traffic statistics. Table TRA0302: motor vehicle flow by road class and region and country in Great Britain
36. Fields N, Cronley C, Mattingly SP, Murphy ER, Miller VJ (2019) "You are really at their mercy": examining the relationship between transportation disadvantage and social exclusion among older adults through the use of innovative technology. Transp Res Rec 2673(7):12–24. https://doi.org/10.1177/0361198119839969
37. Dahlberg L, McKee KJ, Fritzell J, Heap J, Lennartsson C (2020) Trends and gender associations in social exclusion in older adults in Sweden over two decades. Arch Gerontol Geriatr 89:104032. https://doi.org/10.1016/j.archger.2020.104032
38. Van Regenmortel S, De Donder L, Smetcoren A-S, Lambotte D, De Witte N, Verté D (2018) Accumulation of disadvantages: prevalence and categories of old-age social exclusion in Belgium. Soc Indic Res 140(3):1173–1194. https://doi.org/10.1007/s11205-017-1817-8
39. Chung S, Jeon H, Song A, Kim J-H (2019) Developmental trajectories and predictors of social exclusion among older Koreans: exploring the multidimensional nature of social exclusion. Soc Indic Res 144(1):97–112. https://doi.org/10.1007/s11205-018-2045-6
40. Chen Y, Edossa A, Morrow-Howell N (2015) Social exclusion among older adults: a spatial analysis using geographic information system (GIS). Gerontologist 55(Suppl_2):93–93. https://doi.org/10.1093/geront/gnv499.02
41. Tong H, Lai DWL (2016) Social exclusion and health among older Chinese in Shanghai, China. Asia Pacific Journal of Social Work and Development 26(2–3):120–141. https://doi.org/10.1080/02185385.2016.1219272
42. Hajek A, König H-H (2018) The association between obesity and social exclusion in middle-aged and older adults: findings from a nationally representative study in Germany. BMC Geriatr 18(1):258. https://doi.org/10.1186/s12877-018-0946-5
43. Lee S (2020) Social exclusion and subjective well-being among older adults in Europe: findings from the European social survey. J Gerontol Series B 76(2):425–434. https://doi.org/10.1093/geronb/gbaa172
44. Lee J, Cagle JG (2018) Social exclusion factors influencing life satisfaction among older adults. J Poverty Soc Justice 26(1):35–50. https://doi.org/10.1332/175982717X15127351091521
45. Rahman MHU, Singh A (2019) Disability and social cohesion among older adults: a multi-country study. Int J Soc Econ 46(4):485–502. https://doi.org/10.1108/IJSE-05-2018-0230
46. McCausland D, Luus R, McCallion P, Murphy E, McCarron M. (2021) The impact of COVID-19 on the social inclusion of older adults with an intellectual disability during the first wave of the pandemic in Ireland. J Intellect Disabil Res 65(10):879–889. https://doi.org/10.1111/jir.12862

47. Dahlberg L (2020) Ageing in a changing place: a qualitative study of neighbourhood exclusion. Ageing Soc 40(10):2238–2256. https://doi.org/10.1017/S0144686X1900045X
48. Ekoh PC, Chukwu NE, Okoye UO (2021) Aging in rural Nigeria: gendered exclusion of rural older adults and its impact on their perceived life satisfaction in South-East Nigeria. J Women Aging 35:1–13. https://doi.org/10.1080/08952841.2021.1999734
49. Mikulioniene S, Rapolienė G (2020) Perceived incentives and barriers to social participation: the case of older adults living alone in Lithuania. Intersections 6(2):176–194. https://doi.org/10.17356/ieejsp.v6i2.626
50. Burns VF, Leduc JD, St-Denis N, Walsh CA (2020) Finding home after homelessness: older men's experiences in single-site permanent supportive housing. Hous Stud 35(2):290–309. https://doi.org/10.1080/02673037.2019.1598550
51. Zhang J, Yamamoto FJ (2017) The kindness of strangers: exploring interdependencies and shared mobilities of elderly people in rural Japan. Social inclusion 5(4):183–195
52. Restrepo Pineda JE, López Lorduy AC, Arismendy Mejía A (2020) Approach to the process of social stigmatization of LGBT seniors in Antioquia, Colombia. Papeles de población 26(105):219–251. https://doi.org/10.22185/24487147.2020.105.27
53. Barbosa Neves B, Sanders A, Kokanović R (2019) "It's the worst bloody feeling in the world": experiences of loneliness and social isolation among older people living in care homes. J Aging Stud 49:74–84. https://doi.org/10.1016/j.jaging.2019.100785
54. Gebeyaw G, Kotecho MG, Adamek ME (2021) "Homelessness and health problems are not distinct:" the challenges of rural-urban migrant homeless older people in Ethiopia. J Cross Cult Gerontol 36(4):347–368. https://doi.org/10.1007/s10823-021-09443-x
55. He SY, Cheung YH, Tao S (2018) Travel mobility and social participation among older people in a transit metropolis: a socio-spatial-temporal perspective. Transp Res Part A Policy Pract 118:608–626
56. Women's Health East (2019) The unheard story: the impact of gender on social inclusion for older women
57. Burchardt T, Le Grand J, Piachaud D (1999) Social exclusion in Britain 1991–1995. Soc Policy Adm 33(3):227–244. https://doi.org/10.1111/1467-9515.00148
58. Burholt V, Scharf T (2014) Poor health and loneliness in later life: the role of depressive symptoms, social resources, and rural environments. J Gerontol B Psychol Sci Soc Sci 69(2):311–324. https://doi.org/10.1093/geronb/gbt121
59. Essex MJ, Nam S (1987) Marital status and loneliness among older women: the differential importance of close family and friends. J Marriage Fam 49:93–106. https://doi.org/10.2307/352674
60. Yunong H (2012) Family relations and life satisfaction of older people: a comparative study between two different hukous in China. Ageing Soc 32(1):19
61. Jiang D, Fung HH, Lay JC, Ashe MC, Graf P, Hoppmann CA (2019) Everyday solitude, affective experiences, and well-being in old age: the role of culture versus immigration. Aging Ment Health 23(9):1095–1104. https://doi.org/10.1080/13607863.2018.1479836
62. Adeel M, Yeh G-O, Zhang F (2016) Transportation disadvantage and activity participation in the cities of Rawalpindi and Islamabad, Pakistan. Transp Policy 47:1–12. https://doi.org/10.1016/j.tranpol.2015.12.001
63. Al-Rashid MA, Shamsul Harumain YA, Goh HC, Ali Z, Nadeem M, Campisi T (2023) Perceived norms of public transport use as the determinants of older adults' social exclusion: evidence from Pakistan. Cities 137:104264. https://doi.org/10.1016/j.cities.2023.104264
64. Pakistan Bureau of Statistics (2017) 6th Population & housing census 2017. https://www.pbs.gov.pk/content/final-results-census-2017
65. Government of the Punjab (2017) UC Metropolitan Corporation Lahore, Lahore
66. Rana IA, Bhatti SS (2018) Lahore, Pakistan—urbanization challenges and opportunities. Cities 72:348–355. https://doi.org/10.1016/j.cities.2017.09.014
67. Moustakas C (1994) Phenomenological research methods. Sage
68. Dearnley C (2005) A reflection on the use of semi-structured interviews. Nurse Res 13(1). https://doi.org/10.7748/nr2005.07.13.1.19.c5997

69. Ruslin R, Mashuri S, Rasak MSA, Alhabsyi F, Syam H (2022) Semi-structured interview: a methodological reflection on the development of a qualitative research instrument in educational studies. IOSR J Res Method Educ (IOSR-JRME) 12(1):22–29
70. Saunders M, Lewis P, Thornhill A (2009) Research methods for business students. Pearson Education
71. Kuno E (1976) Multi-stage sampling for population estimation. Popul Ecol 18(1):39–56. https://doi.org/10.1007/BF02754081
72. Al-Rashid MA., Rao MN, Ahmad Z. (2020) Using GIS measures to analyze the spatial equity to public parks in Lahore metropolitan. J Res Archit Plan 28(8):8–19
73. Creswell JW (2014) A concise introduction to mixed methods research. Sage.
74. Morse JM (1995) The significance of saturation. Sage, Thousand Oaks, CA
75. Malterud K, Siersma VD, Guassora AD (2016) Sample size in qualitative interview studies: guided by information power. Qual Health Res 26(13):1753–1760. https://doi.org/10.1177/1049732315617444
76. Gibbs G (2018) Analyzing qualitative data, 2nd edn. https://doi.org/10.4135/9781526441867
77. Curl A (2018) The importance of understanding perceptions of accessibility when addressing transport equity: a case study in Greater Nottingham, UK. J Transp Land Use 11(1)
78. Hui V, Habib KN (2014) An investigation of transport-related social exclusion of the at-risk community (homeless people) in Toronto, Canada
79. Kenyon S, Lyons G, Rafferty J (2002) Transport and social exclusion: investigating the possibility of promoting inclusion through virtual mobility. J Transp Geogr 10(3):207–219
80. Hsieh M-O, Leung P (2019) Social exclusion experienced by older adults: factors for designing healthcare settings in Taipei, Taiwan. Soc Work Health Care 58(4):368–381. https://doi.org/10.1080/00981389.2019.1575941
81. Al-Rashid MA, Harumain YAS, Goh HC, Campisi T (2023) Psychosocial barriers to public transport use triggers older adults' social exclusion: empirical evidence from Pakistan. J Urban Plan Dev 149(1):04022050. https://doi.org/10.1061/JUPDDM.UPENG-4097
82. Kizony R, Schreuer N, Rotenberg S, Shach-Pinsly D, Sinoff G, Plaut P (2020) Participation in out-of-home activities among older adults: the role of mobility, attitudes and travel behaviors. J Transp Health 17:100846. https://doi.org/10.1016/j.jth.2020.100846
83. Levasseur M, Généreux M, Bruneau J-F, Vanasse A, Chabot É, Beaulac C, Bédard M-M (2015) Importance of proximity to resources, social support, transportation and neighborhood security for mobility and social participation in older adults: results from a scoping study. BMC Public Health 15(1):503. https://doi.org/10.1186/s12889-015-1824-0
84. Itrat A, Taqui AM, Qazi F, Qidwai W (2007) Family systems: perceptions of elderly patients and their attendents presenting at a university hospital in Karachi, Pakistan. J Pakistan Med Assoc 57(2):106
85. Ali SM, Kiani MFK (2003) Ageing and poverty in Pakistan. https://ideas.repec.org/p/pid/mimaps/200318.html
86. Ashiq U, Asad AZ (2017) The rising old age problem in Pakistan. J Res Soc Pakistan 54(2)
87. Sampson RJ, Morenoff JD, Gannon-Rowley T (2002) Assessing "neighborhood effects": social processes and new directions in research. Ann Rev Sociol 28(1):443–478. https://doi.org/10.1146/annurev.soc.28.110601.141114
88. Bruggencate TT, Luijkx KG, Sturm J (2018) Social needs of older people: a systematic literature review. Ageing Soc 38(9):1745–1770. https://doi.org/10.1017/S0144686X17000150
89. Walsh K, O'Shea E, Scharf T, Shucksmith M (2014). Exploring the impact of informal practices on social exclusion and age-friendliness for older people in rural communities. J Community Appl Soc Psychol 24(1):37–49. https://doi.org/10.1002/casp.2176

Muhammad Ahmad Al-Rashid is a PhD researcher at the Department of Urban and Regional Planning, University of Malaya, Malaysia. He is an expert in sustainable mobility and urban planning. He has published more than 25 articles in reputed Web of Science (WoS) and Scopus Indexed journals. His main areas of interest include urban planning, transport planning, social exclusion, smart cities, and active mobility. He has previously worked in think tanks, the corporate sector, the government, and non-profit organizations. He has a track record of leading community and urban development research in various sectors by applying concrete research methodologies and evidence-based interventions. He holds master's and bachelor's degrees in city and regional planning.

Prof. Tiziana Campisi is serving as an Assistant Professor, Department of Engineering and Architecture, University of Enna Kore, Cittadella Universitaria 94100, Enna (En), Italy. She obtained her degree in Civil and Environmental Engineering at the University of Catania and her specialised degree (cum Laude) in Civil Protection Engineering at the University of Enna Kore where she also obtained her PhD degree. She is an expert in road quality management and sustainable mobility. She has published more than 130 manuscripts in international and indexing journals and conferences. She is responsible for the local unit of the project (PRIN2017) titled "WEAKI TRANSIT"(CUP Code: F74I19001290001), supported by MIUR (Ministry of Education, University and Research) and funded by the PRIN 2017 programme. She serves as Editor in Chief for the Open Transportation Journal Bentham and as a member of several Editorial Boards and she is a reviewer for several international journals and conferences. Her research interest lies on transport supply/demand, sustainable and shared mobility, road safety, vulnerable road users, and microsimulation of road traffic.

Muhammad Nadeem is a postdoctoral researcher at the Graduate School of Urban Innovation, Yokohama National University, Japan, and obtained his PhD degree in urban innovation from the same institute. He is a motivated researcher in the field of urban planning and has published more than 20 articles in international and indexing journals and conferences. His research key areas include sustainable urban planning, transit-oriented development, transportation planning, travel behavior, and smart cities. He earned his master's and bachelor's degrees in city and regional planning from University of Engineering and Technology in Lahore, Pakistan.

Bilal Saghir is an independent researcher in the field of urban planning with a track record of successfully spearheading community research initiatives across diverse sectors. His interest lies in delivering evidence-based research and analysis that provide decision-makers with actionable insights, enabling them to make well-informed, strategic decisions rooted in a comprehensive understanding of market dynamics. Bilal earned his Master's in City and Regional Planning from King Fahd University Saudi Arabia.

Rethinking Urban Planning and Development for the Post-Pandemic Urban Crowd Management

Jiyon Shin, Reazul Ahsan, and Soumaya Ezazaa

Abstract The demand for collective gatherings and recreational space has been rising as a response to the three-year itch of staying indoors and social distancing due to the global COVID-19 pandemic. Crowds are spilling out in the city and rejuvenating places, yet also causing dangerously overcrowded and uncontrolled situations. Planning is also challenged by the temporary use of space and the unpredictable nature of crowds during events, requiring a deep understanding of the crowd flow and flexibility in planning. How can places be prepared to accommodate the sudden influx of crowds safely? This chapter focuses on the case study of the tragic crowd crush in a once-festive alley celebrating Halloween in Itaewon, Seoul, the Republic of Korea, in 2022. What sort of planning methods could have prevented the crowd crush? Many past similar events worldwide reported a significant number of casualties and deaths, especially related to stampedes. This chapter analyzes the tragedy in retrospect in terms of **form, programming, and the roles of stakeholders** while reviewing and drawing lessons learned from relevant crowd-control measures elsewhere that could have been applied to avoid such an incident. Ultimately, this chapter offers insights and guidelines for planning for future crowd-raising occasions in limited spaces.

Keywords Urban planning · Crowd management · Post-pandemic planning · And urban safety

J. Shin
World Bank, City and Metropolitan Planning, University of Utah, Incheon, South Korea

R. Ahsan (✉)
City and Metropolitan Planning Department, University of Utah, Incheon, South Korea
e-mail: reazul.ahsan@utah.edu

S. Ezazaa
Urban Design and Planner, Reichen et Robert & Associés International, Rabat, Morocco

K. M. Nahiduzzaman (ed.), *Making Sense of Planning and Development for the Post-Pandemic Cities*, City Development: Issues and Best Practices,
https://doi.org/10.1007/978-981-97-5481-6_12

1 Introduction

Urban planning and population control are essential for mitigating risks in cities and urban regions. As cities continue to grow and urbanize, they must be constructed and maintained to minimise the risk of harm to their citizens and visitors. Urbanization can pose significant challenges for urban crowd management due to the rapid increase in population density in urban areas. The high concentration of people in cities can lead to overcrowding in certain places, which can exacerbate problems of public safety and security. Managing crowds in urban areas become even more difficult during massive gatherings, such as festivals or sporting events, where vast numbers of people converge on a single location. Especially after the global COVID-19 pandemic has become more controllable with the distribution of vaccines, cities have been lifting bans on massive gatherings and facial masks, collective gatherings are expected to increase. Further complicating the problem of crowd control in cities, the permanence of the layout and infrastructure of urban areas can make it difficult to move large numbers of people quickly and efficiently without proper planning. Ensuring the safety of urban places is further challenged by the temporary use of space and the unpredictable nature of crowds during events, requiring a deep understanding of the crowd flow and flexibility in planning. How can places be prepared to safely accommodate the sudden influx of crowds when the spatial form is fixed and limited?

Considering that in modern days, major crowd events occur on a larger scale with higher densities and as crowd tragedies occur all over the world, it is time for urban professionals to think of a new approach to urban policy, design, and research that gives whole new techniques and insights [1]. For this, two different disciplines come into play. Urban planning is the purposeful and methodical design of urban places and structures, considering the community's requirements and safety. Conversely, crowd management is the design and execution of techniques to manage and regulate the movement and conduct of huge crowds in public settings. Urban planning and crowd management directly impact the safety and well-being of the urban people. Including risk minimization as a primary design and implementation aspect for both is crucial.

Therefore, city planners and authorities need to develop effective strategies for crowd management that can cope with the increasing demands of urbanization. This includes implementing technologies and infrastructure improvements that can help manage and direct people efficiently and training emergency services and other first responders to quickly respond to situations that may arise. Also, it is important to critically assess past mistakes, then incorporate various methods to mitigate the damage, which can include contemporary technologies and data analysis.

This paper focuses on South Korea's unfortunate and large-scale crowd crush that occurred in the narrow back alley of a popular area in Seoul, South Korea in October 2022. The crowd crush is South Korea's deadliest disaster since the sinking of the MV Sewol in 2014, and it is the city's deadliest mass death incident since the Sampoong Department Store collapse in 1995 [2]. It is the country's bloodiest crowd crush, surpassing the 1959 event at the Pusan Municipal Stadium that killed 67 people [3]. After weeks of inquiry, a special police team decided that a lack of safety procedures and other lapses contributed to the high number of deaths. This tragedy has been selected as a case study as it alerted the world that even in a city considered highly developed, modern, and safe, such a disaster can occur, forcing planners to reconsider urban planning from a different perspective and to incorporate a variety of crowd control methods to make crowded urban spaces safer.

2 Literature Review on Crowd Management

One of the critical components of crowd management is first **understanding the crowd** in mass gatherings. Some studies focused on the characteristics of the crowd. In a study about the crowd management of outdoor music festivals, the type of performance, crowd demographics, and consumption of drugs and alcohol were identified as major factors influencing crowd behavior and safety [4]. Fruin [5] further adds to the crowd management literature that the force and pressure of the crowd, the information upon which the crowd reacts, space (form), and time (the duration of the crowding and event scheduling) can largely determine the outcome of a crowd disaster.

Next, preparing the event's space configuration (form) by assessing its capacity is an essential step for crowd management. In addition, it is also important to understand the movement pathways of the crowd and its flow directions because the space configuration impacts these. By doing so, organizers can identify spatial obstacles that block the flow of people and remove them accordingly while strategically finding spatial solutions to divide and direct the crowd to avoid overcrowding. In the annual Hajj mecca event, a crowd crush happened twice, once in 1992, causing 1,426 deaths, and later in 2006, with 346 deaths after onsite improvement and enhanced crowd management [6]. The crowd's high density is said to be the main cause of the crushing disaster, especially by doubling the number of pilgrims between 1982 and 2010. To tackle this problem, on-site improvement has been implemented by redesigning and widening the Jamarat Bridge, which is crossed by millions of pilgrims during the Hajj ritual stoning of the devil [7, 8]. The improved spatial management of the site of the Hajj, although it could not entirely prevent casualties, remains a good example proving the possibility of decreasing the disasters of such similar tragic events in the future [9], and simultaneously enhancing the experience of visitors.

On the other hand, a multitude of crowd management research focuses on the actual management and **programming** (as opposed to the non-form-related) of the event while pointing out that inadequate emergency planning, insufficient training and experience with security, and understaffing are the primary and common attributes of crowd disasters [4, 10, 11, 12] emphasize that 90% of crowd management is the planning before the event. The remaining 10% of the efforts focus on execution. In their work [12], understanding the crowd, location, and time of the event beforehand while cooperating and strategizing with multiple institutions and picking out on-site security personnel are all part of the pre-planning phase,executing the plan involved constant monitoring, controlling flows, preventing accidents, while ensuring communication throughout the event.

Moreover, crowd management is a collaborative practice involving numerous **stakeholders** [12]. The **stakeholders** involved in a mass gathering event include both the attendees of the event and the staff responsible for their safety (police, firemen, medical services). In the pre-planning stage, the information about the nature of the group attending the event and the capacity and responsibilities of the staff to ensure safety are crucial elements to consider for planning mass gatherings. For both the pre-planning stage and the actual event, full communication coordination among stakeholders should be provided, which includes staff, local police, fire, nearby businesses and attendees, and medical emergency services.

Lastly, in both the planning and execution phase, **technology** can be incorporated to strengthen security [7, 13]. Big data and IoT sensors can provide sufficient coverage at various spots throughout a city or venue through a mobile-operator pilot program. The system can uncover early opportunities for safety enhancements and enable stakeholders to make substantial improvements to how events, transportation networks, and venues are handled to ensure safe entrance and exit and minimize hazards.

In sum, planning for crowd management involves multiple aspects but can be roughly categorized into first understanding the crowd and planning for safe crowd circulation and emergencies (programming) while considering and maneuvering the spatial characteristics (form), properly collaborating with relevant stakeholders, and incorporating technology to assist the process before and during the event. Planning strategies related to these factors can help avoid disasters in mass gathering events and also in generally crowded areas in the city.

3 The Gap and Challenges in the Itaewon Halloween Event

The translation of the factors related to crowd management in the case of the Halloween event in Itaewon can help understand the planning gaps and challenges to tackle for future similar events in the city of Seoul.

3.1 Case Context: Itaewon Crowd Crush

Seoul is one of the OECD cities with the highest population density. In the midst of the global Covid-19 outbreak, Seoul has implemented stringent safeguards, particularly for events and festivals where large crowds could spread the virus. Life gradually returned to normal after the pandemic, and restrictions were eased, allowing people to resume enjoying outdoor activities and events. After nearly two years, the famed district of Itaewon collected a crucial number of people to celebrate Halloween in

October 2022. Unfortunately, this cyclic event ended tragically with the death of 156 people and injuries in 170 due to a stampede in a small alley [14] (Figs. 1, 2).

Site Area of Itaewon and the section detail explain the crowd crush in October 2022

In light of the tragic event of the district of Itaewon, crowd management in urban planning and design for dense cities such as Seoul, generally and in limited spaces specifically, is ultimately needed to mitigate the disasters of similar events in the future.

The next section examines the relevant regulations and the reality of crowd management in terms of programming, form, and stakeholders in the case of the Itaewon crowd crush.

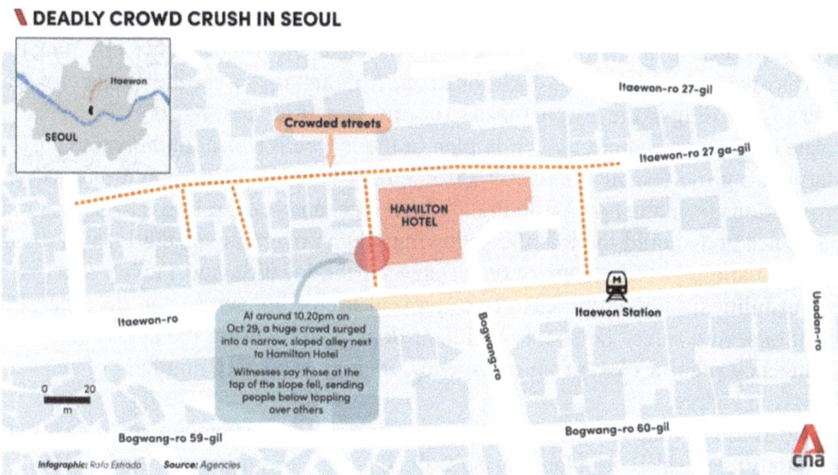

Fig. 1 The locational map where the urban crowd management failed and led to the deadly crowd crush [15]

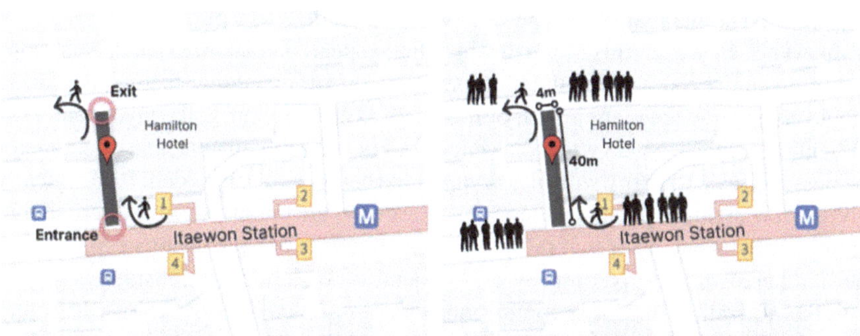

Fig. 2 Thematic description of the location, crowd density, and accessibility during the event

4 Programming for Mass Gatherings in South Korea (Republic of Korea)

In the wake of the Halloween disaster in South Korea, the Chief Police Officer stated that "there are no manuals for planning for mass gatherings with no specific organizer in charge." However, some manuals and relevant regulations that could have been flexibly applied to prevent or mitigate the disaster have existed, which were updated time and time again.

The upper law (Framework Act on the Management of Disasters and Safety) defines that celebrations that need collaborative safety measures are applied to "local festivals" hosted by local authorities or private entities (Article 66-11 in the April 2022 version). The conditions of such local festivals can fall into one of the following categories: local festivals that anticipate minimum 1,000 people at one time during the event or local festivals with high risk factors such as the use of highly inflammable or explosive materials or the location (e.g., a mountain or on the surface of the water). According to the law, in the case of private entities hosting such festivals, the private entity must submit an action plan to the respective local authority to ensure safety during the festival (Said Act, Article 66-1-③). The Plan should incorporate consultations from the Fire Department, the Police, and other relevant organizations in advance and should include the following: (1) an overview of the local festival; (2) information on the organizer or manager of the location and facility and one's responsibilities; (3) safety measures to prevent fire or human-related disasters; (4) a plan to secure safety personnel and their deployment during the event; and (5) a contingency plan in case of an emergency, including contact numbers of relevant officials in charge.

Two Manuals following the law (Framework Act on the Management of Disasters and Safety) were also circulated prior to the disaster. The first is the *Safety Manual for Unsystematized Mass Gatherings* (Korea National Police Service 2014[1]) which covers various events, such as festivals, performances, and athletic games—whether the event is public or private, indoors or outdoors, domestic or international, and profit or non-profit. It is mostly for the internal use of the police and relevant event organizers. The manual prototype was first created in 2005 and was limited to for-profit events. However, this Manual, which first appeared in 2006, expanded its scope to non-profit events. It has detailed information on various threat factors organizers should consider when planning for events, such as crowd control, the layout and exit-and-entrance plans, a plan when the venue goes beyond its capacity, and planning for the possibility that nearby subways or bus stops become overcrowded. It specifically advises that having separate exits and entrances ways are preferred, while there should be more exits. Also, it states if the exit way or stairways are narrow or steep, limiting the number of people passing by at the same time should be considered (p. 31) with the utilization of crowd control managers in particular entry points. It further suggests closing down nearby roads for better pedestrian flow. It also includes plans

[1] Korean title: Dajung-Unjib Haengsa Ancheon-Gwallee Manual (다중운집행사안전관리매뉴얼).

to stop escalators in nearby subways while managing the crowd at subway entrances to avoid potential disasters due to sudden influxes of people. Throughout the event, a ready hotline for emergencies and constant communication should be available.

The second Manual that follows the law (Framework Act on the Management of Disasters and Safety) is the regularly updated *Local Festival Area Safety Management Manual*,[2] published in 2005 (Ministry of the Interior and Safety). This Manual instructs how events, regardless of the event organizer, should plan for safety throughout the event. While its recent update was completed in 2021, this Manual was initially made in response to a series of stampedes during a music concert in Gyeongbuk Sangjoo Stadium in 2005 (11 deaths and 162 injured) and during the opening of an amusement park in Seoul in 2006 (35 injured). In this Manual, guidelines before, during, and after the event are included, while also highlighting particular (1) locational risks in mountain areas, water surfaces (sea, river, lake), (2) material-based risks, such as fire, inflammable materials, snow and ice, and (3) time-related risks, such as when events are held during the night. Although there are no guidelines specifically concerning sporadic events held separately by different private entities sharing the same narrow alleyway, the manual states venues with slopes should be reviewed by authorities if it is appropriate, that small and narrow areas should have a proper enter-and-exit plan, that safety guards that control the capacity should be present, and that any large-scale event must have an emergency communication system with relevant organizations in place.

In sum, these laws and manuals have existed, giving practical guidelines to ensure safety during any event with large crowds with specific attention given to slopy narrow areas, stating the necessity of having crowd control managers and managing entrances and exits, along with roadside control in connection with nearby mass transit areas. The implementation and application of the regulations may have been what was lacking.

5 Form

In the case of the Itaewon Halloween event, one of the main causes of casualties and fatalities was directly influenced by crowd density in comparison to the limited space (**form**). While the back alley of Hamilton Hotel was a popular hangout for visitors for offering a variety of international cuisine and bar and club options, it dramatically attracted more crowds during Halloween weekends because of its festivity with people celebrating the night in flashy Halloween costumes. In terms of form, the problem area in the Itaewon alleyway was narrow and steep: 3.2 m in width and 40 m long with a nearly 10% slope. The catastrophe occurred in this short path connecting the neighborhood to one of the main subway stations on the main boulevard. The concentration of an immense throng in this steep lane impacted the capacity, which exceeded the limit and caused a crush.

---

[2] Korean title: Jeeyeok-Chukje-Jang Ahnjuhn-Gwallee Manual (지역축제장안전관리매뉴얼).

Studies explain that the capacity for people per square meter is up to two point five to three people, and from four and above, the crowd can become unpredictable and at high risk [10]. The crowd can start to "float" with seven people per square meter, with individuals experiencing difficulty breathing with extreme body pressure due to the crowd and also being unable to move in any particular direction of one's will. However, in the case of Itaewon, there were up to 12 people per square meter, subjecting each individual to immense pressure tantamount to 224 kg to 560 kg—equal to four to ten times one's body weight [16]. Therefore, once some people could not endure the pressure anymore and began to fall in the middle of the narrow and steep alleyway, the crowd from the upper slope continued to flow downwards, causing serial collapses of bodies falling over bodies for ten minutes. Since there were no emergency route plans, crowd circulation plans, or any crowd control plans, the alleyway became uncontrollably crowded with people inundating through both ends, making it impossible to either exit or enter.

According to the media and investigation reports, from the mid-point to the lower end of the alleyway became narrower because of illegal expansions of buildings on the side, eventually worsening the bottleneck. These illegally expanded alleyway buildings were not registered on the official building registry. Furthermore, the main road leading to the alleyway was crowded by obstacles, such as the subway ventilation box structure, and some street vendors using the space to sell clothes or street food. In spite of a previous area study on Itaewon commissioned by the government five years ago that recommended that pedestrian roads should be widened by undergrounding the adjacent subway ventilation box structure on the road nearby, no changes were made [17].

6 Stakeholders

According to the Framework Act on the Management of Disasters and Safety (Article 4: Responsibilities of the State and Other Entities), it is "the State and local governments that are responsible for protecting lives, bodies, and property of people from disasters or various other accidents," and that they "shall endeavor to prevent disasters or various other accidents and to mitigate damage therefrom, and shall formulate and implement plans to promptly deal with and recover from the damage." Also, The head of a disaster management agency "…shall formulate and implement a safety management plan relating to his/her duties, and cooperate with a Special Metropolitan City…" having jurisdiction over his/her agency in performing disaster and safety management affairs. In other words, in general, the district authority in which the event takes place should be in charge of forming and implementing an overall safety management plan.

Whereas in the same Act, Article 5 (Responsibilities of Citizens) states that "People shall fully cooperate with the State and local governments in performing disaster and safety management affairs, and endeavor to protect buildings and facilities they own or use from any disaster or various other accidents." This provides a basis that citizens who are direct stakeholders in the event, even if they are not the major host or organizer, should be fully cooperative to protect their "buildings and facilities." However, does the law also include the alleyways "outside" of their buildings?

The foreword in the Safety Manual (Korea National Police Service 2014), which is a more detailed version of the law, elaborates on the changing trends of how private events with a considerable number of participants are now equipping themselves with safety measures that private–public cooperative safety measures are on the rise, instead of the police-centric top- down safety management. The Manual ultimately highlights that it is, in principle, the private businesses' responsibility to prepare themselves with the proper safety measures prior to the event. In this perspective, the law may be translated as businesses, along with the police and local authorities, are also partly responsible for considering beyond the interior of their buildings to ensure safety.

Nevertheless, unlike "local festivals" as stated in the law and manuals, the weekend of Halloween was no large-scale concert or event hosted by the local authorities where the "one in charge" is clear. At least three major stakeholders were involved in the Itaewon disaster: the local authority (Gu-district office), the police, and the businesses. Like any other year, in preparation for the weekend of Halloween, the Yongsan Gu-district office evidently had an emergency meeting before Halloween with limited members of the local police present. This year (2022), the issues on the agenda were focused on preventing the spread of Covid-19, drug or sex-related crime, keeping the streets clean, and preventing illegal parking and facility- related accidents [18]. It is noticeable that there were no elements of traffic or crowd control in the meeting. Also, contrary to other cross-cutting emergency meetings with 40 participants that included 119 emergency personnel in preparation for Halloween held two years prior to the year of the tragedy, this year's meeting was much smaller in scale [19]. Since this was not considered a "local festival" organized by the Gu, a detailed safety management plan was not mandatory either.

The police also respectively created their response plan prior to Halloween, yet there was more of a focus on policing illegal activities and Covid-19 control than focusing on crowd control, unlike in previous years. After there were repeated calls from citizens an hour and also 30 min prior to the disaster, the police repeatedly requested more police deployment, yet the police control tower did not respond to the request, losing another opportunity to mitigate the disaster.

The Itaewon Merchant Association also met with the police prior to Halloween. However, according to the meeting minutes of the police, to tell the police to either reduce their deployment or be less stringent on policing not to discourage their visitors from enjoying the first mask-free event [20]. Businesses in the alleyway were solely focused on gladly greeting their customers with great expectations after suffering from years of deficit due to the pandemic. At most, they were concerned about the safety within their buildings.

The crux of the matter in this tragedy was that there was no single primary entity in charge of the security of the alleyway filled with visitors. Thus every relevant entity could evade the responsibility to ensure the collective safety in an area commonly used by all other local businesses. There was the law and manuals, but no single entity was willing to plan for proper crowd control as they were not "mandatorily" obliged by law to do so. Security and safety in the streets are not directly translated into profits, so businesses may also lack the incentive to take the initiative. Yet, when there is a lack of security and safety, the results can be devastating as this incident unravels.

7 Discussion: The Gap and Challenges in the Itaewon Halloween Event

In the case of the Itaewon Halloween catastrophe, safeguard measures composed of form, programming, and stakeholder cooperation were mishandled or simply missing. There was no sole organizer responsible for orchestrating the programming for crowd control before or during the event; thus, no action resembling holistic programming was done. There was no measure to control the foot traffic in the narrow alleyway. Planning for narrow streets to be exit-only ways for the tentative space with appropriate police lines or privately hired security staff controlling the crowd were not present; there was no hotline or clear communication method among relevant stakeholders to notify each other of any emergencies, no extra measures to remove the obstructions in the narrow alleyway or reduce the bulk of the illegal buildings blocking the already narrow road. Needless to say, former advice on widening adjacent roads was also ignored.

While there was general miscoordination of staff responsible for safety before and during the disaster, language was an additional barrier to communication during the disaster. This event generally attracted the younger population in addition to many foreign visitors that were found as victims among the crowd. Providing equal crowd communication through on-site media, message text, and other means for both categories of attendees, locals and foreigners, could have helped to ensure collective safety.

In response to the Halloween disaster, the Seoul Metropolitan Council legislated an ordinance on December 30, 2022 that ensures the safety of events with no single or specific organizer in a particular place with at least 1,000 visitors in one place. The Mayor will be in charge of establishing a comprehensive safety plan which stipulates collaboration with the police when needed. It is yet too early to evaluate the effectiveness of such a measure, yet it can be considered a step in the right direction to prevent such tragedies in the future.

Also, Seoul, known worldwide as a smart city, may plan in advance for events to use its communication infrastructures and urban big data information to handle large gatherings [21]. According to the Smart Seoul Data of Things Project,[3] there are real-time smart crowd sensors on nearby roads in Itaewon. Yet, there is no indication of sensors near the location as shown in Fig. 2. Expanding such infrastructure in

[3] https://data.seoul.go.kr/dataVisual/seoul/guide.do.

places with foreseeable crowding (as every year, the Itaewon Hamilton district was well known to get very crowded) in preparation of mass crowding and linking them with emergency hotlines while planning in advance can be helpful for effective crowd management (Fig. 3).

The site analysis reveals that the Itaewon incident site has a single entrance and no traffic diversion measures (Fig. 2). This was identified as one of the causes for the occurrence. With a better urban design concept and emergency pedestrian traffic diversion, any urban event with a large crowd can be easily managed. The following diagram depicts a potential urban design and plan alternative that would reduce the likelihood of such urban incidents (Fig. 4).

Diverting pedestrian traffic is always considered a successful urban crowd management tool. Such planning and design techniques have been used in Singapore, London Olympics, and Mecca Hajj pilgrim.

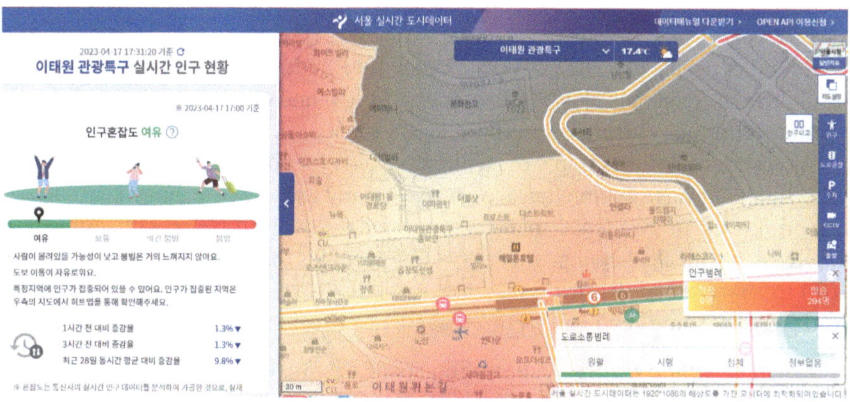

Fig. 3 The real-time site analysis of Itewon district and the incident site [22]

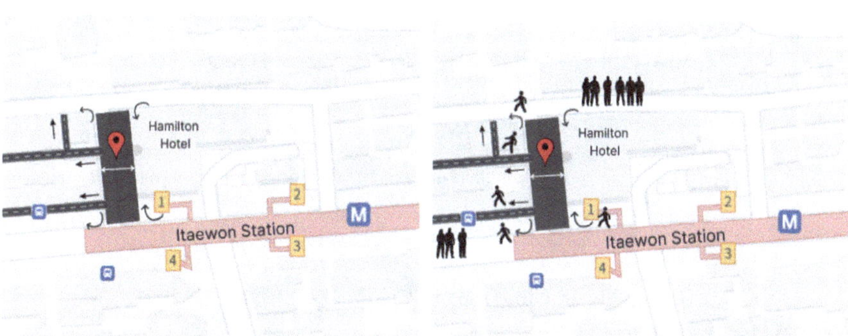

Fig. 4 Alternative methods for managing urban crowds include widening entrances and exits and redirecting pedestrian traffic

The following section introduces some global examples of urban crowd management that can provide guidelines to prevent crowd disasters in the future.

8 Global Examples of Urban Crowd Management

Advanced planning and upgrading of urban accessibility have become a technique for managing urban crowds, particularly during public events. Numerous major events throughout the world use technology. At the same time, accessibility management through urban design has proven to be a successful toll in reducing crowd crush in public events. One of the successful planning projects to control urban crowd management is the 'Jamarat' bridge in Mecca. The Jamarat bridge is designed to divers the pilgrims in an outflow to control the crowd and reduce the risks [6]. The following diagram shows the crowd management techniques of Jamatara bridge (Fig. 5).

Few other events like 2012 Olympic Games in London [24], and New Years's Eve Fesitivities in Sydney [11] used urban traffic follow and open space traffic direction methods to minimize the crowd crus in urban events.

Another successful example of an effective plan for crowd management is Singapore. Singapore has a comprehensive urban planning and crowd management strategy that prioritizes its residents' and visitors' safety and well-being [25]. Singapore employed physical planning, transportation planning, and the Internet of Things (IoT) to analyze and distribute the crowd to a low-crowded zone for urban crowd management. Some key features of Singapore's crowd management plan include:

1. ***Efficient public transportation***: Singapore has a well-developed public transportation system that includes buses, trains, and taxis. This helps regulate people's flow and minimize overcrowding in public spaces.

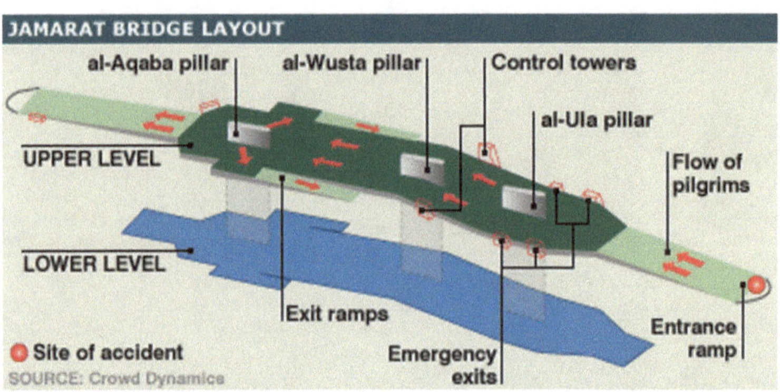

Fig. 5 The Jamarat bridge a planning approach to urban event crowd management in Mecca [23]

2. ***Designated event areas***: Singapore has designated event areas for large-scale events, such as concerts and festivals. These areas are designed to accommodate large crowds and are equipped with appropriate infrastructure, such as restrooms and food and beverage concessions.
3. ***Intelligent traffic management systems***: Singapore has implemented an intelligent traffic management system that helps regulate traffic flow and minimize congestion during peak hours and large events.
4. ***Emergency response systems***: Singapore has a well-coordinated emergency response system with a network of hospitals, ambulance services, and fire departments. This helps ensure that immediate medical assistance is available in an emergency.
5. ***Crowd control measures***: Singapore uses a range of crowd control measures, such as road closures, traffic diversions, and crowd management personnel, to minimize the risk of accidents and other incidents during large events [24].

9 Conclusion

By having a greater understanding of how people move about a region, innovative crowd management technologies for cities and huge events will improve stakeholder experiences and production efficiency in the short and long run. Cities and site managers should consider how crowd control services could be useful in their particular context and the difficulties they can help resolve. Many cities today use various physical planning measures to deflect the urban populace, particularly during various public events. The most prevalent physical planning aspects include.

Designing public spaces such as parks, plazas, and pedestrian zones facilitates people's flow and reduces crowding in specific areas. This can be accomplished by providing multiple entrances and exits, utilizing barriers or physical elements to direct pedestrian flow, and creating comfortable seating and rest areas to disperse crowds.

Traffic management can also play an important role in reducing urban congestion. This can be accomplished by implementing traffic control measures such as traffic lights, roundabouts, and pedestrian crossings. By effectively managing traffic flow, reducing the population density in certain areas is possible.

Effective use of technology such as real-time traffic monitoring systems and crowd monitoring can assist in identifying high-congestion areas and enabling swift responses to prevent overcrowding. It can also aid in identifying problem areas and facilitate the effective management of crowds during events or emergencies.

Promoting alternative modes of transportation, such as walking, cycling, and public transportation, can help reduce the number of vehicles on the road and, in turn, traffic congestion and crowding.

Event management, event organizers can reduce attendance by meticulously arranging and coordinating the event. This includes the event's date, time, and location, along with the quantity of tickets sold. Additionally, effectively plan with event attendees to assist them in determining their travel and arrival times. This can help to reduce congestion by spreading out the arrival and departure of masses. Adding more transportation options for attendees will also reduce rush-hour congestion. This includes enhanced bus and train service, in addition to shuttle buses to and from the event. Importantly, organizers can select facilities with sufficient space to facilitate the anticipated number of attendees. Security personnel with extensive training could monitor the crowd and assure its safety. This can aid in the prevention of throng crushes and stampedes.

As the world recovers from the pandemic, it is evident that managing urban crowds will continue to be a top priority for communities everywhere. Effective crowd control measures will require technological breakthroughs, improved communication, and enhanced collaboration between the government, the private sector, and the community. Equally important is that these programmes prioritise the health and safety of all individuals while simultaneously promoting accessibility, inclusivity, and equity.

Moving forward, city planners and officials must examine the lessons learned from the epidemic as well as the changing demands and expectations of the residents. By focusing on sustainable and adaptable urban population management solutions, we can help guarantee that our cities remain dynamic, resilient, and safe places to live, work, and visit.

References

1. Sengupta RR, Bennett DA (2003) Agent-based modelling environment for spatial decision support. Int J Geogr Inf Sci 17(2):157–180. https://doi.org/10.1080/713811747
2. Choe S-H (2022) The Itaewon tragedy is the worst peacetime disaster in South Korea since the Sewol ferry sank in 2014, killing more than 300 people. The New York Times. https://www.nytimes.com/live/2022/10/29/world/korea-halloween-stampede-itaewon#91cdc8ff-08c1-5125-828e-055161c8a4c4
3. Deskreport (2022) Chronology of major stampedes in S. Korea. Yonhalnew Agency. https://en.yna.co.kr/view/AEN20221030005500315
4. Johansson A, Batty M, Hayashi K, Al Bar O, Marcozzi D, Memish ZA (2012) Crowd and environmental management during mass gatherings. Lancet Infect Dis 12(2):150–156. https://doi.org/10.1016/S1473-3099(11)70287-0
5. Fruin JJ (2002) The causes and prevention of crowd disasters
6. Aldossari M, Aljoudi A, Celentano D (2019) Health issues in the Hajj pilgrimage: a literature review. J East Mediterr Health 25(10). https://doi.org/10.26719/2019.25.10.744
7. Ali T, Altamimi AB (2019) Crowd management in open spaces. ArXiv: abs/1904.12625
8. Helbin D, Johansson A, Al-Abideen HZ (2007) Dynamics of crowd disasters: an empirical study. Phys Rev. https://doi.org/10.1103/PhysRevE.75.046109
9. Soomaroo L, Murray V (2012) Disasters at mass gatherings: lessons from history. PloS Curr 2. https://doi.org/10.1371/currents.RRN1301
10. Al-Kodmany KM (2013) Crowd management and urban design: new scientific approaches. Urban Des Int 18:282–295
11. Patty A (2023) Crowd crush in The Rocks during an otherwise safe city celebration. The Daily Morning Herald. https://www.smh.com.au/national/nsw/crowd-crush-in-the-rocks-during-an-otherwise-safe-city-celebration-20230101-p5c9p0.html

12. Martella C, Li J, Conrado C, Vermeeren, A. (2017) On current crowd management practices and the need for increased situation awareness, prediction, and intervention. Saf Sci 91:381–393. https://doi.org/10.1016/j.ssci.2016.09.006
13. Wijermans N, Conrado C, van Steen M, Martella C, Li J (2016) A landscape of crowd-management support: an integrative approach. Saf Sci 86:142–164. https://doi.org/10.1016/j.ssci.2016.02.027
14. Blair G, Henley J (2022) At least 153 killed in crowd crush during Halloween festivities in Seoul. The Guardian. https://www.theguardian.com/world/2022/oct/29/dozens-of-people-crushed-by-large-crowd-at-halloween-festivities-in-south-korea
15. Druy J (2022) Commentary: what caused the Halloween crowd crush in South Korea. Channelnewsasia CNA. https://www.channelnewsasia.com/commentary/seoul-crowd-crush-death-trap-itaewon-halloween-how-it-happened-3036291
16. JTBC (Producer) (2023) 넘어진 지점부터 10m 걸쳐 끼임 560kg 압력에 질식사. https://news.jtbc.co.kr/article/article.aspx?news_id=NB12110991
17. Jae-wook J (2022) [단독]'이태원 보행로 넓혀야'…5년前 정부보고서만 이행됐어도 [단독]'이태원 보행로 넓혀야'…5년前 정부보고서만 이행됐어도 'We need to widen the Itaewon pedestrian path'... Even if only the government report from 5 years ago was implemented. naver.com. https://www.edaily.co.kr/news/read?newsId=03129126632521456&mediaCodeNo=257
18. Min-hyeok J (2022) "할로윈 데이" 앞두고… 이태원 있는 용산구 "안전확보에 만전" Ahead of 'Halloween Day'... Yongsan-gu, located in Itaewon, "Fully committed to securing safety". Ahnjeon News. https://www.safetynews.co.kr/news/articleView.html?idxno=216531
19. Kim S-s (2022) Two years ago, 40 ward offices, police, and merchants held a Halloween countermeasure meeting, but there was no this year. Hankyeore. https://www.hani.co.kr/arti/area/capital/1065646.html
20. MBN News (Producer) (2022) 간담회 문건엔 "상인회가 경찰 배치 자제 요청"…상인회 "사실 과 달라". https://www.youtube.com/watch?v=G5ZxmwIDA0s
21. Sharma D, Bhondekar AP, Shukla AK, Ghanshyam C (2018) A review on technological advancements in crowd management. J Ambient Intell Humaniz Comput 9(3):485–495. https://doi.org/10.1007/s12652-016-0432-x
22. Kyung-don N (2022) Site of incident. The Korea Herald. Daily, October 30. https://www.koreaherald.com/view.php?ud=20221030000272
23. Still K (2005) Crowd safety and crowd risk analysis. https://www.gkstill.com/CV/Projects/Jamarat.html
24. Ancliffe S (2017) Crowd planning for Glastonbury and London 2012 Olympics. Movement strategies a GHD documentary. https://www.movementstrategies.com/news/crowd-planning-for-public-safety
25. GSMA (2016) GSMA smart cities guide: crowd management. Barcelona. https://www.gsma.com/iot/resources/gsma-smart-cities-guide-crowd-management-3/

Jiyon Shin is an Adjunct Professor at the University of Utah, Asia Campus in South Korea, and she also serves as a consultant at the World Bank Korea Green Growth Trust Fund. She has a PhD in Urban Planning from Seoul National University and a Master's degree in urban planning from New York University. She has experience practicing urban planning from the Seoul Metropolitan Government and is interested in research regarding inclusive and sustainable urban planning practices.

Reazul Ahsan is an Associate Professor in the University of Utah's Department of City and Metropolitan Planning and was appointed as a program coordinator of the Urban Ecology Programme at the University of Utah Asia Campus in Songdo, South Korea. Dr. Ahsan's research interests include smart urbanism, climate-adaptive urban design, urban resilience, and environmental sustainability. He received his PhD in Urban and Regional Planning from the University of South Australia in 2013 and worked as a research fellow at MIT for the Malaysian Sustainable City Programme (2015-2016). Reazul Ahsan is a seasoned specialist, having worked in higher education and research in Australia, Malaysia, Japan, Thailand, the United Kingdom, and the United States for more than decades.

Soumaya Ezazaa is an Urbanist and Architect who currently works as a Project Coordinator at CARTA—REICHEN ET ROBERT ASSOCIES ARCHITECTES URBANISTES in Rabat, Morocco. She completed her Master's degree in urban planning and Design from the University of Seoul in 2021. Her key research interest is in human—and community-centered design, which focuses on new ways to approach urban problems through urban design, urban planning, and policy-making.

Rethinking Urban Development and Built Environment Attributes in the Post-Pandemic World: A Case of High-Density Hong Kong

Kazi Humayun Kabir and Md. Ayatullah Khan

Abstract This chapter is intended for a review of the present urban planning and built environment scenarios in Hong Kong and to analyse the challenges that Hong Kong faces in tackling the pandemic with its urban planning and built environment. Lastly, this study came up with a few suggestions for how urban development and the built environment in Hong Kong could be adjusted after a pandemic. The objectives of the study were met through a meticulous review of journal databases like Google Scholar, PubMed, and Science Direct, as well as several reports from government and non-government organisations. The thematic analysis developed four themes, including urban development planning scenarios, built environment dynamics, COVID-19 transmission linked with urban design and built environment dynamics, and lessons learned and policy rethinking options in the post-pandemic era in Hong Kong. Urban development and built environment in Hong Kong include high-density development, zoning regulations, transport networks, compact urban form, mixed-use developments, skyscrapers and high-rise buildings, pedestrian-friendly infrastructure, and green building practices. Moreover, regions with higher population density, dense and high-rise structures, fewer sky views and fewer green spaces, and a higher private residential, commercial, and transport land use density were regarded as high-risk zones in several phases of COVID-19 transmission in Hong Kong. Compact urban development with better access to services and public health infrastructure, redesigning the older high-rise buildings using contactless smart technologies, smart mobility restrictions implementation, and more green space allocations could be useful policy options to become a pandemic-resilient city in Hong Kong.

K. H. Kabir (✉)
Department of Urban Planning and Design, The University of Hong Kong, Hong Kong, China
e-mail: kazikabir@connect.hku.hk; khkabir@ds.ku.ac.bd

K. H. Kabir · Md. A. Khan
Development Studies Discipline, Khulna University, Khulna, Bangladesh

Md. A. Khan
Department of Geography, Hong Kong Baptist University, Hong Kong, China
e-mail: ayatullah@life.hkbu.edu.hk

K. M. Nahiduzzaman (ed.), *Making Sense of Planning and Development for the Post-Pandemic Cities*, City Development: Issues and Best Practices,
https://doi.org/10.1007/978-981-97-5481-6_13

Keywords COVID-19 pandemic · High-density city · High-rise building · Compact city development · Smart technologies · Smart mobility · Hong Kong

1 Introduction

The global urban population faced unparalleled hardships because of the COVID-19 outbreak. The effects of the outbreak have extended far beyond the realm of public health, changing the underlying characteristics of urbanization and the built environment [1, 2]. There are far-reaching consequences for city life, infrastructure, and design as a result of the outbreak, which have been reluctantly accepted by most nations. Cities have been forced to reevaluate their current urban development policies due to a number of factors, including the rapid increase in remote employment, shifts in transportation preferences, and a focus on healthcare and well-being [1, 3, 4]. However, prospects for innovation and resilience have emerged as an outbreak has revealed limitations and highlighted loopholes in contemporary urban areas.

Evidence revealed that the majority of cities with a dense population that featured prevalent growth in urbanization and high-rise dwellings were more susceptible to the spread of COVID-19 and that their urban planning and built environment were considered to be one of the most crucial attributes for the speedy dissemination of COVID-19 [5, 6, 7, 8, 9]. The built environment, by influencing how people lead their daily lives and communicate with one another, can have far-reaching effects on the temporal and spatial trends of disease dissemination. The geographical structure and operational areas of an urban area greatly influence individual mobility and interactions with others, which are intimately connected with the expansion of transmissible diseases. To give just one example, the spread of dengue fever has been linked to people's increased mobility [10]. Moreover, the transport networks of the industrialized and high-density metropolitan societies are considered to be one of the great influencers when simulating the spread of diseases that are transmissible [11, 12].

Like others, Hong Kong has also a substantial impact on the trends in the spread of transmissible diseases like COVID-19 because of its urban setting, huge population density, small living spaces, and busy metropolitan areas. Additionally, decisions made during urban planning, such as those regarding land use patterns, building designs, transportation networks, and the administration of public spaces, have a significant impact on how likely it is for diseases to spread inside the city which is also supported by the earlier studies [5, 7, 13]. To maintain the well-being and resilience of Hong Kong in the face of these threats, it is essential to rethink urban development policies and built environment dynamics, and to do so, an in-depth analysis on how current urban development and built environment dynamics policies and practices influence the transmission of COVID-19 around the city is crucial. Moreover, the lessons that learned from the COVID-19 transmission in Hong Kong to identify the policy rethinking options for the development of pandemic resilient Hong Kong city are also important to explore. However, there is lack of a comprehensive study on all these issues in the existing literature. For example, earlier studies just assessed the risk of COVID-19 spread in Hong Kong with the built environment [5, 7, 13, 9, 14, 15]; the effects of the built environment on urban vibrancy under COVID-19 in Hong Kong [16]; the impact of urban geometry and socio-demographic characteristics on COVID-19 in Hong Kong [17]; etc. To fill the literature gap, this study therefore aimed to evaluate the existing urban planning and built environment scenarios of Hong Kong and to assess the challenges faced by Hong Kong to tackle the pandemic due to its urban design and built environment. Finally, this study proposed some rethinking options for urban development and built environment attributes in the post-pandemic world in Hong Kong. The findings from this study will hopefully add to the continuing discussion about urban revitalization following a pandemic, especially from the perspective of Hong Kong. It is hoped that the results will be used to improve urban settings in terms of their livability, sustainability, and agility in the face of upcoming healthcare emergencies, as well as to guide policy considerations.

2 Theoretical and Conceptual Framework

The city's urban development plans and the features of its built environment have played a significant role in defining Hong Kong's distinct landscape. The concept of compact and high-density development lies at the heart of urban planning in Hong Kong. Due to the lack of available land, high-rise structures dominate the city's environment. This technique maximises the utilisation of available land while enabling the housing of a sizable population in a limited area [18, 19]. The government also advocates the necessity of including green spaces and public parks into urban development in order to enhance the quality of life for inhabitants and decrease the negative consequences of high-density living [20, 21]. Easily accessible public transportation is yet another attractive feature of the city. Together, the Mass Transit Railway (MTR), buses, and ferries make up Hong Kong's robust and dependable public transportation system. Through the use of this multimodal network, the use of private vehicles is reduced, traffic is made less congested, and ecologically beneficial transportation options are promoted [22, 23, 24]. In recent years, the government has also placed a strong emphasis on green building techniques and sustainable development. The Hong Kong Green Building Council is one initiative that actively promotes the use of sustainable building materials and methods [25, 26]. The use of renewable energy sources like solar panels and the installation of energy-efficient technologies in new construction support the city's sustainability goals [27, 28]. Overall, Hong Kong has been striving hard to develop into a "smart city," utilising cutting-edge technology to enhance the productivity, sustainability, and quality of life of the city's residents [29, 30, 31].

However, the dynamics of Hong Kong's urban built environment and the urban development strategies that guide its growth have had a major impact on the spread of the COVID-19 virus. The city's high density of living and compact urban planning has been both a benefit and an obstacle to controlling the outbreak. One could argue that the density of the built environment has aided in the quick introduction of social isolation and lockdowns because it is simpler to regulate a compact area. The government's capacity to swiftly seal off public areas and limit access has been useful in containing transmission to fewer territories. In contrast, the likelihood of neighbourhood transmission has risen because of the close association of residential and commercial structures and because of the congestion of public transit. Limitations on the availability of open areas and the difficulty of maintaining physical distance in densely populated regions have also hampered the successful implementation of preventative measures. To lessen the effects of the built environment's dynamics on COVID-19 transmission, the government has stepped up its testing, contact tracing, and vaccination initiatives.

In order to increase pandemic resilience, sustainable urban planning and design should be encouraged. The transmission of infectious diseases can be slowed down by increasing greenery, enhancing ventilation and sky views in buildings, and using flexible and adaptive construction in public areas. Additionally, promoting a pandemic-resilient Hong Kong city in the future may benefit from investments in digital infrastructure, such as a contactless entry in shared spaces, smart mobility, and the use of data-driven technology that ensures privacy (Fig. 1).

3 Materials and Methods

This study was based on a thorough review of the association among COVID-19 transmission, urban development planning and built environment features in Hong Kong, using journal databases like Google Scholar, PubMed, and Science Direct, as well as several reports from government and non-government organisations. In this study, both the exclusion and inclusion criteria were followed. Firstly, the main sources of data for this study were publications from journals with empirical data and different publications from government and non-government organisations. Review articles, books, book chapters, book series, and conference proceedings were not utilised. The second requirement was that the studies use quantitative or mixed methods. The selection of only quantitative and mixed-method studies was based on the assumption that these types of studies are less likely to be skewed and more likely to provide results that can be used in other situations [32]. A mixed-methods study was the best option because it was built on both quantitative and qualitative findings. Quantitative studies also give people a solid basis for making well-informed decisions [33, 34]. Thirdly, only articles that originally appeared in English were considered. This was done to avoid confusion and make interpretation easier for international readers. Fourth, a research period of 14 years (from 2010 to 2023) was chosen. This is enough time to study the relationship between urban growth and built environment characteristics in Hong Kong after the pandemic. Lastly, only papers about Hong Kong were chosen because the main goal of the study is to look at the assess the challenges faced by Hong Kong to tackle the pandemic due to its urban design and built environment as well as to find out policy options to tackle the challenges in the

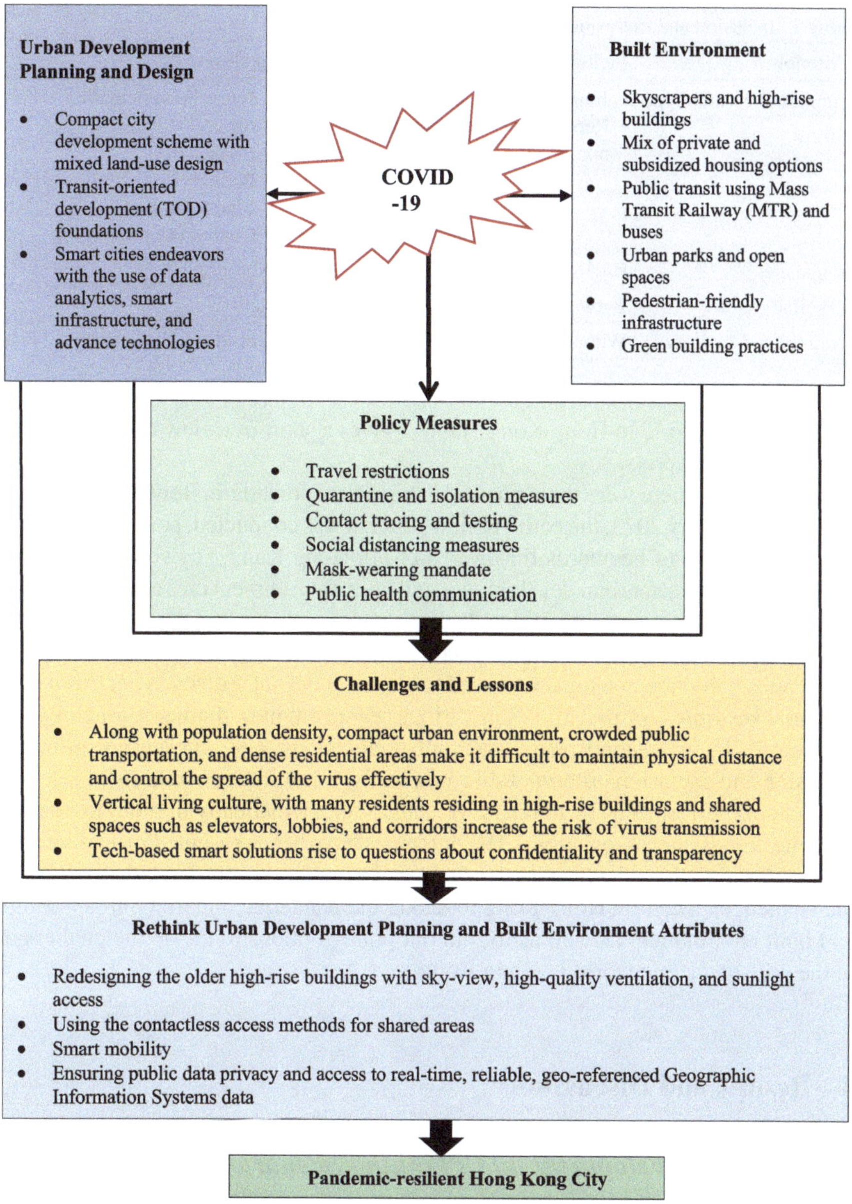

Fig. 1 Theoretical and conceptual framework on rethinking urban development and built environment attributes in the post-pandemic world in Hong Kong

Table 1 Inclusion and exclusion criteria

Criterion	Inclusion	Exclusion
Literature/research type	• Journal (Research article) • Reports from government and non-governmental organizations	• Journals (systematic review), • Book series • Book • Chapter in Book • Conference proceeding
Language	English	Non-English
Timeline	Between 2010 to 2023	<2010
Country and territory	Within Hong Kong	Outside of Hong Kong

post-pandemic world in Hong Kong. Table 1 gives a short overview of the inclusion and exclusion criteria.

In this paper, there were three phases to the review procedure. Between December 2022 and February 2023, the entire review process was conducted, commencing with the identification of keywords for the search. In Hong Kong, keywords relevant to the linkages between urban development and built environment characteristics in the post-pandemic era were used (Table 2). Keywords were chosen based on previously completed or related studies and a thesaurus. The formulation of various keywords applicable to the research topic was used to identify relevant articles. After evaluating potential keywords for the articles in various search engines, duplicate articles were eliminated. In the second phase, relevant publications were chosen according to inclusion and exclusion criteria (Table 1). Finally, a qualitative synthesis of selected articles was performed. The findings were analysed thematically in order to identify patterns, trends, and associations. The outcomes have been presented in a systematic way, addressing the objectives of the study and providing an in-depth description of the challenges faced by Hong Kong to tackle the pandemic due to its urban design and built environment as well as to find out policy options to tackle the challenges in the post-pandemic world in Hong Kong.

4 Results and Discussion

4.1 Urban Development and Planning Scenarios in Hong Kong

A variety of urban planning and development strategies have been taken into account and executed over the course of time in order to meet the changing requirements of Hong Kong. For instance, to make the most of the available space and reduce the likelihood of urban sprawl, Hong Kong has adopted a compact city development scheme. Taking this strategy means prioritizing infrastructural efficiency and mixed

Table 2 Key words and search terms

Database	Key words and search terms
Science Direct	{(urban development design OR city growth planning) AND (built environment OR city landscape) AND (challenges OR hardships) AND (policy OR scheme OR strategy) AND (post-pandemic OR COVID-19) AND (Hong Kong)}
PubMed	(association [Title word][MeSH Terms] OR relationship [All Fields]) AND (urban development design [Title word] [MeSH Terns] OR city growth planning [Title word][MeSH Terms]) AND (built environment [Title word][MeSH Terms] OR city landscape [Title word][MeSH Terms]) AND (challenges [Title word][MeSH Terms] OR hardships [All Fields]) AND (policy [Title word][MeSH Terms] OR strategy [All Fields]) AND (post-pandemic [Title word][MeSH Terms] OR COVID-19 [All Fields]) AND (Hong Kong [Title word] OR hongkong [All Fields])
Google Scholar and Google	First search: Urban development planning and built environment attributes in Hong Kong Second search: Challenges faced by Hong Kong to tackle the pandemic due to its urban design and built environment Third search: Future policy options for rethinking the urban development planning and built environment dynamics in the post-pandemic world in Hong Kong

land-use design to handle the high population densities. The objective is to design livable, long-lasting cities where people can easily access an extensive variety of services by walking [35, 36, 37]. In addition, the public transit system in Hong Kong is highly advanced and comprehensive. Urban planning has adopted transit-oriented development (TOD) foundations, which prioritize the development of compact, walkable neighborhoods in close proximity to transit hubs. Sustainable urban mobility is fostered by this strategy, which prioritizes public transportation above individual car use [38, 39, 15]. Moreover, the Hong Kong government has recognized additional development sites, including the New Territories and the East Lantau Metropolis, to handle future population expansion and reduce demand for accommodation. It is possible to build large-scale cities in these regions, complete with all the supporting infrastructure and transportation networks, as well as a variety of housing options, retail establishments, and entertainment venues [40, 41, 42]. Furthermore, technology for smart cities and endeavors has been widely adopted in Hong Kong to improve the city's productivity, communication, and quality of life. The use of data analytics, smart infrastructure, and technological advances are all examples of approaches that are used to enhance urban services, the management of transportation, and resource distribution [29, 44, 45, 46, 47, 48, 49, 31].

4.2 Built Environment Dynamics in Hong Kong

Various initiatives have been launched by Hong Kong to improve the city's built environment scenarios. To direct the evolution of Hong Kong's built environment, the government, for example, has enacted extended urban planning guidelines. In order to maximise land utilisation and encourage sustainable development, zoning restrictions, density control regulations, and land use planning are all part of the process [50, 51, 52, 53, 54]. Additionally, Hong Kong has embarked on massive land reclamation operations to increase its geographical size in light of its scarcity of land. More space for living, working, and recreation has been made possible because of land reclamation, which has reduced congestion in already developed regions [55]. Further, in order to house its expanding population, the city has welcomed high-rise buildings. Improved dwelling space and more effective land use are both results of vertical development. To guarantee the stability and security of high-rise buildings, the government has developed building laws and guidance [56, 57]. In addition, the city of Hong Kong has made a concerted effort to promote eco-friendly practises and environmentally responsible design. It involves encouraging the utilisation of energy from renewable sources, green building practises, and green rooftops. In order to lessen the impact on the environment, the city actively promotes the use of sustainable building practises and resources [58, 59, 26]. Furthermore, Hong Kong's robust and reliable transport network was created to sustain the city's ever-changing built environment. The city's extensive network of public transit, which includes the Mass Transit Railway (MTR), buses, and ferries, enables decreased reliance on private vehicles and increased promotion of environmentally friendly modes of transportation [39, 60]. Also, the government has launched urban regeneration initiatives to improve the condition of the existing built environment and bring revitalization to decaying neighborhoods. Programmes like the aim to make neighborhoods more attractive places to live by investing in necessary infrastructure upgrades, protecting historical landmarks, and encouraging the construction of mixed-use buildings that minimise environmental impact [61, 62]. With an eye towards improving the quality of life in the city, Hong Kong has also invested heavily in the development of open spaces for the public, parks, waterfront promenades, and recreational places [63, 64].

4.3 COVID-19 Transmission Linked with Urban Design and Built Environment Dynamics in Hong Kong

Despite successfully controlling COVID-19 transmission through smart measures such as universal masking, social distancing, testing of all symptomatic and high-risk groups for isolation of confirmed cases in healthcare facilities, and quarantine of contacts without city lockdown or border closure [65], the urban design and built environment dynamics of Hong Kong have had a significant role in COVID-19 transmission in the city. For instance, results from a study indicate that confirmed cases tend to congregate in areas with a greater risk profile, including those with close proximity to public transportation, a multitude of skyscrapers, a great deal of commercial enterprises, and a wide range of mixed land use [7]. According to another study, restaurants and public markets are the two built environments that challenge physical distancing with close interactions and drive the number of COVID-19 confirmed cases in Hong Kong during various phases of the outbreak [14]. Further research revealed that COVID-19 waves in Hong Kong city are correlated with residential surroundings and urban operations, correspondingly [66]. Several clusters of the COVID-19 virus have been found to have spread vertically in Hong Kong's high-rise residential buildings. Old high-rises with limited living space and cheap real estate were particularly vulnerable to transmission. The virus was thought to be spreading mostly through shared drains in the buildings. Apartments on the upper and lower floors of a building could be infected if their drainpipes are located inside the building. Instead, the danger of infection was decreased in buildings where the drain stacks were placed outside [67]. Additional analysis has shown that the COVID-19 threat exists primarily in the central or downtown area of the city compared to the sub-urban area on Hong Kong. Results showed that various aspects of the built environment, including proximity to public transportation, private residential density, the height of buildings, density of population, commercial density, green space density, and sky view, all have positive or negative connections with COVID-19 incidence based on the spatial patterns. Results also showed a positive correlation between private residential density and both the incidence rate (the number of confirmed cases per 1000 people) and the venue density (the number of venues or buildings visited by the confirmed cases), suggesting that geographic regions with greater private residential density are likely to have a greater COVID-19 risk. However, both the incidence rate and the number of venues are inversely related to population density. The density of transport facilities and the height of buildings are positively related to the incidence rate, while the diversity of land uses is inversely related to the incidence rate. Greater densities of green areas are correlated with higher incidence rates and larger numbers of venues, suggesting that confirmed cases are more likely to frequent these areas. The explanation is that people are more likely to engage in multiple outdoor activities when there are green spaces nearby, and some of the confirmed cases may be found among them. In addition, there is an inverse correlation between sky view and venue density, which indicates that safer places tend to have a higher-quality sky view. The findings imply that exposure to sunlight reduces the likelihood of contracting COVID-19 [5]. In another study, the building coverage ratio and population density were found to have a significant impact on the infection rate in urban neighborhoods at various scales in Hong Kong [9]. Evidence from another study linked COVID-19 incidences to factors including building structure and transportation connectivity. The influence of urban design on COVID-19 incidence is also growing in Hong Kong [17]. The study also showed that regions with higher population density, dense and high-rise structures, fewer sky views and fewer green spaces, and a higher private residential, commercial, and transport land use density were regarded as high-risk zones in several phases of COVID-19 transmission in Hong Kong. Additionally, locations that were commonly frequented, such as marketplaces, shopping centres, and restaurants, were major sites of COVID-19 transmission, and some local individuals who worked there were affected [13].

4.4 Lessons Learned and Future Policy Options for Rethinking the Urban Development and Built Environment Attributes in the Post-Pandemic World in Hong Kong

This study includes several lessons learned and future policy options for rethinking the urban development and built environment attributes in the post-pandemic world in Hong Kong. Table 3 provides the overview.

5 Conclusion

This study has analysed the difficulties Hong Kong has as a result of its urban planning and built environment in coping with the COVID-19 outbreak and presented a complete evaluation of existing urban planning and built environment scenes in the city. The goals of this study were accomplished through an extensive review of research sources and publications from government and non-government organisations. Urban development planning scenarios built environment dynamics, the impact of COVID-19 transmission on urban design and the built environment, and lessons learned and policy recommendations for the post-pandemic period in Hong Kong were identified as the four primary issues from the thematic analysis. High-density development, zoning regulations, transportation networks, compact urban form, mixed-use developments, skyscrapers and high-rise buildings, pedestrian-friendly infrastructure, and green building practises are just some of the aspects of Hong Kong's urban development and built environment that were highlighted in the study. Moreover, at several stages of COVID-19 transmission in Hong Kong, places with a high population density, dense and tall structures, limited sky views and green spaces, and a higher land use density for residential, commercial, and transport

Table 3 Lessons learned and future policy options for rethinking the urban development and built environment attributes in the post-pandemic world in Hong Kong

Areas	Lesson learned	Policy options
Urban development planning and design	• Virus transmission relies on a number of other factors besides population density in Hong Kong • Inadequate green and open spaces prevent certain areas from meeting the needs of their residents for outdoor exercise and recreation and from meeting their need for social distance • Higher private residential, commercial, and transport land use densities were also regarded as high-risk zones in several phases of COVID-19 transmission	• High-density locations may be less at risk of pandemics if they have better access to services and public health infrastructure [68, 69] • Urban planners should persist in advocating for compact urban development because of the several advantages it offers [68, 69] • Public parks and green spaces need to be granted more space [68, 69]

(continued)

Table 3 (continued)

Areas	Lesson learned	Policy options
Built environment dynamics	• High-rise buildings density and fewer sky views enhanced the risk of COVID-19 transmission • Old high-rises with limited living space and cheap real estate were particularly vulnerable to transmission • Danger of infection was higher in buildings where the drainpipes were located inside the building • Greater access to public transport served as a risk factor that contributed to the higher transmission of the virus	• Redesigning the older buildings with modified drainage, heating, ventilation, and air conditioning systems [70] • Design concepts for high-rise buildings should emphasise the use of contactless access methods, such as lifts and door/window locks controlled by smartphones and automatic door/window opening with facial recognition [70] • A higher-quality sky view in a building constructed with ultraviolet filter glass and a transparent solar panel could allow sunlight to enter the building, thereby promoting the development of indoor plants and increasing vitamin D levels in humans [70] • Implementation of smart mobility restrictions depending on the likelihood of transmission through various means of transportation may control the dissemination of the virus [68, 69] • Getting people to switch to walking and biking is a great way to encourage more healthy transportation [69]
Smart city perspectives	• Identification and isolation of infected individuals, reduction of human-to-human contacts in service delivery, etc. are just a few examples of how smart solutions helped establish more effective and efficient response and recovery procedures • Transmission has been controlled through technological means, but this has given rise to questions about confidentiality and transparency	• Accessibility to real-time, reliable, geo-referenced Geographic Information Systems data is crucial for effective response and recovery efforts, and permissions must be granted to the public in order to do it securely [68, 69] • Disregarded technology that threatens personal security and set protocols to share public information [68, 69] • Blended methods are better for stopping the pandemic, handling issues related to privacy, making coordination and information sharing easier, and stopping the propagation of misleading misinformation [68, 69]

objectives were recognised as high-risk zones. Several recommendations for modifying Hong Kong's urban development and built environment characteristics after a pandemic are offered based on the study's findings. Among these are the adoption of smart mobility limits, the allocation of additional green spaces, the use of contactless smart technology in the redesign of existing high-rise structures, and the promotion of compact urban growth with improved access to services and public health facilities. Moreover, Hong Kong should ensure secure access to reliable geo-referenced Geographic Information Systems data and data privacy to become an effective smart city. Hong Kong can take steps towards becoming a pandemic-resilient metropolis by implementing these policy changes. To reiterate, this chapter is illuminating because it sheds light on how urban planning, the built environment, and pandemic response interact in Hong Kong. To better address future public health emergencies, cities will need to adapt and reinforce their urban infrastructure, which is where the findings and recommendations given here come in. The results of this research are meant to help government stakeholders, urban planners, and decision-makers in Hong Kong and other cities with similar problems design safer, more sustainable cities for the post-pandemic age. To tackle the upcoming outbreak in the world, Hong Kong has the opportunity to become a role model for resilient cities by rethinking urban development and built environment dynamics in a way that benefits the community as a whole.

References

1. Ahsan MM (2020) Strategic decisions on urban built environment to pandemics in Turkey: lessons from COVID-19. J Urban Manage 9(3):281–285. https://doi.org/10.1016/j.jum.2020.07.001
2. Fatmi MR (2020) COVID-19 impact on urban mobility. J Urban Manage 9(3):270–275. https://doi.org/10.1016/j.jum.2020.08.002
3. Gaisie E, Oppong-Yeboah NY, Cobbinah PB (2022) Geographies of infections: built environment and COVID-19 pandemic in metropolitan Melbourne. Sustain Cities Soc 81:103838. https://doi.org/10.1016/j.scs.2022.103838
4. Li B, Peng Y, He H et al (2021) Built environment and early infection of COVID-19 in urban districts: a case study of Huangzhou. Sustain Cities Soc 66:102685. https://doi.org/10.1016/j.scs.2020.102685
5. Huang J, Kwan MP, Kan Z et al (2020) Investigating the relationship between the built environment and relative risk of COVID-19 in Hong Kong. ISPRS Int J Geo Inf 9(11):624. https://doi.org/10.3390/ijgi9110624
6. Irandoost K, Alizadeh H, Yousefi Z et al (2023) Spatial analysis of population density and its effects during the Covid-19 pandemic in Sanandaj, Iran. J Asian Archit Building Eng 22(2):635–642. https://doi.org/10.1080/13467581.2022.2047983
7. Kan Z, Kwan MP, Wong et al (2021) Identifying the space-time patterns of COVID-19 risk and their associations with different built environment features in Hong Kong. Sci Total Environ 772:145379. https://doi.org/10.1016/j.scitotenv.2021.145379
8. Tong H, Li M, Kang J (2022) Relationships between building attributes and COVID-19 infection in London. Build Environ 225:109581. https://doi.org/10.1016/j.buildenv.2022.109581

9. Xu Y, Guo C, Yang J et al (2023) Modelling impact of high-rise, high-density built environment on COVID-19 risks: empirical results from a case study of two Chinese cities. Int J Environ Res Public Health 20(2):1422. https://doi.org/10.1016/j.buildenv.2020.107472
10. Vazquez-Prokopec GM, Kitron U, Montgomery B et al (2010) Quantifying the spatial dimension of dengue virus epidemic spread within a tropical urban environment. PLoS Negl Trop Dis 4(12):e920. https://doi.org/10.1371/journal.pntd.0000920
11. Mei S, Chen B, Zhu Y et al (2015) Simulating city-level airborne infectious diseases. Comput Environ Urban Syst 51:97–105. https://doi.org/10.1016/j.compenvurbsys.2014.12.002
12. Mpolya EA, Yashima K, Ohtsuki H et al (2014) Epidemic dynamics of a vector-borne disease on a villages-and-city star network with commuters. J Theor Biol 343:120–126. https://doi.org/10.1016/j.jtbi.2013.11.024
13. Kan Z, Kwan MP, Huang J et al (2021) Comparing the space-time patterns of high-risk areas in different waves of COVID-19 in Hong Kong. Trans GIS 25(6):2982–3001. https://doi.org/10.1111/tgis.12800
14. Yip TL, Huang Y, Liang C (2021) Built environment and the metropolitan pandemic: analysis of the COVID-19 spread in Hong Kong. Build Environ 188:107471. https://doi.org/10.1016/j.buildenv.2020.107471
15. Yu Z, Zhu X, Liu X (2022) Characterizing metro stations via urban function: thematic evidence from transit-oriented development (TOD) in Hong Kong. J Transp Geogr 99:103299. https://doi.org/10.1016/j.jtrangeo.2022.103299
16. Xiao L, Liu J (2023) Exploring non-linear built environment effects on urban vibrancy under COVID-19: the case of Hong Kong. Appl Geogr 155:102960. https://doi.org/10.1016/j.apgeog.2023.102960
17. Kwok CYT, Wong MS, Chan KL et al (2021) Spatial analysis of the impact of urban geometry and socio-demographic characteristics on COVID-19, a study in Hong Kong. Sci Total Environ 764:144455. https://doi.org/10.1016/j.scitotenv.2020.144455
18. Hou J, Song Y (2018) Compact development in Hong Kong. In: Lin Z, Gámez LSJ (eds) Vertical urbanism, 1st edn. Routledge, London, pp 170–188
19. Lau S, Wang J, Giridharan R et al (2005) High-density, high-rise and multiple and intensive land use in Hong Kong: a future city form for the new millennium. In: Jenks M, Dempsey N (eds) Future forms and design for sustainable cities, 1st edn. Routledge, London, pp 153–166
20. Lam KC, Ng SL, Hui WC et al (2005) Environmental quality of urban parks and open spaces in Hong Kong. Environ Monit Assess 111:55–73. https://doi.org/10.1007/s10661-005-8039-2
21. Xue F, Gou Z, Lau S (2017) The green open space development model and associated use behaviors in dense urban settings: lessons from Hong Kong and Singapore. Urban Design Int 22:287–302. https://doi.org/10.1057/s41289-017-0049-5
22. Lo HK, Tang S, Wang DZ (2008) Managing the accessibility on mass public transit: the case of Hong Kong. J Transp Land Use 1(2):23–49
23. Su J, Sze NN (2022) Safety of walking trips accessing to public transportation: a Bayesian spatial model in Hong Kong. Travel Behav Soc 29:125–135. https://doi.org/10.1016/j.tbs.2022.06.003
24. Tang S, Lo HK (2008) The impact of public transport policy on the viability and sustainability of mass railway transit—the Hong Kong experience. Transp Res Part A Policy Pract 42(4):563–576. https://doi.org/10.1016/j.tra.2008.01.022
25. Iwan A, Poon KK (2018) The role of governments and green building councils in cities' transformation to become sustainable: case studies of Hong Kong (East) and Vancouver (West). Sustain City 67
26. Wadu Mesthrige J, Kwong HY (2018) Criteria and barriers for the application of green building features in Hong Kong. Smart Sustain Built Environ 7(3/4):251–276. https://doi.org/10.1108/SASBE-02-2018-0004
27. Jia J, Lee WL (2018) The rising energy efficiency of office buildings in Hong Kong. Energy Build 166:296–304. https://doi.org/10.1016/j.enbuild.2018.01.062
28. Sang X, Pan W, Kumaraswamy MM (2014) Informing energy-efficient building envelope design decisions for Hong Kong. Energy Procedia 62:123–131. https://doi.org/10.1016/j.egypro.2014.12.373

29. Govada SS, Rodgers T, Cheng L et al (2020) Smart environment for smart and sustainable Hong Kong. In: Smart environment for smart cities, pp 57–90. https://doi.org/10.1007/978-981-13-6822-6_2
30. Li X, Nam KM, Khoo CK (2020) Smart-city vision and strategy in Hong Kong. In: Joo Y, Tan T (eds) Smart cities in Asia. Edward Elgar Publishing, Cheltenham, United Kingdom, pp 38–60. https://doi.org/10.4337/9781788972888.00010
31. Office of the Government Chief Information Officer (2019) Smart city development in Hong Kong. IET Smart Cities 1(1):23–27. https://doi.org/10.1049/iet-smc.2019.0036
32. Carr LT (1994) The strengths and weaknesses of quantitative and qualitative research: what method for nursing? J Adv Nurs 20(4):716–721. https://doi.org/10.1046/j.1365-2648.1994.20040716.x
33. Bryman A (2016) Social research methods. Oxford University Press, Oxford, United Kingdom
34. Rasinger SM (2013) Quantitative research in linguistics: an introduction. A&C Black.
35. Tian Y, Jim CY (2011) Factors influencing the spatial pattern of sky gardens in the compact city of Hong Kong. Landsc Urban Plan 101(4):299–309. https://doi.org/10.1016/j.landurbplan.2011.02.035
36. Tian Y, Jim CY (2012) Development potential of sky gardens in the compact city of Hong Kong. Urban For Urban Green 11(3):223–233. https://doi.org/10.1016/j.ufug.2012.03.003
37. Tian Y, Jim CY, Tao Y (2012) Challenges and strategies for greening the compact city of Hong Kong. J Urban Plann Dev 138(2):101–109. https://ascelibrary.org/doi/abs/10.1061/(ASCE)UP.1943-5444.0000076
38. Al-Kodmany K, Xue Q, Sun C (2022) Reconfiguring vertical urbanism: the example of Tall Buildings and Transit-Oriented Development (TB-TOD) in Hong Kong. Buildings 12(2):197. https://doi.org/10.3390/buildings12020197
39. Lu Y, Gou Z, Xiao Y et al (2018) Do transit-oriented developments (TODs) and established urban neighborhoods have similar walking levels in Hong Kong? Int J Environ Res Public Health 15(3):555. https://doi.org/10.3390/ijerph15030555
40. Gang W, Wang S, Gao D et al (2015) Performance assessment of district cooling systems for a new development district at planning stage. Appl Energy 140:33–43. https://doi.org/10.1016/j.apenergy.2014.11.014
41. He P, Xue J, Shen GQ et al (2023) The impact of neighborhood layout heterogeneity on carbon emissions in high-density urban areas: a case study of new development areas in Hong Kong. Energy Buildings 287:113002. https://doi.org/10.1016/j.enbuild.2023.113002
42. Kwok YT, Schoetter R, de Munck C et al (2021) High-resolution mesoscale simulation of the microclimatic effects of urban development in the past, present, and future Hong Kong. Urban Climate 37:100850. https://doi.org/10.1016/j.uclim.2021.100850
43. Wang AQ, Chan EH, Yeung SC et al (2017) Urban fringe land use transitions in Hong Kong: from new towns to new development areas. Procedia Eng 198:707–719. https://doi.org/10.1016/j.proeng.2017.07.122
44. Govada SS, Spruijt W, Rodgers T (2017) Assessing Hong Kong as a smart city. In: Smart economy in smart cities. International collaborative research: Ottawa, St. Louis, Stuttgart, Bologna, Cape Town, Nairobi, Dakar, Lagos, New Delhi, Varanasi, Vijayawada, Kozhikode, Hong Kong, pp 199–228. https://doi.org/10.1007/978-981-10-1610-3_8
45. Hartley K (2023) Public perceptions about smart cities: governance and quality-of-life in Hong Kong. Soc Indic Res 166(3):731–753. https://doi.org/10.1007/s11205-023-03087-9
46. Ip KK, Cheng KCC (2022) Smart city development in Hong Kong: an ethical analysis. China Perspect 3:21–30
47. Lam PT, Yang W (2020) Factors influencing the consideration of Public-Private Partnerships (PPP) for smart city projects: evidence from Hong Kong. Cities 99:102606. https://doi.org/10.1016/j.cities.2020.102606
48. Leung KY, Lee HY (2023) Implementing the smart city: who has a say? Some insights from Hong Kong. Int J Urban Sci 27(sup1):124–148. https://doi.org/10.1080/12265934.2021.1997634

49. Ma R, Lam PT (2019) Investigating the barriers faced by stakeholders in open data development: a study on Hong Kong as a "smart city." Cities 92:36–46. https://doi.org/10.1016/j.cities.2019.03.009
50. Guo J, Sun B, Qin Z et al (2017) A study of plot ratio/building height restrictions in high density cities using 3D spatial analysis technology: a case in Hong Kong. Habitat Int 65:13–31. https://doi.org/10.1016/j.habitatint.2017.04.012
51. Liang J, Jiayu CHEN, De TONG et al (2022) Planning control over rural land transformation in Hong Kong: a remote sensing analysis of spatio-temporal land use change patterns. Land Use Policy 119:106159. https://doi.org/10.1016/j.landusepol.2022.106159
52. Wang H, Zhang X, Skitmore M (2015) Implications for sustainable land use in high-density cities: evidence from Hong Kong. Habitat Int 50:23–34. https://doi.org/10.1016/j.habitatint.2015.07.010
53. Yu KH, Hui ECM (2017) An empirical analysis of Hong Kong's planning control decisions for residential development. Habitat Int 63:89–102. https://doi.org/10.1016/j.habitatint.2017.03.014
54. Yu KH, Hui ECM (2019) Pogodzinsk land-use planning, certainty and flexibility: a study of planning control decisions on residential development in Hong Kong. Habitat Int 83:85–98. https://doi.org/10.1016/j.habitatint.2018.11.008
55. Lai LW, Lu WW, Lorne FT (2014) A catallactic framework of government land reclamation: the case of Hong Kong and Shenzhen. Habitat Int 44:62–71. https://doi.org/10.1016/j.habitatint.2014.04.013
56. Lau SS, Zhang Q (2015) Genesis of a vertical city in Hong Kong. Int J High-Rise Build 4(2):117–125
57. Peng S, Maing M (2021) Influential factors of age-friendly neighborhood open space under high-density high-rise housing context in hot weather: a case study of public housing in Hong Kong. Cities 115:103231. https://doi.org/10.1016/j.cities.2021.103231
58. Khoo CK, Li X, Huang J (2022) Green behaviors and green buildings: a post-occupancy evaluation of public housing estates in Hong Kong. Sustainability 14(16):9862. https://doi.org/10.3390/su14169862
59. Tam VW, Hao JL, Zeng SX (2012) What affects implementation of green buildings? An empirical study in Hong Kong. Int J Strateg Prop Manag 16(2):115–125. https://doi.org/10.3846/1648715X.2011.645559
60. Wadu Mesthrige J, Maqsood T (2022) Transport infrastructure, accessibility and residential property values: evidence from Hong Kong. Built Environ Project Asset Manage 12(2):163–179. https://doi.org/10.1108/BEPAM-01-2021-0019
61. Yau Y (2012) Homeowner involvement, land readjustment, and sustainable urban regeneration in Hong Kong. J Urban Technol 19(1):3–22. https://doi.org/10.1080/10630732.2012.626705
62. Yung EHK, Zhang Q, Chan EH (2017) Underlying social factors for evaluating heritage conservation in urban renewal districts. Habitat Int 66:135–148. https://doi.org/10.1016/j.habitatint.2017.06.004
63. Tang BS (2017) Is the distribution of public open space in Hong Kong equitable, why not? Landsc Urban Plan 161:80–89. https://doi.org/10.1016/j.landurbplan.2017.01.004
64. Yung EH, Conejos S, Chan EH (2016) Public open spaces planning for the elderly: the case of dense urban renewal districts in Hong Kong. Land Use Policy 59:1–11. https://doi.org/10.1016/j.landusepol.2016.08.022
65. Wong SC, Au AKW, Lo JYC et al (2022) Evolution and control of COVID-19 epidemic in Hong Kong. Viruses 14(11):2519. https://doi.org/10.3390/v14112519
66. Yu Z, Liu X (2022) Spatial variations of the third and fourth COVID-19 waves in Hong Kong: a comparative study using built environment and socio-demographic characteristics. Environ Plan B Urban Analytics City Sci 50:1144–1160. https://doi.org/10.1177/23998083221107019
67. Zhao P (2023) Analysis of COVID-19 clusters involving vertical transmission in residential buildings in Hong Kong. Build Simul 16:701–711. https://doi.org/10.1007/s12273-022-0929-5
68. Afrin S, Chowdhury FJ, Rahman MM (2021) COVID-19 pandemic: rethinking strategies for resilient urban design, perceptions, and planning. Front. Sustain. Cities 3:668263. https://doi.org/10.3389/frsc.2021.668263

69. Sharifi A, Khavarian-Garmsir AR (2020) The COVID-19 pandemic: impacts on cities and major lessons for urban planning, design, and management. Sci Total Environ 749:142391. https://doi.org/10.1016/j.scitotenv.2020.142391
70. Navaratnam S, Nguyen K, Selvaranjan K et al (2022) Designing post COVID-19 buildings: approaches for achieving healthy buildings. Buildings 12(1):74. https://doi.org/10.3390/buildings12010074
71. Raj Sapru R (2012) Sustainable urban growth: is Hong Kong a model for China? Business for Social Responsibility. https://www.bsr.org/en/blog/sustainable-urban-growth-is-hong-kong-a-model-for-china

Kazi Humayun Kabir is a faculty member in the Development Studies Discipline, Khulna University, Bangladesh. In 2009, He started his teaching career as lecturer of Urban and Rural Planning Discipline in the same university. Currently, he is pursuing (has submitted the draft thesis for examination) his PhD at The University of Hong Kong since 2019. Prior to joining the academia, he was involved in several projects of development and planning home and abroad. His research interests include GIS and Remote Sensing applications in Built Environment, Poverty and Inequality, Disaster Studies and Sustainable Development, Climate Change, Big Data and Machine Learning, Smart City and Spatially Integrated Social Science.

Md. Ayatullah Khan is a Doctoral Researcher at the Department of Geography, Hong Kong Baptist University, Hong Kong. He holds both master's and bachelor's degrees in Development Studies from Khulna University, Bangladesh, with an intense focus on research, monitoring, and evaluation. As part of his professional career, he has lectured at United International University, Dhaka, Bangladesh, as a part-time faculty member. In addition, he held the position of researcher at the Institute of Informatics and Development (IID) in Dhaka, Bangladesh. Also, he has had hands-on experience serving as a humanitarian and development activist at CARE Bangladesh. Furthermore, he has been engaging in multiple research collaborations with faculty members from Khulna University, Bangladesh. His research interests, but are not limited to, include vulnerability assessment, climate change, climatic events and disasters, sustainable development, flood management, city sustainability, and urban development.

Conclusion—Navigating Changes: Influence of the Disruptive Forces on Transformational Urban Planning Conventions

Kh Md Nahiduzzaman

Abstract The current migration, housing, and nature of employments have experienced substantial changes as a result of climate change, evolving global economic dynamics, and the pandemic. These causes have compelled individuals, corporations, and policymakers to adapt to changing circumstances. Climate change has led to a rise in extreme weather occurrences, resulting in people moving from affected places to more affordable and safer areas.The impact of worldwide inflation and increasing living expenses has influenced individual's housing choices, prompting a significant number of people to search for residences in areas beyond the typical cordon of citie's residential districts. The pandemic has reinforced the prevalence of remote work arrangements, resulting in the adoption of virtual and hybrid work. This has left a deep impact on lifestyles, organizational structures, and urban land use. This book is an attempt to unearth the relationship between urban planning, housing choices, land uses, mobility trends, and technological progressions "during" and the "post" pandemic era. It emphasizes the need for flexibility and inclusiveness in planning to re-address sub-urbanization trends, increasing need for affordable housing, and sustainable transportation modes. As cities adapt, the primary focus must be on establishing resilient, accessible, and environmentally sustainable urban settings that can effectively respond to changing needs and future disruptions, such as pandemics.

Keywords Post-pandemic cities · Urban transformation · Change dynamics · Urban Resilience

K. M. Nahiduzzaman (✉)
Faculty of Applied Science, The University of British Columbia (UBC) Okanagan, 1137 Alumni Ave, Kelowna, BC V1V 1V7, Canada
e-mail: Kh.Nahiduzzaman@ubc.ca; KhMd.Nahiduzzaman@um6p.ma

K. M. Nahiduzzaman (ed.), *Making Sense of Planning and Development for the Post-Pandemic Cities*, City Development: Issues and Best Practices,
https://doi.org/10.1007/978-981-97-5481-6_14

1 Prelude

The contemporary landscape of migration, housing, and work patterns have experienced significant transformations, mainly due to the combination of climate change, shifting global economic dynamics, and the extensive effects of the pandemic [1, 2]. These complex factors have caused a significant change in how society behaves, forcing individuals, contemporary businesses, and policymakers to adapt to "new" realities.

The impact of climate change, as indicated by the increasing frequency of extreme weather events such as wildfires, floods, droughts, cyclones, etc., has caused significant migrations [3]. People from regions greatly impacted, like British Columbia in Canada, have sought shelter in places with more affordable living conditions and less exposure to extreme weather events. This is evident in the noticeable migration to other provinces such as Alberta, as documented by Haddad and McAvay [4] and [5]. This migration trend highlights the strong urge to move, driven by both financial and environmental factors. Moreover, worldwide monetary inflation, characterized by escalating interest rates and elevated living expenses, has had a cascading impact on housing decisions and preferences. The increased expenses associated with mortgages and property price have caused individuals to reassess their housing choices, leading many to prefer residences located away from the main districts of the city [6]. The increasing popularity of hybrid work arrangements has made it easier to reconsider the significance of relocating themselves because of the virtual connectivity with the employers. Even it is affordable and convenient if the employees are in need to travel a long distance for a few days of the month if they need to work in-person.

The epidemic had a significant and far-reaching impact and played a pivotal role in reinforcing remote work dynamics. The imperative and subsequent triumph of hybrid or totally virtual work arrangements have resulted in a general embrace, while their continuation has become a popular norm. This transformation has had a profound impact on both individual lifestyles and organizational structures, greatly impacting changes in urban land uses and housing preferences. In essence, these major elements have combined caused transformation in migration, housing preferences, and work patterns. The interconnection of these transformations highlights the intricate interplay between environmental issues, financial factors, and the transforming influence of the pandemic on the individual lifestyles, work, and plan within the urban settings [7].

2 Impacts of Climate Change and Pandemic

The consequences of climate change-induced disruptions, specifically the increased frequency of wildfires, floods, and droughts, along with economic downturn, have had significant effects. This is evident in the relocation of families and individuals from provinces like British Columbia, which experience recurring floods, droughts, and heatwaves, to cities such as Calgary and Edmonton. These cities offer relatively more stable weather conditions, lower living expenses, and stable job prospects [8]. This migration pattern has led to a combination of social, economic, and environmental consequences, shaping the dynamics of these cities in different ways. From a social perspective, this movement demonstrates the significant impact that climate-induced events have on the lives and choices of the people affected. From an economic perspective, the (compulsory) movement of people from regions with higher costs of living to areas with lower expenses highlights the financial difficulties. Put simply, this shift towards economically wealthy regions is the effort to maintain livelihoods, handle living expenses, and provide economic stability in the midst of climate-related difficulties. From an environmental perspective, the rise in migration across different regions leads to changes in how populations are distributed, which in turn affects how land is used and resources are allocated. Evidently, the COVID-19 epidemic worsened the already vulnerable living conditions of migrants affected by climate change and financial difficulties, particularly by increasing the difficulties of finding suitable jobs. The pandemic-imposed limitations, encompassing lockdowns, social distancing protocols, and corporate shutdowns, resulted in a substantial reduction in job prospects across diverse industries. Despite the increase in work opportunities following the COVID pandemic, the newly demanded skillsets are preventing many job seekers from being re-hired [9].

3 Pandemic-Induced Work Changes

The COVID-19 pandemic has significantly changed the traditional work arrangements, leading to a widespread and quick acceptance and implementation of hybrid and fully virtual work setups. This shift in work habits originated from the need for remote work models to guarantee the uninterrupted operation of businesses throughout the limitations imposed by the epidemic. The advantages of these hybrid or virtual work options have been pivotal. They have established a mutually beneficial arrangement, advantageous to both employees and employers. Remote employment greatly improved work-life balance for individuals by providing them with increased flexibility. By eliminating the daily drive, workers have acquired increased autonomy over their schedules, allowing them to more effectively manage their personal and professional obligations. Furthermore, employees observed a rise in productivity and job satisfaction due to the remote work environment, which offered a more comfortable and flexible setting for task completion.

The hybrid work approach has proven to be advantageous for organizations in several ways. It decreased expenses associated with office maintenance and enabled access to a wider range of skilled individuals, regardless of geographical limitations. Moreover, remote work arrangements have enhanced the ability of businesses to withstand and recover from unexpected disruptions, hence proving the effectiveness of remote work in maintaining uninterrupted corporate operations. The positive experiences of both individuals and businesses throughout the epidemic resulted in the continuation of these remote work arrangements beyond the pandemic. Several firms recognized the potential for increased productivity and work satisfaction among their employees. Consequently, they integrated hybrid work models into their long-term goals, which influenced organizational structures by promoting a more versatile and responsive attitude to work. The transformation mentioned has had a substantial influence on both individual lifestyles and organizational structures, fundamentally reshaping the way work is seen and carried out [10]. It has not only transformed how employees handle their work-life equilibrium but also impacted how organizations view their operational dynamics and the organizational structure needed to facilitate such adaptable work models.

4 Post-COVID Global Monetary Inflation and Choice for Housing Locations

The proliferation of global monetary inflation, predominantly manifested through escalated interest rates, has significantly impacted the prices of mortgages and properties. The increase in interest rates, due to global economic changes, has increased the financial strain on people looking for housing, resulting in higher mortgage payments and higher property prices. The economic hardship has influenced individuals to choose alternate accommodations, so greatly influencing their tastes and choices. The increasing financial burden caused by heightened mortgage and property expenses has led many to reconsider their housing choices. The increased costs associated with conventional housing options in urban areas have resulted in a significant change in people's preferences. Many residents are increasingly choosing to live in dwellings that are located outside of the city boundaries in order to reduce financial burden.

5 COVID, Hybrid Employments and Housing Choice Dynamics

The emergence of hybrid work arrangements and the rapid growth of online based commerce (i.e., e-commerce), triggered by the pandemic and ongoing to this day, have sparked a significant transformation in various facets of everyday life, notably in housing, transportation, and urban planning. The emergence and extensive implementation of hybrid work arrangements, together with the rapid growth of e-commerce, have profoundly started to transform the dynamics of residential locational choices. With the increasing ability for individuals to work remotely or have flexible work hours, they are now re-evaluating their residential preferences [7,

11]. The allure of urban cores, hitherto predominantly shaped by their closeness to areas of employment, is increasingly tending to diminish. Suburban, rural territories, and small city regions are getting increasingly popular due to the availability of larger living spaces with lower costs. The shift in work arrangements has led people to choose larger homes or properties in less urbanized areas, prioritizing comfort, space, and calmness above city-centric locales. Furthermore, the adoption of mixed employment models has greatly transformed daily mobility patterns. Commuting has experienced a significant transition, as fewer people tend to depend on public transportation or travel to centralized places or commercial and business districts on a daily basis. The decreasing need to go to conventional employment has led to a reduction in daily commutes, resulting in less traffic congestion and altering the demand for transportation [12, 13].

The availability of work-from-home choices has led to a reconsideration of conventional views on residential areas and urban development. The trend towards decentralized work and living preferences has prompted a re-assessment of the significance of spatial proximity to workplaces, therefore impacting land use, housing designs, and urban planning strategies to cope with this changing lifestyle. The rise of hybrid employment arrangements, which have been expedited by the effects of the pandemic, has significantly influenced the way people choose their accommodation. The current work environment enables individuals to work from distant regions and has significantly reduced the importance of living close to cities. With the widespread acceptance of remote work, individuals are no longer limited to residing close to their employers. Consequently, people have been increasingly looking for houses in areas that are distant from the popular urban districts. This changing pattern corresponds to the demand for more cost-effective housing alternatives, allowing individuals to manage lower housing expenses while still being able to work remotely or in a hybrid work setup. The increased flexibility in employment models has freed individuals from the "contemporary" limitations of residing in urban areas, resulting in a notable change in housing choices and preferences. The convergence of economic demands and the adaptability of remote employment has catalyzed a significant transformation in the way individuals perceive and choose housing, ultimately reconfiguring the dynamics of what we call "urban" living.

6 Concluding Thoughts

This book, featuring contributions from leading researchers and practitioners worldwide, attempts to offer a thorough examination of the complex and transformative relationship between urban planning, housing preferences, transformative land uses, mobility patterns, and technological advancements during and after the pandemic. As cities adapt to shifting dynamics, some critical observations have re-surfaced.

Efforts in urban planning, management, and development aimed at attaining resilience need to adjust to the changing preferences for dwelling locations. The landscape during and after the pandemic has experienced a notable "shift" towards sub-urbanization and moving out from the cities. This trend has been influenced by factors such as remote employment with an urge for affordable and livable residential areas. Therefore, professionals involved in planning and related fields must prioritize "flexibility" and "inclusiveness" in their thoughts, choice of planning methods and their rightful execution.

The land uses that are developing as a result of the pandemic have favorable prospects for increased availability of inexpensive housing options. Cities may achieve accessible, affordable, and environmentally sustainable dwelling options by re-purposing underutilized spaces and adopting new design solutions. Nevertheless, it is important to make collaborative endeavors to tackle the challenges related to affordability, especially considering the continuous rise in land and construction expenses that have been further increased due to global monetary inflation.

The effects of transformative decisions about the location of housing and the adoption of hybrid and remote work arrangements are fundamentally changing the way people move and the modes of transportation they choose to use. With the increasing popularity of remote work, the way people commute is changing. This is causing a discernible shift in the use of public transit and an increasing need for alternate transportation modes, including ride sharing and carpooling, and infrastructure for active transportation e.g., walking and biking. Therefore, it is crucial for the researchers, practitioners and policy makers to prioritize feasible investment options for flexible infrastructure that facilitate multiple transportation modes and encourage sustainable transportation solutions that are both financially accessible and environmentally sustainable.

Given the transformative dynamics following the pandemic, together with the increasing reliance on technological breakthroughs, the future of land use pattern in cities is expected to undergo substantial changes. E-commerce, as a powerful force, is continuously disrupting traditional retail models and the way people contemporarily shop. As a result, urban environments are likely to be transformed to accommodate fulfillment facilities, last-mile delivery hubs, and the changing mobility patterns of shoppers. This evolution highlights the necessity of flexible land use regulations (bylaws) that support the integration of various development options and promote resilient urban settings.

Essentially, the fate of cities depends on their capacity to readily accept while being creative to establish urban environments that are accessible, environmentally friendly, and pleasant to live in. Urban planners, architects, designers, engineers, economists, sociologists, and other professionals may construct cities that satisfy the evolving demands of their inhabitants and promote economic development and environmental stewardship on account of inclusivity, affordability, and resilience.

References

1. Nahiduzzaman KM (2021) COVID-19 and change dynamics in the transformational cities. J Urban Manag 10(2)
2. Nahiduzzaman KM, Lai SK (2020) What does the global pandemic COVID-19 teach us? some reflections. J Urban Manag 9(3)
3. Punia A, Singh RP, Chauhan NS (2023) Drivers of climate migration. In: Singh P, Ao B, Yadav A (eds) Global climate change and environmental refugees. Springer, Cham. https://doi.org/10.1007/978-3-031-24833-7_4
4. Haddad M, McAvay H (2024) Are international and internal migration distinct phenomena? In: Recchi E, Safi M (eds) Handbook of human mobility and migration. Edward Elgar Publishing. https://doi.org/10.4337/9781839105784.00011
5. Kaur M, Selim S, Nahiduzzaman KM, Hewage K, Sadiq R (2023) An investigation of the current situation of BC floodplain mapping in British Columbia using a framework: a fuzzy rule-based approach. J Urban Plan Dev
6. Abbas N, Macchiarelli C (2024) Housing affordability remains stretched amid higher interest rate environment. Prospective home buyers face high prices and elevated borrowing costs, while homeowners refrain from listing their properties. IMF, Monetary Policy. https://www.imf.org/en/Blogs/Articles/2024/01/11/housing-affordability-remains-stretched-amid-higher-interest-rate-environment
7. Nahiduzzaman KM, Holland M, Sikder S, Shaw P, Hewage K, Sadiq R (2021) Urban Transformation towards a Smart City: An E-commerce induced Path-Dependent Analysis. J Urban Plan Dev 147(1)
8. Ballard J (2024) Almost 70,000 people left B.C. last year—mostly for Alberta. CBC News. https://www.cbc.ca/news/canada/british-columbia/bc-70-thousand-people-exodus-1.7159382
9. Gurchiek K (2021) Report: pandemic drives workers' desire to learn new job skills. https://www.shrm.org/topics-tools/news/organizational-employee-development/report-pandemic-drives-workers-desire-to-learn-new-job-skills
10. Linda R, Louise V, Lisa R, Diana H (2021) Remote work: effects on Nordic people, places and planning 2021–2024. Stockholm: Nordregio
11. Nahiduzzaman KM, Aldosary AS, Ishak M (2019) Framework analysis of e-commerce induced shift in the spatial structure of a city. J Urban Plan Dev 145(3)
12. Campisi T, Nahiduzzaman KM, Nikiforiadis A, Stamatiadis N, Basbas S (2022) Will the effects of COVID-19 on commuting and daily activities of the university students be maintained? evidence from a small town in Sicily. Sustainability 14(10)
13. Campisi T, Nahiduzzaman KM, Al-Rashid M, Tesoriere G (2022) COVID-19 and impacts of the evolving city models on mobility. AIP Conf Proc

Prof. Kh Md Nahiduzzaman is Nordic trained Canadian urban planner who received PhD from the Royal Institute of Technology (KTH), Sweden, and MPhil from Norwegian University of Science and Technology (NTNU), Norway. Prior to joining Mohammed VI Polytechnic University (UM6P), Morocco, Prof. Nahiduzzaman held faculty positions at the University of British Columbia Okanagan (Canada), King Fahd University of Petroleum & Minerals (Saudi Arabia), and the Department of Urban Planning and the Environment in the Royal Institute of Technology (Sweden). His core research interest is urban resilience through the lenses of digital twins, urban transformational planning, smart cities, and climate risks and vulnerability assessment. He is a recognized researcher in the field. He is heavily engaged on collaborative research projects with the key governments and industrial partners, and (indigenous) communities across British Columbia, Canada and Morocco. So far, he has secured reputed research grants worth more than $3 million from the prestigious national and international funding agencies. He has published more than 100 scientific manuscripts in the top ranked journals, books and other reputed proceedings. Many of his works such as non-structural flood mitigation, floodplain maps and land use planning strategies in BC have direct impact on the provincial policies. He is an "Executive" and "Book Review" Editor of the Journal of Urban Management, Editor-in-Chief of "City Development: Issues and Best Practices" journal, and Editorial Board Member of the Journal of Urban Planning and Development. He is also seating on many national and international scientific committees, fora and think-tanks. Prof. Nahiduzzaman is the editor of the book "Making Sense of Planning and Development for the post-pandemic Cities" by Springer. He is also a co-chair of "Smart and Connected Cities" track in the prestigious US-Africa Frontiers program to be held on February 18-20, 2025, in Kigali, Rwanda.